FINITELY
AXIOMATIZABLE
THEORIES

SIBERIAN SCHOOL OF ALGEBRA AND LOGIC

Countable Boolean Algebras and Decidability • *Sergei S. Goncharov*

Definability and Computability • *Yuri L. Ershov*

Finitely Axiomatizable Theories • *Mikhail G. Peretyat'kin*

Right-Ordered Groups • *Valeriĭ M. Kopytov and Nikolaĭ Ya. Medvedev*

FINITELY AXIOMATIZABLE THEORIES

Mikhail G. Peretyat'kin

Institute of Pure and Applied Mathematics
Kazakh Academy of Sciences
Almaty, Kazakhstan

Consultants Bureau • New York, London, and Moscow

Library of Congress Cataloging-in-Publication Data

On file

Sci
QA
9.6
.P47
1997

Siberian School of Algebra and Logic is a simultaneous translation
of the book series Sibirskaya Shkola Algebry i Logiki,
which is published in and translated by
Scientific Books (RIMIBE NGU)
2, Pirogova Street, Novosibirsk 630090, Russia

ISBN 0-306-11062-8

©1997 Consultants Bureau, New York
A Division of Plenum Publishing Corporation
233 Spring Street, New York, N.Y. 10013

http://www.plenum.com

10 9 8 7 6 5 4 3 2 1

Printed in the United States of America

To the memory of W. Hanf

Preface

Finitely axiomatizable theories are a classical subject of mathematical logic. They attract the attention of researchers not only as an object of investigation but also as an effective method of studying various problems. For instance, the well-known result of Church on the algorithmic undecidability of predicate logic was based on a regular method of constructing finitely axiomatizable theories. This method was modified by Trakhtenbrot in order to prove the undecidability of the class of finite models. To prove the undecidability of elementary theories of various classes of models, Ershov created a universal method of relative elementary definability, which has its origins in the above-mentioned method.

What can be expressed by a single formula of classical predicate logic? Posed in this general setting, the question incorporates many particular problems on the expressibility of finitely axiomatizable theories which arose as a result of the successful development of model theory in the 1950's. Despite its timeliness, the problem of expressibility of finitely axiomatizable theories did not have a satisfactory solution for a long time. Some results were obtained in this direction, but they were fragmented and did not present the entire picture. Thus, even in the late 1970's the above question remained open.

As the first attempt to determine the expressibility of finitely axiomatizable theories in general, we recall the result of Kleene which asserts
that any recursively axiomatizable theory of a finite signature is interpreted in a finitely axiomatizable theory of a signature containing a single
auxiliary binary predicate. As a result of the active development of model
theory, algorithm theory, and other branches of logic, a number of key
questions concerning finitely axiomatizable theories were formulated. The
Hanf problem on coincidence of recursive isomophism types for the Lindenbaum algebras of finitely axiomatizable theories and those of recursively
axiomatizable theories and the Vaught–Morley problem on the existence
of a complete uncountably categorical finitely axiomatizable theory (which
was divided by Shelah into two cases: the categorical variant and the noncategorical variant in countable cardinality) exert the primary influence
on further development along this direction.

Only during the past decade was the problem of expressibility of formulas of predicate logic given an exhaustive solution. The author [42]
constructed an example of a complete uncountably categorical finitely
axiomatizable theory. The example itself gave a positive answer to the
well-known Vaught–Morley ptoblem. Together with the profound ideas of
Hanf on constructing formulas with prescribed semantic properties, this
example formed the basis for a solution to the general problem on the expressibility of finitely axiomatizable theories of predicate logic. The author
suggested that a universal construction be formulated, which showed that
a majority of the natural model-theoretic properties of expressibility coincide for finitely axiomatizable theories and for recursively axiomatizable
theories. This result enables us to regularly reduce a lot of questions on
finitely axiomatizable theories to the considerably easier case of recursively
axiomatizable theories.

The book follows the same direction as the Hanf problem and the
countably noncategorical variant of the Vaught–Morley problem. A lot of
positive information on the expressibility of finitely axiomatizable theories
was obtained in this direction. Thus, a more precise title of the book
would be *Expressibility of Finitely Axiomatizable Theories and Their
Applications in Logic (Positive Aspects)*.

Notice that the countably categorical variant of the Vaught–Morley
problem was negatively solved by Zil'ber [65]. The results of Zil'ber were
strengthened and generalized to a wider class of superstable theories by
Lachlan, Cherlin, and Harrington. These results have not been included
in this book due to the great difference in notions and methods used.

The book is based on a single, involved construction, but the contents
are not reduced only to this construction and its applications.

First, the main construction has a simpler predecessor, which is also presented in the book. Thus, the reader may approach the difficult goal of understanding the universal construction by a successive approximation of ideas that are presented initially in a simpler variant.

Second, the book contains self-contained proofs for universal construction as well as for some results that are very interesting in themselves. These include, most importantly, a separate chapter containing a solution to the Vaught–Morley problem and a separate chapter with an intermediate construction which solves the Hanf problem. At present, the proofs of these results have been considerably simplified compared to the original proofs. Due to this, compact and self-contained exposition may be of interest to specialists in logic.

Presently, no monograph devoted to the expressibility of finitely axiomatizable theories has been written. This book summarizes the investigations in the field at a time when considerable progress has been achieved. It treats systematically all positive results obtained concerning the expressibility of finitely axiomatizable theories and states a number of new natural questions in this field, thus providing prospects for further development of the theory

ACKNOWLEDGEMENT

I would like to thank my teacher, Yu. L. Ershov, who directed my initial steps in science. I would like to acknowledge my colleagues S. S. Goncharov, V. L. Selivanov, E. A. Palyutin, S. D. Denisov, A. S. Morozov, and D. E. Pal'chunov for their useful discussions.

I am also grateful to A. V. Kravchenko who carefully checked the text of this book.

I express thanks to my editor Tamara Rozhkovskaya for the extensive editorial work she performed in transforming my manuscript into a fine book.

Mikhail G. Peretyat'kin

Alma Ata, Kazakhstan

July, 1996

Contents

Introduction

We begin with a brief summary of necessary notions and facts. Except for some facts from the theory of Boolean algebras, model theory, and the theory of algorithms, no preliminary knowledge is required for reading this chapter.

All theories are considered in the classical first-order predicate logic with equality. As for general notions, we follow [3, 13, 56, 60]. Throughout the book, we use the following notation: $|X|$ is the cardinality of a set X, $\mathcal{P}(X)$ is the set of all subsets of a set X, $|\mathfrak{M}|$ is the universe of a model $\mathfrak{M}$, the notation $\mathfrak{M} \upharpoonright X$ means the restriction of the model $\mathfrak{M}$ to the set X, $IM(\lambda, T)$ is the number of isomorphism types for models of the theory T of cardinality λ, $J_\alpha(\beta)$ is the cardinal-valued function defined as follows:

$$J_0(\beta) = \beta$$
$$J_{\alpha+1}(\beta) = 2^{J_\alpha(\beta)}$$
$$J_\gamma(\beta) = \sup\left\{J_\delta(\beta),\ \delta < \gamma\right\} \text{ if } \gamma \text{ is a limit ordinal}$$

where α stands for an ordinal and β denotes a cardinal.

In the notation of a signature, superscripts indicate the numbers of places of the corresponding symbols. We deal with only those signatures for which Gödel numberings of formulas exist. Such signatures are called *enumerable*. A finite signature is called *rich* if it contains at least one predicate symbol or functional symbol of the number of places $n \geqslant 2$ or two unary functional symbols. Let σ denote a signature. We denote by $FL(\sigma)$ the set of all such formulas of the signature σ and by $FL_n(\sigma)$ the set of all formulas in n free variables $x_1, \ldots, x_n$. The set of all sentences of the signature σ is denoted by $SL(\sigma)$. If $\mathfrak{M}$ is a model and $\mathfrak{M}^*$ is its complete enrichment by constants, then $AD(\mathfrak{M}^*)$ stands for the set of closed atomic formulas that are true in $\mathfrak{M}^*$ and $FD(\mathfrak{M}^*)$ is the set of all closed formulas that are true in $\mathfrak{M}^*$.

If T_1 and T_2 are theories of signatures σ_1 and σ_2, then $T_1 \oplus T_2$ denotes the *direct sum* of T_1 and T_2.

A theory T is called a *model-complete theory* if $\mathfrak{M} \subseteq \mathfrak{N} \Leftrightarrow \mathfrak{M} \preccurlyeq \mathfrak{N}$ for any models $\mathfrak{M}$ and $\mathfrak{N}$ of the theory T, which is equivalent to the $\exists$-reducibility and $\forall$-reducibility of formulas in T.

We denote by GRE the extension of the graph theory in the signature $\sigma = \{\Gamma^2\}$ that is defined by the following axioms:

$$(\forall x)^\neg \Gamma(x, x), (\forall x, y)(\Gamma(x, y) \leftrightarrow \Gamma(y, x)), (\exists x, y)\Gamma(x, y),$$
$$(\exists x, y)(x \neq y \& {}^\neg\Gamma(x, y))$$

We denote by SI the theory of the signature $\sigma = \{\lhd^2, c\}$ whose axioms assert that $\lhd$ is the successor relation, c has no $\lhd$-predecessor, the remaining elements have $\lhd$-predecessors, and all cycles are inhibited. This theory is ω_1-categorical and complete.

0.1. Axiomatizable Theories
and Recursively Enumerable Indices

The finitely axiomatizable theories form a subclass of the class of recursively axiomatizable theories. The latter can be defined as the class of recursively enumerable theories or as the class of recursive-enumerably axiomatizable theories. Following traditions, we often write "axiomatizable theory" instead of "recursively axiomatizable theory" for brevity. In fact, such theories are defined by recursively enumerable systems of axioms. The key idea of the book is to investigate expressive possibilities of finitely axiomatizable theories by comparison with the case of recursively axiomatizable theories, which have been well studied.

Using the standard Post numbering W_n, $n \in \mathbb{N}$, of recursively enumerable sets, we construct an effective numbering of the class of axiomatizable theories as follows. Let σ be an enumerable signature. We fix a Gödel numbering Φ_i, $i \in \mathbb{N}$, of all sentences of the signature σ. If a theory T of the signature σ is given by the set of axioms $\{\Phi_i \mid i \in W_m\}$, then we say that m is a *recursively enumerable index* of the theory T. The notion of an index can be introduced in another way. For the same set of axioms $\{\Phi_i \mid i \in W_m\}$ we define a theory T' of the signature $\sigma' \subseteq \sigma$ containing only those symbols of σ that occur in formulas of the sequence Φ_i, $i \in W_m$. The number m is called a *weak recursively enumerable index* of the theory T'.

0.2. Enumerated Models

Let $\mathfrak{M}$ be an at most countable model. By a *numbering* of the model $\mathfrak{M}$ we mean a mapping $\nu : \mathbb{N} \overset{\text{onto}}{\to} |\mathfrak{M}|$, where $\mathbb{N}$ is the set of natural numbers. If $\mathfrak{M}$ is a model of an enumerable signature σ and ν is a numbering of $\mathfrak{M}$, then $\mathfrak{M}_\nu$ stands for the model of the signature $\sigma_1 = \sigma \cup \{c_i \mid i \in \mathbb{N}\}$, where the new constant symbol c_i has the value $\nu(i)$ in the model $\mathfrak{M}_\nu$. An enumerated model $\langle \mathfrak{M}, \nu \rangle$ is called *constructive* if there is an algorithm of verification of the truth of quantifier-free formulas in $\mathfrak{M}_\nu$, and *strongly constructive* if the first-order theory $FD(\mathfrak{M}_\nu)$ is decidable. A model $\mathfrak{M}$ is called (*strongly*) *constructivizable* if it admits a (strong) constructivization.

Two numberings ν_1 and ν_2 of a model $\mathfrak{M}$ are called *autoequivalent* if the enumerated models $\langle \mathfrak{M}, \nu_1 \rangle$ and $\langle \mathfrak{M}, \nu_2 \rangle$ are constructively isomorphic, i.e., if there exists an automorphism μ of the model $\mathfrak{M}$ and a general recursive function $f(x)$ such that $\mu\nu_1(x) = \nu_2 f(x)$ for all $x \in \mathbb{N}$.

A strongly constructivizable model $\mathfrak{M}$ is called *autostable* if any two strong constructivizations of $\mathfrak{M}$ are autoequivalent. In general, a model $\mathfrak{M}$ may have exactly $d \in \{0, 1, \omega\}$ pairwise nonautoequivalent strong constructivizations. The number $d = d(\mathfrak{M})$ is called the *algorithmic dimension* of the model $\mathfrak{M}$ with respect to strong constructivizations.

Let T be a model-complete recursively enumerable theory. Then any constructivization ν of any model $\mathfrak{M}$ of the theory T is strong. Conversely, if a model $\mathfrak{M}$ of the theory T is not strongly constructivizable, then it is not constructivizable.

An enumerated algebra $\langle \mathcal{B}, \nu \rangle$ of an enumerable signature σ is called *positively enumerated* if, in $\mathcal{B}$, the operations are uniformly represented by recursive functions on numbers and the equality predicate is recursively enumerable with respect to the numbering ν.

0.3. Lindenbaum Algebras

Let $\mathcal{L}_n(T)$ denote the Lindenbaum algebra of a theory T over formulas in n free variables $x_1, \ldots, x_n$. The algebra $\mathcal{L}_0(T)$ is referred to as the *Lindenbaum algebra* of the theory T and is denoted by $\mathcal{L}(T)$. The algebra $\mathcal{L}(T)$ presents the structure of all completions of the theory T. In particular, there is a natural one-to-one correspondence between all completions T^* of the theory T and all elements of the Stone space St $(\mathcal{L}(T))$, i.e., all ultrafilters $\mathcal{F}$ of $\mathcal{L}(T)$.

Let T be an axiomatizable theory of an enumerable signature σ. For the Lindenbaum algebra $\mathcal{L}(T)$ there is a natural numbering γ induced by the Gödel numbering of the set of sentences $SL(\sigma)$. We call γ the *Gödel numbering* of the algebra $\mathcal{L}(T)$. Since the theory T is axiomatizable, the pair $(\mathcal{L}(T), \gamma)$ is a positively enumerable Boolean algebra. Sometimes, we speak about recursive properties of the algebra $\mathcal{L}(T)$ without specifying the numbering γ.

The following lemma can be proved in a standard way.

Lemma 0.3.1. *Let T_0 and T_1 be theories of enumerable signatures and let $\mu : \mathcal{L}(T_0) \to \mathcal{L}(T_1)$ be a recursive isomorphism between the Lindenbaum algebras $\mathcal{L}(T_0)$ and $\mathcal{L}(T_1)$ of the theories T_0 and T_1. If complete extensions T_0^* and T_1^* of the theories T_0 and T_1 correspond to each other under μ, then the following assertions hold:*

(a) *the theories T_0^* and T_1^* have the same 1-degree of algorithmic complexity,*

(b) *the theory T_1^* is decidable if and only if the theory T_0^* is decidable.*

Let T be a theory of a signature σ. A set $S \subseteq SL(\sigma)$ is called the set of *generators* of the Lindenbaum algebra $\mathcal{L}(T)$ if, in T, every sentence $\Phi \in SL(\sigma)$ is equivalent to some Boolean combination of sentences of S.

Lemma 0.3.2. *Let T be a theory of a signature σ. A set $S \subseteq SL(\sigma)$ is a set of generators of the Lindenbaum algebra $\mathcal{L}(T)$ if and only if for any subset $S' \subseteq S$ the theory defined by the set of sentences*

$$T[S'] = T \cup S' \cup \{\neg\Psi \mid \Psi \in S \backslash S'\}$$

is complete or is inconsistent.

PROOF. The necessity is obvious. Let us prove the sufficiency. We assume that the theory $T[S']$ is complete or is inconsistent for any $S' \subset S$. Let Φ be a sentence of the signature σ. We enumerate the set S and represent it in the form $\{\Psi_i \mid i \in \mathbb{N}\}$. Consider the set $\mathcal{D}$ of all sequences of the form

$$\langle \alpha_0, \alpha_1, \ldots, \alpha_{s-1} \rangle, \quad s < \omega, \; \alpha_i \in \{0, 1\}$$

such that Φ is not provable and is not disprovable in $T \cup \{\Psi_k^{\alpha_k} \mid k < s\}$; here and in what follows, we use the ordinary notation $\Psi^0 = \neg \Psi$ and $\Psi^1 = \Psi$.

The set $\mathcal{D}$, being a tree, cannot be infinite; otherwise, there is a sequence of length ω such that each of its initial segments belongs to $\mathcal{D}$. Hence the sequence defines a set $S' \subseteq S$ such that Φ is not provable and is not disprovable in the theory $T[S']$, which contradicts the assumption. Thus, the set $\mathcal{D}$ is finite. Therefore, it is possible to choose a natural number k which is greater than the length of each sequence of $\mathcal{D}$. We denote by Ψ the disjunction of those elementary conjunctions of the form $C(\alpha) = \Psi_0^{\alpha_0} \& \Psi_1^{\alpha_1} \& \ldots \& \Psi_{k-1}^{\alpha_{k-1}}$ for which the formula Φ is provable in the theory $T \cup \{C(\alpha)\}$. It is easy to see that the initial formula Φ is equivalent to the constructed formula Ψ in T. $\square$

0.4. Model-Theoretic Properties

We specify the notion of a model-theoretic property, which is necessary for the definition of the semantic similarity of theories.

Two theories T_0 and T_1 of signatures σ_0 and σ_1 are called *isomorphic* (in symbols, $T_0 \approx T_1$) if T_1 is obtained from T_0 by a finite number of renamings of signature symbols and by addition and elimination of those symbols that can be first-order definably expressed in terms of other signature symbols. If the signatures σ_0 and σ_1 are disjoint, then the theories T_0 and T_1 are isomorphic if and only if T_0 and T_1 are subtheories of a common theory T of the signature $\sigma = \sigma_0 \cup \sigma_1$; moreover, $T_i = T \upharpoonright \sigma_i$, $i = 1, 2$, and, in the theory T, all signature symbols are first-order definable with respect to σ_0 and with respect to σ_1 as well.

In the case of isomorphisms, signature symbols of each theory are represented by formulas of another theory, so that these theories are indistinguishable from the point of view of any of their properties that are definable on the basis of the structure of first-order definable relations.

Therefore, in the majority of cases of model-theoretic applications, isomorphic theories are regarded as copies of the same theory.

By a *model-theoretic* (semantic) property we mean a class p of complete theories of enumerable signatures which satisfies the closedness condition with respect to the first-order definable equivalence of the form $T_0 \approx T_1 \Rightarrow (T_0 \in p \Leftrightarrow T_1 \in p)$ for complete theories T_0 and T_1 of any enumerable signatures. We denote by ML the set of all possible model-theoretic properties. An arbitrary subset $L \subseteq ML$ is called a *list of model-theoretic properties* or simply a *list*.

0.5. Semantic Similarity of Theories

The definition of the similarity of two theories is based on a natural division of properties of the theory T into two groups. The first group contains properties concerning the structure of the Lindenbaum algebra $\mathcal{L}(T)$, such as consistency, completeness, existence of a decidable completion, and so on. All these properties are uniquely defined by the recursive isomorphism type of the Lindenbaum algebra of T. The second group contains model-theoretic properties of various completions T^* of the theory T, in particular, the algorithmic properties such as stability, existence of a prime model, strong constructivizability, and so on.

Thus, a theory T can be characterized by the triple $(\mathcal{L}(T), \gamma, f)$, where $(\mathcal{L}(T), \gamma)$ is the corresponding Lindenbaum algebra with the Gödel numbering γ and f is the mapping from the Stone space St $(\mathcal{L}(T))$ into the power-set $\mathcal{P}(L) = \{E \mid E \subseteq L\}$ that is defined as follows. For a completion T^* in St $(\mathcal{L}(T))$ we set $f(T^*) = \{p \in L \mid T^* \text{ has the property } p\}$. The above triple is called the *generalized Lindenbaum algebra of the theory T over the list L* or simply the *Lindenbaum L-algebra of the theory T*.

We can introduce a special class in order to represent the isomorphism type of Lindenbaum L-algebras. To this end, we consider triples of the form $\tau = (\mathcal{B}, \nu, h)$, where $(\mathcal{B}, \nu)$ is a positively enumerated Boolean algebra and h is a mapping from the Stone space St $(\mathcal{B})$ into the power-set $\mathcal{P}(L)$. Such triples are called *abstract semantic L-types* or simply *L-types*.

Two abstract L-types $\tau = (\mathcal{B}, \nu, h)$ and $\tau' = (\mathcal{B}', \nu', h')$ are called *equivalent* if there is an isomorphism $\lambda : \mathcal{B} \to \mathcal{B}'$ between Boolean algebras $\mathcal{B}$ and $\mathcal{B}'$ which is recursive with respect to the numberings ν, ν' and $h(\mathcal{F}) = h'(\mathcal{F}')$ for any ultrafilter $\mathcal{F} \in$ St $(\mathcal{B})$ and the corresponding ultrafilter $\mathcal{F}' = \lambda(\mathcal{F}) \in$ St $(\mathcal{B}')$.

We associate abstract L-types with theories. Let T be an axiomatizable theory with the Lindenbaum L-algebra $(\mathcal{L}(T), \gamma, f)$ and let $\tau = (\mathcal{B}, \nu, h)$ be an abstract semantic L-type. We say that the *theory T has type τ* or the *type τ is realized in the theory T* if the corresponding triples $(\mathcal{B}, \nu, h)$ and $(\mathcal{L}(T), \gamma, f)$ are equivalent.

Let T_1 and T_2 be axiomatizable theories and let L be a list of model-theoretic properties. We say that T_1 and T_2 are *semantically similar* with respect to the list L (denoted by $T_1 \equiv_L T_2$) if T_1 and T_2 have the same abstract type.

In other words, theories T_1 and T_2 are semantically similar with respect to a list L if there is a recursive isomorphism $\mu : \mathcal{L}(T_1) \to \mathcal{L}(T_2)$ between the Lindenbaum algebras $\mathcal{L}(T_1)$ and $\mathcal{L}(T_2)$ such that any completion T_1^* of the theory T_1 and the corresponding completion $T_2^* = \mu(T_1^*)$ of the theory T_2 have the same description within the framework of the list L. In this case, we say that the *isomorphism μ preserves all the properties from the list L*.

0.6. Main Theorem

The problem of characterization of the expressive possibilities of finitely axiomatizable theories can be stated as the problem of description of abstract L-types $(\mathcal{B}, \nu, h)$ that are realized in finitely axiomatizable theories. In such a formulation the problem is difficult. Only conditions for the first two components are well known. Namely, the pair $(\mathcal{B}, \nu)$ can be an arbitrary positively enumerable Boolean algebra [25, 52, 46], but the conditions on the function h turn out to be complex. Furthermore, the conditions depend on the specific character of properties from the list L and their interdependence.

The problem is not simplified even in the case of axiomatizable theories. However, there is an indirect approach to the above problem. Instead of studying an initial problem, we compare the possibilities for abstract L-types realized in axiomatizable theories and in finitely axiomatizable theories as well. Then a global problem is to find a list L of model-theoretic properties that is as large as possible and the possibilities that these classes coincide. This approach is applied to the problem on expressive possibilities of finitely axiomatizable theories. The result is formulated as the following general theorem.

Theorem 0.6.1. *Let T be a recursively axiomatizable theory without finite models and let σ be a finite rich signature. Then for a recursively enumerable index of T it is possible to construct effectively a finitely axiomatizable model-complete theory $F = \mathbb{F}_\sigma(T)$ of the signature σ and a recursive isomorphism $\mu : \mathcal{L}(T) \to \mathcal{L}(F)$ between the Lindenbaum algebras $\mathcal{L}(T)$ and $\mathcal{L}(F)$ such that any completion T^* of the theory T and the corresponding completion $F^* = \mu(T^*)$ of the theory F have the same description within the framework of the following list of model-theoretic properties:*

(a) *stability, superstability, ω-stability, stability in cardinality α,*

(b) *the existence of a prime model and the quantity of its algorithmic dimension (with respect to strong constructivizations); the number of atomic models of cardinality $\alpha > \omega$,*

(c) *the number of countable minimal models (Jónsson models) and the quantities of their algorithmic dimensions; the number of minimal models of cardinality $\alpha > \omega$,*

(d) *the existence of a countable strongly constructivizable homogeneous model; the existence of a countable strongly constructivizable α^+-homogeneous model; the existence of an α^+-homogeneous model of cardinality $\alpha \geqslant \omega$,*

(e) *the existence and strong constructivizability of a countable saturated model; the existence of a saturated model of cardinality $\alpha > \omega$,*

(f) *the existence of a model with first-order definable elements and its strong constructivizability; the existence a model with almost first-order definable (algebraic) elements and the quantity of its algorithmic dimension,*

(g) *the number of countable rigid models (with a unique trivial automorphism) and the quantities of their algorithmic dimensions; the number of rigid models of cardinality $\alpha > \omega$,*

(h) *nonmaximality of the spectrum function.*

For the sake of brevity, we write $\mathbb{F}(T)$ instead of $\mathbb{F}_\sigma(T)$ if the choice of a signature σ is unimportant or is clear from the context. The list of model-theoretic properties mentioned in Theorem 0.6.1 will be called *universal* and will be denoted by MQL or the MQL-list. Theorem 0.6.1 is proved in Chapters 4–6.

From Theorem 0.6.1 we obtain the following general claim.

Theorem 0.6.2. *Let the existence of a theory with infinite models and given properties formulated within the framework of the MQL-list be proved in the class of axiomatizable theories. Then there exists a finitely axiomatizable theory with the same properties.*

Theorem 0.6.1 is an assertion of the most general character and has different applications. However, its simplified versions (direct corollaries) are often more convenient for concrete purposes. As an example, we present a variant of Theorem 0.6.1 under the completeness condition.

Theorem 0.6.3. *Let T be a complete decidable theory without finite models. Then there exists a complete, model-complete, and finitely axiomatizable theory F of a finite rich signature σ such that the theories T and F have the same description within the framework of properties from the MQL-list.*

REMARK 0.6.1. Theorem 0.6.1 is applicable to the case in which T has finite models. Starting from the theory T, it is possible to construct a finitely axiomatizable theory $F = \mathbb{F}_\sigma(T)$ and a recursive isomorphism μ. In addition, the MQL-properties of each complete extension T^* of the theory T with infinite models can be transferred to the corresponding completion F^* of the theory F. If T is a theory of a finite model, then the corresponding theory F^* has only infinite models and, within the framework of the MQL-list, has the same properties as the theory $T^* \oplus SI$, where SI is the ω_1-categorical successor theory, defined above.

REMARK 0.6.2. Theorem 0.6.1 remains valid in the case of weak recursively enumerable indices of a theory T. In this case, it is possible to make the construction from Theorem 0.6.1 totally effective. Taking the maximally large enumerable signature (with an infinite number of symbols of each number of places) and using weak recursively enumerable indices, we can define a common numbering for all possible axiomatizable theories of any enumerable signatures. Then the main construction $T \mapsto \mathbb{F}_\sigma(T)$ becomes an effective operator acting from the class of all axiomatizable theories into the class of finitely axiomatizable theories of a given finite rich signature.

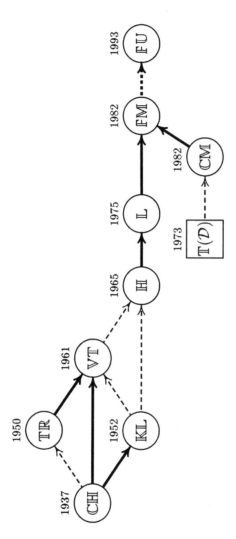

Fig. 0.7.1. Constructions of finitely axiomatizable theories:

direct reinforcement of a result

succession and ideological progress

reinforcement with loss of details

axiomatizable construction

0.7. Constructions of Finitely Axiomatizable Theories

In this section, we consider a number of constructions of finitely axiomatizable theories and a construction of axiomatizable theories. Each construction represents a general method of constructing theories of some class and provides a construction of a series of theories depending on one or several parameters. We can manage properties of the obtained theory by the choice of parameters. For parameters we can take natural numbers, axiomatizable theories, and recursively enumerable trees. The strength of the construction is defined by the collection of controllable properties and the simplicity of the dependence of the properties of the obtained theory on the choice of parameters.

The general scheme (cf. Fig. 0.7.1) demonstrates connections between constructions and the dates of their creations. Arrows are directed toward strong constructions.

We note that $\mathbb{F}\mathbb{U}$ is the full name of the universal construction (cf. Theorem 0.6.1). It is also denoted by $\mathbb{F}(T)$ in concrete applications. Every construction allows us to construct a finitely axiomatizable theory of any given finite rich signature σ.

We note that the construction $\mathbb{T}(\mathcal{D})$ yields axiomatizable theories of a signature with infinite number of unary predicates. It is included in the scheme for a complete exposition. We will denote by K a creative set.

We begin with the *Church construction* [5].

Theorem 0.7.1. *For a natural number n, it is possible to construct effectively a finitely axiomatizable theory $F = \mathbb{C}\mathbb{H}(n)$ such that*

(a) *for $n \in K$, the theory F is inconsistent,*

(b) *for $n \notin K$, the theory F is consistent.*

The *Trakhtenbrot construction* [62] establishes a similar relation for finite models.

Theorem 0.7.2. *For a natural number n, it is possible to construct effectively a finitely axiomatizable theory $F = \mathbb{T}\mathbb{R}(n)$ such that*

(a) *for $n \in K$, the theory F has a finite model,*

(b) *for $n \notin K$, the theory F has no finite models.*

The *Vaught construction* [63] generalizes the Church construction and the Trakhtenbrot one.

Theorem 0.7.3. *For a natural number* n, *it is possible to construct effectively a finitely axiomatizable theory* $F = \mathbb{V}\mathbb{T}(n)$ *and a special halt-sentence* Φ *such that*

(a) *for* $n \in K$, *the relation* $\vdash (F \to \Phi)$ *holds; moreover, the theory* F *has a model* $\mathfrak{M}$ *which is finite and unique up to an isomorphism,*

(b) *for* $n \notin K$, *the relation* $\neg \vdash (F \to \Phi)$ *holds; moreover, the theory* F *is hereditarily and essentially undecidable and has no recursively enumerable models.*

A creative set can be thought of as the halt problem for a universal Turing machine. Then the halt-sentence gives signals about the halt. The input of properties of the halt-sentence to the statement of the theorem turns out to be useful for applications. For example, the Vaught result implies the Church theorem and the Trakhtenbrot theorem. To justify this assertion, it suffices to set

$$\mathbb{C}\mathbb{H}(n) = \neg(\mathbb{V}\mathbb{T}(n) \,\&\, \neg\Phi), \quad \mathbb{T}\mathbb{R}(n) = \mathbb{V}\mathbb{T}(n) \,\&\, \Phi$$

The Kleene [30] result has a significant place among the pioneering constructions. It demonstrates the expressive possibilities of finitely axiomatizable theories.

Theorem 0.7.4. *Let* T *be an axiomatizable theory of a finite signature* σ *without finite models. Then for a recursively enumerable index of* T *it is possible to construct effectively a finitely axiomatizable theory* $F = \mathbb{K}\mathbb{L}(T)$ *of the signature* $\sigma' = \sigma \cup \{R^2\}$ *such that* $T = F \upharpoonright \sigma$.

Theorem 0.7.4 [a variant]. *Let* T *be an axiomatizable theory. Then for a recursively enumerable index of* T *it is possible to construct effectively a finitely axiomatizable theory* $F = \mathbb{K}\mathbb{L}^*(T)$ *and an interpretation* I *of the theory* T *in the theory* F.

In the Kleene construction, the binary predicate R is used to define a finitely axiomatizable fragment of the Peano arithmetic, in which the system of axioms of the theory T is enumerated. The Gödel coding of formulas and a special mechanism of determining the truth of formulas in T are also applied. We note that $\mathbb{K}\mathbb{L}$ is the first construction with an arbitrary axiomatizable theory taken as a parameter.

We now proceed to the *Hanf construction*. We fix a finitely axiomatizable decidable theory H with the atomless Lindenbaum algebra $\mathcal{L}(H)$.

As such a theory, we can take the standard successor theory with the signature $\{\lhd^2, c\}$ containing a unique initial element c (cycles are admissible).

Theorem 0.7.5 [24]. *For any recursively axiomatizable theory T, there exists a finitely axiomatizable theory $F = \mathbb{H}(T)$ such that the Lindenbaum algebras $\mathcal{L}(F)$ and $\mathcal{L}(T \oplus H)$ of the theories F and $T \oplus H$ are recursively isomorphic: $\mathcal{L}(F) \cong \mathcal{L}(T \oplus H)$.*

The following facts are direct consequences of Theorem 0.7.5:

(a) the existence of an undecidable finitely axiomatizable theory whose undecidability degree is strictly less than the creative degree;

(b) the existence of an undecidable finitely axiomatizable theory such that it is neither hereditarily undecidable nor essentially undecidable;

(c) the existence of a decidable finitely axiomatizable theory such that the corresponding decision algorithm cannot be described by a primitive-recursive function in the standard Gödel numbering of formulas.

We complete the characterization of constructions.

Owing to Theorem 0.7.5, the natural question arises whether a similar assertion remains valid without the term H.

The Hanf Problem. *Is it true that for any recursively axiomatizable theory T there is a finitely axiomatizable theory F such that there exists a recursive isomorphism between the Lindenbaum algebras $\mathcal{L}(F)$ and $\mathcal{L}(T)$?*

The second key problem is presented by the well-known Vaught–Morley problem stated in the early 1960s.

The Vaught–Morley Problem. *Does a complete finitely axiomatizable uncountably categorical theory exist?*

The Hanf problem and the Vaught–Morley problem were included in the well-known list of 102 problems by Friedman [15]. These key problems had a beneficial effect on solving a general problem on the expression of finitely axiomatizable theories.

To explain the significance of the subject, we introduce the notion of a basic theory for a construction. By a *basic theory* we mean the simplest theory on the basis of which we construct the framework for functioning of the computational mechanism (as a Turing machine) which provides the required properties of the finitely axiomatizable theory. In this sense, we would say that the constructions by Church, Trakhtrenbrot, Vaught, and

Hanf are based on a finitely axiomatizable fragment of the ordinary successor theory and its variants. In the constructions listed, cyclic formations of the basic theory lead to cyclic formations in models of the obtained theory, which prevents one from obtaining stronger results. The advances by Hanf [25] became possible because for a basic theory the complete finitely axiomatizable superstable theory constructed by Makowsky [33] in 1974 was used.

By the early 1980s, the author constructed an example of an uncountably categorical, complete, and finitely axiomatizable theory. The use of this theory allows one to force constructions and, as a result, obtain an exhaustive answer to the general question concerning the expressive possibilities of finitely axiomatizable theories.

Chapter 1

Quasisuccessor Theory of Rank 2

In this chapter, we describe the complete, uncountably categorical, and finitely axiomatizable theory which is denoted by QS and is called the *quasisuccessor theory* because it looks like an ordinary infinitely axiomatizable successor theory. We emphasize that the example of such a theory itself yields a solution of the Vaught–Morley problem (cf. Introduction). Owing to its properties, the quasisuccessor theory QS is an ideal basic theory for the construction of finitely axiomatizable theories, and it will be used as such in Chapter 3, below.

The theory QS presented corresponds to the description in [53]. This description is simpler than the first example [42] of the theory QS', which was concerned with the Vaught–Morley problem and was published in 1980. We note that the theory QS turns out to be simpler than the theory QS' because the Morley rank of QS is less than that of QS'. Namely, for the theory QS the Morley rank of the formula $(x = x)$ is equal to 2; therefore, models of the theory QS are, in some sense, two-dimensional. Therefore, the theory QS can be referred to as the *quasisuccessor theory of rank* 2. Furthermore, the Morley rank of the theory QS, being equal to the upper bound of the ranks of formulas in a single variable, has the value 3. For comparison, models of the theory QS' are three-dimensional and the Morley rank of the theory QS' is equal to 4.

The quasisuccessor theory QS is an ordinary successor relation without endpoints and cycles, and classes by some equivalence play the role of elements. The internal combinatorial mechanisms of the theory, based on a finite number of predicates and described by a finite set of axioms, suppress any cycles but do not prevent isomorphisms between models from the point of view of the successor relation on the classes. It is precisely these properties that explain the choice of the theory QS for the basic theory in constructions of finitely axiomatizable theories.

We specify some notions. In this chapter, a binary relation R is called a *successor relation* (or a *succession relation*) if any element of the universe has exactly one R-successor and exactly one R-predecessor and the R-cycles of length 1 are absent, i.e., if $\neg R(x, x)$ for all x. The assertion that a predicate P is a successor relation on classes by some equivalence has the same meaning.

1.1. Axiomatics and the Simplest Properties

The signature σ of the theory QS has the form

$$\sigma = \{\lhd^2, \sim^2, M^2, H^2, \approx^2, Q^2, R^2, S^2, D^1, U^1, V^1, \varPi^2, \varSigma^2\}$$

where the superscripts indicate the numbers of places of predicates.

Axioms of the theory QS

B.1. $\sim$ is an equivalence relation, $\lhd$ is a successor relation on $\sim$-classes.

B.2. H is an equivalence relation, M is a successor relation on H-classes.

B.3. Every $\sim$-class intersects every H-class.

B.4. $x \approx y \leftrightarrow x \sim y \,\&\, H(x, y)$.

B.5. In every $\sim$-class, the predicate D distinguishes exactly one $\approx$-class.

B.6. In every H-class, the predicate D distinguishes exactly one $\approx$-class.

B.7. $x \sim y \,\&\, M(x, y) \,\&\, y \lhd z \,\&\, H(y, z) \to \left(D(x) \leftrightarrow D(z)\right)$.

B.8. Every $\approx$-class contains exactly two elements.

B.9. $Q(x, y) \leftrightarrow x \approx y \,\&\, x \neq y$.

B.10. R is a successor relation.

B.11. S is a successor relation.

B.12. $R(x, y) \to x \lhd y \;\&\; H(x, y)$.

B.13. $S(x, y) \to x \sim y \;\&\; M(x, y)$.

B.14. $Q(x, u) \;\&\; Q(y, v)$ implies $R(x, y) \leftrightarrow R(u, v)$.

B.15. $Q(x, u) \;\&\; Q(y, v)$ implies $S(x, y) \leftrightarrow S(u, v)$.

B.16. $R(x, u) \;\&\; R(y, v)$ implies $S(x, y) \leftrightarrow S(u, v)$.

B.17. $D(x) \leftrightarrow U(x) \vee V(x)$.

B.18. $\neg U(x) \vee \neg V(x)$.

B.19. $Q(x, y)$ implies $U(x) \leftrightarrow V(y)$.

B.20. $S(x, y) \;\&\; R(y, z)$ implies $U(x) \leftrightarrow U(z)$.

B.21. The relation $\Pi(x, y) \vee \Sigma(x, y)$ is equivalent to the conjunction of the relations $x \not\approx y$, $(\exists u)[D(u) \;\&\; u \sim x \;\&\; H(u, y)]$, $(\exists v)[D(v) \;\&\; v \sim y \;\&\; H(v, x)]$.

B.22. $\neg \Pi(x, y) \vee \neg \Sigma(x, y)$.

B.23. $\Pi(x, y) \leftrightarrow \Sigma(y, x)$.

B.24. $Q(x', x'')$ implies $\Pi(x', y) \leftrightarrow \Sigma(x'', y)$.

B.25. $S(x, x')$, $R(y, y')$, $x \not\approx y$, $x' \not\approx y'$ implies $\Pi(x, y) \leftrightarrow \Pi(x', y')$.

B.26. $D(x) \;\&\; S(x, y) \;\&\; R(x, z)$ implies $\Pi(y, z)$.

We give some simple consequences of the axioms of the theory QS. For the sake of convenience, we continue the enumeration of the axioms.

Axioms B.1–B.6 imply that the predicate $\approx$ is an equivalence relation and the predicate D is well defined on $\approx$-classes.

C.27. $x \approx y \to \big(D(x) \leftrightarrow D(y) \big)$.

In addition, Axioms B.17 and B.19 yield the following relation:

C.28. $D(x) \leftrightarrow (\exists y)[x \approx y \;\&\; U(y)]$.

Axioms B.1–B.7 describe the connection between the $\lhd$-successor and the M-successor by means of $\approx$-classes satisfying D. Using them, it is possible to prove the following relations:

C.29. $D(x) \;\&\; D(y) \to \big(x \sim y \leftrightarrow H(x, y) \big)$.

C.30. $D(x) \;\&\; D(y) \to \big(x \lhd y \leftrightarrow M(x, y) \big)$.

Axiom B.9 provides the symmetry of the predicate Q, which, together with Axiom B.8, defines Q as a successor relation with 2-cycles. Therefore, for any element the Q-successor coincides with the Q-predecessor. Moreover, from the axioms it is easy to obtain the following relation:

C.31. $x \approx y \leftrightarrow x = y \vee Q(x, y)$.

From B.23, for Π and Σ we obtain the following version of B.24 with respect to the second argument:

C.32. $Q(y', y'')$ implies $\Pi(x, y') \leftrightarrow \Sigma(x, y'')$.

Similarly, we can obtain versions of B.25 replacing Π by Σ and interchanging R and S.

1.2. The Impossibility of Cycles

In this chapter, the symbol $\vdash$ means the provability in the theory QS. Formal tree-like proofs are often applied. We begin with the main lemma on quasisuccession.

Lemma 1.2.1. $\lhd$-*Cycles are impossible in the theory* QS.

PROOF. Assume the contrary. In some model $\mathfrak{N}$ of the theory QS there is a $\lhd$-cycle of length $s \geqslant 2$:

$$[a_0]_\sim \lhd [a_1]_\sim \lhd \ldots \lhd [a_{s-1}]_\sim \lhd [a_0]_\sim, \quad a_i \not\sim a_j, \quad i < j < s \quad (1.2.1)$$

By B.5, we can assume that a_i, $i < s$, satisfy $D(x)$. By C.29 and C.30,

$$[a_0]_H M [a_1]_H M \ldots M [a_{s-1}]_H M [a_0]_H, \neg H(a_i, a_j), \quad i < j < s \quad (1.2.2)$$

Using C.28, we find an element b_0^0 satisfying the relations

$$b_0^0 \approx a_0, \quad U(b_0^0) \quad (1.2.3)$$

Using B.10–B.16 and starting from the element b_0^0, we construct a double "net" (with possible repetitions) of elements b_i^j, c_i^j, $i, j \in \mathbb{Z}$, such that

$$Q(b_i^j, c_i^j), R(b_i^j, b_{i+1}^j), R(c_i^j, c_{i+1}^j), S(b_i^j, b_i^{j+1}), S(c_i^j, c_i^{j+1}), i, j \in \mathbb{Z} \quad (1.2.4)$$

Taking (1.2.3) for the initial condition and using the cycles (1.2.1), (1.2.2), by B.12 and B.13, we obtain the inclusions

$$b_i^j \in [a_k]_\sim \text{ for } i \equiv k \,(\text{mod}\, s), \quad b_i^j \in [a_k]_H \text{ for } j \equiv k \,(\text{mod}\, s)$$

As a result, for all $i, j, k, l \in \mathbb{Z}$ we have

$$b_i^j \approx b_k^l \Leftrightarrow i \equiv k \,(\text{mod}\, s) \,\&\, j \equiv l \,(\text{mod}\, s) \quad (1.2.5)$$

We begin again from (1.2.3) and use Axiom B.20 repeatedly in order to obtain $U(b_i^i)$, $i \in \mathbb{Z}$. By B.17, we also obtain $D(b_i^i)$ for all i.

By (1.2.4), Axiom B.26 for $x = b_0^0$ yields

$$\Pi(b_0^1, b_1^0) \tag{1.2.6}$$

The same axiom for $x = b_1^1$ yields $\Pi(b_1^2, b_2^1)$. Applying B.25 several times, we find $\Pi(b_1^i, b_i^1)$, $i = 2, 3, \ldots, s$, in particular, $\Pi(b_1^s, b_s^1)$.

Using a formal tree-like proof, we find

$$31 \frac{b_0^0 \approx b_s^s \qquad \dfrac{\dfrac{U(b_s^s)}{U(b_0^0),\ \neg V(b_s^s)} 18}{b_0^0 = b_s^s \vee Q(b_0^0, b_s^0),\ \neg Q(b_0^0, b_s^s)} 19}{b_0^0 = b_s^s} \tag{1.2.7}$$

We consider b_0^0 and b_s^0. From (1.2.5), it follows that $b_0^0 \approx b_s^0$. By C.31,

$$b_0^0 = b_s^0 \vee Q(b_0^0, b_s^0) \tag{1.2.8}$$

In accordance with (1.2.8), the further proof is divided into two cases.

CASE 1: $b_0^0 = b_s^0$. By the above relations and the uniqueness of successors and predecessors for Q, R, and S, we have

$$(1.2.7) \frac{\dfrac{\dfrac{b_0^0 = b_s^0}{b_s^s = b_s^0 \quad b_0^0 = b_s^0}}{b_1^s = b_1^0,\ b_0^0 = b_s^1,\ \Pi(b_1^s, b_s^1)}}{\dfrac{\Pi(b_1^0, b_0^1)}{\Sigma(b_0^1, b_1^0)} 23}$$

CASE 2: $Q(b_0^0, b_s^0)$. Based on the same properties, we have

$$(1.2.7) \frac{\dfrac{\dfrac{Q(b_0^0, b_s^0),\ Q(c_s^0, b_s^0)}{b_0^0 = c_s^0} \quad Q(b_0^0, b_s^0),\ Q(b_0^0, c_0^0)}{b_s^s = c_s^0 \quad b_s^0 = c_0^0}}{b_1^s = c_1^0, b_s^0 = c_0^1, \Pi(b_1^s, b_s^1)}}{\dfrac{\dfrac{\Pi(c_1^0, c_0^1)}{\Pi(b_1^0, b_0^1)} 24, 32}{\Sigma(b_0^1, b_1^0)} 23}$$

Thus, in any case, the assumption (1.2.1) contradicts (1.2.6) in view of Axiom B.22. □

Lemma 1.2.2. *M-cycles are impossible in the theory QS.*

PROOF. Assume the contrary: in some model $\mathfrak{N}$ of the theory QS there is an M-cycle of length $s \geqslant 2$:

$$[a_0]_H \ M \ [a_1]_H \ M \ldots M \ [a_{s-1}]_H \ M \ [a_0]_H$$

By Axiom B.6, we can assume that the representatives a_i, $i < s$, are D-elements. Using C.30, we can construct the $\triangleleft$-cycle, but this is impossible in view of Lemma 1.2.1.   □

The previous lemmas imply the following lemma.

Lemma 1.2.3. *The theory QS has no finite models.*

1.3. Informal Description of Models

The consistency of the theory QS will be proved with the help of some explicit analytic description of models of the theory QS in Sec. 1.5 below. In this section, we informally describe models of the theory QS, giving geometric illustrations of characteristic details.

By Axioms B.1–B.7 and the fact that $\triangleleft$-cycles and M-cycles are absent, an arbitrary model $\mathfrak{N}$ of the theory QS is a plane (two-dimensional) structure represented in the form of an orthogonal covering of the M-successor by the $\triangleleft$-successor (cf. Fig. 1.3.1). In addition, every $\sim$-class intersects every H-class, and any such intersection consists of exactly two elements. The predicate D distinguishes the *diagonal* of the model, which consists of U-elements and V-elements. By C.29 and C.30, the diagonal establishes a one-to-one correspondence between the set of $\sim$-classes and the set of H-classes. This correspondence is, in some sense, an "isomorphism" between $\triangleleft$-successors and M-successors. Therefore, in any model $\mathfrak{N}$ of the theory QS, the number of $\triangleleft$-chains coincides with the number of M-chains. This number is called the *length* of the model $\mathfrak{N}$ and is denoted by $\mathrm{Len}(\mathfrak{N})$. By a *block* we mean the intersection of a $\triangleleft$-chain and an M-chain. The aforesaid implies that a model of length k consists of k^2 blocks, exactly k blocks of which are intersected by the diagonal.

The minimal nonempty set of elements of a model is called the *net* if it is closed under R-connections and S-connections. By Axiom B.16, every net has a regular structure with simple square cells. By Axioms B.12 and B.13 and the fact that $\triangleleft$-cycles and M-cycles are absent, all nets have strictly plane form without cycles. If σ is a net, then, by Axioms

B.14 and B.15, the set σ' of elements that are Q-connected with σ is also a net. Since Q-connections are symmetric, σ and σ' must be considered simultaneously. It is natural to call them a pair of nets. As shown in Fig. 1.3.1, every pair of nets is a block. There are two types of pairs of nets, depending on the presence of the truth points of unary predicates. In diagonal blocks, pairs of nets have type U–V, whereas the remaining pairs of nets have type $\varnothing$–$\varnothing$.

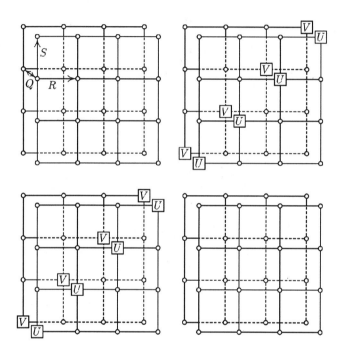

Fig. 1.3.1. Structure of a quasisuccessor model:

$\langle \lhd, \sim \rangle$-successor is directed from left to right, $\langle M, H \rangle$-successor is directed from bottom to top, $\sim$-classes are represented by pairs of vertical lines, H-classes are represented by pairs of horizontal lines.

Two classes $[a]_{\approx}$ and $[b]_{\approx}$ of a model $\mathfrak{N}$ are called *mirror* if, on the sequence (a, b), the formula $\zeta(x, y) = \Pi(x, y) \vee \Sigma(x, y)$ is true. This formula is equivalent to the right-hand side (denoted by $\zeta'(x, y)$) of Axiom B.21. In view of the special form of the formula $\zeta'(x, y)$, the mirror relation is symmetric, antireflexive, and for every class $[a]_{\approx}$ that does not lie on the diagonal there is exactly one class $[b]_\approx$ such that $[a]_\approx$ and $[b]_\approx$ are mirror classes. Moreover, mirror classes are symmetrically located with respect to the diagonal. We also note that the mirror relation between $\approx$-classes induces the mirror relation between blocks. A block is mirror to itself if and only if it is diagonal.

We discuss the possibilities of prescribing the Π and Σ in the model shown schematically in Fig. 1.3.1. At first, we consider the classes $[a]_\approx$ and $[b]_\approx$ that belong to different blocks and are mirror ones. It suffices to define the value of one of the predicates Π and Σ on the pair (a, b). By Axioms B.21–B.25, both predicates can be uniquely extended to all other points of blocks generated by a and b. In the case of a diagonal block, Axiom B.26 gives the initial value of the predicate Π in a neighborhood of the diagonal. Therefore, both predicates Π and Σ are uniquely defined. Having given values of the predicates Π and Σ on all pairs of mirror blocks, we obtain their values on the whole model because the predicates Π and Σ are identically false on other pairs of blocks in view of Axiom B.21. The key combinatorial idea of the theory QS lies in the predicates Π and Σ and the corresponding Axioms B.21–B.26. These predicates present a special mechanism of the suppression of cycles. It appears in the proof of Lemma 1.2.1. The mechanism is based on a specific asymmetric distribution of the above predicates on elements of mirror $\approx$-classes. However, the asymmetry of the action of these predicates is restricted by limits of each separate block, and it disappears in connections between pairs of nets of the type $\varnothing$–$\varnothing$, which allows us to construct isomorphisms between models of the theory QS having equal length.

1.4. Coordinates

A model as a whole is a two-dimensional structure. Therefore, the location of each $\approx$-class is determined by two coordinates, one of which is the $\sim$-class of the $\lhd$-successor and the other of which is characterized by the H-class, which is also reduced to the $\sim$-class of the $\lhd$-successor by transformation of the diagonal D. The two coordinates above characterize elements of the model up to an inclusion of these coordinates in $\approx$-classes. To give a complete characterization, it is necessary to distinguish two elements within an $\approx$-class. To this end, we introduce an additional third coordinate having two values. This coordinate will be called the sign coordinate.

We pass to a strict definition of coordinates. Let $\mathfrak{N}$ be an arbitrary model of the quasisuccessor theory. We consider the quotient model $\mathfrak{N}/_\sim$ of the signature $\sigma' = \{\triangleleft, \sim\}$, which, by Lemma 1.2.1, is an ordinary successor relation without endpoints and cycles.

We fix a discrete linear order $<$ on the quotient model $\mathfrak{N}/_\sim$, which extends the $\triangleleft$-successor. It suffices to put in order all $\triangleleft$-chains of the model $\mathfrak{N}/_\sim$ and, on each of these chains, to introduce an order in accordance with the successor relation. Let σa denote the net generated by a in the model $\mathfrak{N}$ and let $\sigma \mathfrak{N}$ denote the set of all nets of $\mathfrak{N}$ consisting of nondiagonal blocks. Using $<$, we define the mappings

$$f \colon \sigma \mathfrak{N} \to \{-1,\, 1\} \qquad (1.4.1)$$

such that

(a) $f(\sigma a) = f(\sigma b)$ if $[a]_\sim < [b]_\sim$ and $\Pi(a, b)$,

(b) $f(\sigma a) \neq f(\sigma b)$ if $Q(a, b)$.

Having the pair of elements (a, b) of mirror classes from different blocks such that $[a]_\sim < [b]_\sim$ and $\Pi(a, b)$, we can set $f(\sigma a) = f(\sigma b) = k$ for $k \in \{-1,\, 1\}$ in accordance with (a). By (b), f is uniquely defined on the nets that are Q-connected with σa and σb. Thus, to define f we have λ possibilities of the choice, where λ is the number of unordered pairs of mirror nondiagonal blocks of the model considered.

For an arbitrary element a of the model $\mathfrak{N}$, we introduce the coordinates as follows:

$r_1(a) = [a]_\sim$,
$r_2(a) = [x]_\sim$ if $H(a, x)$ & $D(x)$,
$s(a) = 1$ if $\sigma a \cap U \neq \varnothing$ or $[\sigma a \cap (U \cup V) = \varnothing$ & $f(\sigma a) = 1]$,
$s(a) = -1$ if $\sigma a \cap V \neq \varnothing$ or $[\sigma a \cap (U \cup V) = \varnothing$ & $f(\sigma a) = -1]$.

It is easy to verify that the values of the coordinate $r_2(a)$ are independent of the choice of the representative of x. Hence the coordinates are well defined. We note that, in the definition of the sign coordinate, the function (1.4.1) acts as the choice function only on pairs of nets of type $\varnothing$–$\varnothing$. In diagonal blocks, the sign coordinate is defined from the values of the unary predicates U and V.

It is possible to show that for an arbitrary definition of the sign coordinate in the model $\mathfrak{N}$ of the quasisuccessor theory, the following properties hold:

$$Q(x, y) \Rightarrow s(x) \neq s(y) \qquad (1.4.2)$$
$$R(x, y) \Rightarrow s(x) = s(y) \qquad (1.4.3)$$

$$S(x, y) \Rightarrow s(x) = s(y) \tag{1.4.4}$$

$$U(x) \Rightarrow s(x) = 1 \tag{1.4.5}$$

$$V(x) \Rightarrow s(x) = -1 \tag{1.4.6}$$

$$\text{if } \Pi(x, y), \text{ then } s(x) = s(y) \Leftrightarrow [x]_\sim < [y]_\sim \tag{1.4.7}$$

$$\text{if } \Sigma(x, y), \text{ then } s(x) = s(y) \Leftrightarrow [y]_\sim < [x]_\sim \tag{1.4.8}$$

We show the complete representation of the universe of an arbitrary model of the quasisuccessor theory with respect to the coordinates introduced.

Lemma 1.4.1. *Let the sign coordinate be defined in a model $\mathfrak{N}$ of the quasisuccessor theory. For arbitrary elements x and y of the model $\mathfrak{N}$, the following relation holds:*

$$x = y \Leftrightarrow r_1(x) = r_1(y) \ \& \ r_2(x) = r_2(y) \ \& \ s(x) = s(y)$$

PROOF. The implication $\Rightarrow$ is obvious. By (1.4.2), the implication $\Leftarrow$ is a consequence of the following auxiliary relation:

$$r_1(x) = r_1(y), \ r_2(x) = r_2(y) \ \vdash \ x = y \lor Q(x, y)$$

We formalize the last relation and prove it:

$$x \sim y, \ H(x, u) \ \& \ D(u), \ H(y, v) \ \& \ D(v), \ u \sim v \ \vdash \ x = y \lor Q(x, y)$$

$$31 \, \cfrac{2 \, \cfrac{H(x, u), \ H(y, v), \ \cfrac{u \sim v, \ D(u), \ D(v)}{H(u, v)} \, 29}{\cfrac{\cfrac{H(x, y), \ x \sim y}{x \approx y} \, 4}{}}}{x = y \lor Q(x, y)}$$

$\square$

Lemma 1.4.2. *Let the sign coordinate be defined in a model $\mathfrak{N}$ of the quasisuccessor theory. For any elements $a_1, a_2 \in |\mathfrak{N}|$ and number $t \in \{-1, 1\}$ there exists an element $x \in |\mathfrak{N}|$ with the coordinates*

$$r_1(x) = [a_1]_\sim, r_2(x) = [a_2]_\sim, s(x) = t$$

PROOF. We find successively the auxiliary elements u, v, x', x'' such that $a_2 \sim u \ \& \ D(u)$, $a_1 \sim v \ \& \ H(u, v)$, $x' \approx v$, $x'' \approx v$, $x' \neq x''$. The existence of such elements is guaranteed by Axioms B.5, B.3, and B.8. Denote by x any of the elements x', x''. The following chain of formal deductions shows that one of the two elements x' and x'' is the required one:

$$4\frac{x \approx v}{\genfrac{}{}{0pt}{}{}{1\frac{x \sim v,\, a_1 \sim v}{\genfrac{}{}{0pt}{}{}{\frac{x \sim a_1}{r_1(x') = r_1(x'') = [a_1]_\sim}}}}}$$

$$4\frac{x \approx v}{\genfrac{}{}{0pt}{}{}{2\frac{H(x,v),\, H(u,v)}{\frac{H(x,u),\, D(u),\, a_2 \sim u}{r_2(x') = r_2(x'') = [a_2]_\sim}}}}$$

$$9\frac{\dfrac{4\dfrac{x' \approx v,\ x'' \approx v}{x' \neq x'',\, x' \approx x''}}{Q(x',x'')}}{\dfrac{s(x') \neq s(x'')}{s(x') = t \ \lor\ s(x'') = t}}\quad(1.4.2)$$

$\square$

Corollary 1.4.1. *The cardinality of a model $\mathfrak{N}$ of the quasi-successor theory is equal to the cardinality of the quotient model $\mathfrak{N}/\sim$.*

PROOF. The assertion follows from Lemmas 1.4.1 and 1.4.2 because a model of the quasisuccessor theory is infinite. $\square$

1.5. Connection Between Predicates and Coordinates

The introduced coordinates agree well with the language of the first-order theory QS. The following lemma describes some direct "analytic" representations of all predicates of this language in terms of coordinates.

Lemma 1.5.1. *For any elements x and y of a model $\mathfrak{N}$ of the theory QS with some sign coordinate, the following relations hold:*

(a) $x \lhd y \Leftrightarrow r_1(x) \lhd r_1(y)$,

(b) $x \sim y \Leftrightarrow r_1(x) = r_1(y)$,

(c) $M(x,y) \Leftrightarrow r_2(x) \lhd r_2(y)$,

(d) $H(x,y) \Leftrightarrow r_2(x) = r_2(y)$,

(e) $x \approx y \Leftrightarrow r_1(x) = r_1(y) \ \& \ r_2(x) = r_2(y)$,

(f) $Q(x, y) \Leftrightarrow r_1(x) = r_1(y) \ \& \ r_2(x) = r_2(y) \ \& \ s(x) \neq s(y)$,

(g) $R(x, y) \Leftrightarrow r_1(x) \lhd r_1(y) \ \& \ r_2(x) = r_2(y) \ \& \ s(x) = s(y)$,

(h) $S(x, y) \Leftrightarrow r_1(x) = r_1(y) \ \& \ r_2(x) \lhd r_2(y) \ \& \ s(x) = s(y)$,

(i) $D(x) \Leftrightarrow r_1(x) = r_2(x)$,

(j) $U(x) \Leftrightarrow r_1(x) = r_2(x) \ \& \ s(x) = 1$,

(k) $V(x) \Leftrightarrow r_1(x) = r_2(x) \ \& \ s(x) = -1$,

(l) $\Pi(x, y) \Leftrightarrow r_1(x) \neq r_1(y) \ \& \ r_1(x) = r_2(y) \ \& \ r_2(x) = r_1(y)$
$\& \ \big(s(x) = s(y) \Leftrightarrow r_1(x) < r_1(y) \big)$,

(m) $\Sigma(x, y) \Leftrightarrow r_1(x) \neq r_1(y) \ \& \ r_1(x) = r_2(y) \ \& \ r_2(x) = r_1(y)$
$\& \ \big(s(x) = s(y) \Leftrightarrow r_1(y) < r_1(x) \big)$.

PROOF. Relations (a) and (b) follow from the definition of the first coordinate, and relations (c) and (d) follow from Axioms B.1–B.6 and relations C.29 and C.30. Relation (e) is obtained from (b), (d), and Axiom B.4. Relation (f) is a consequence of (e), Axiom B.9, and (1.4.2). The implication $\Rightarrow$ in (g) is valid by Axiom B.12 and relations (a), (d), and (1.4.3). The implication $\Leftarrow$ in (g) can be obtained by using the implication $\Rightarrow$ in (g) and the uniqueness of an R-successor and Lemma 1.4.1. The implication $\Rightarrow$ in (h) follows from Axiom B.13, relations (b) and (c), and (1.4.4). The inverse implication in (h) holds in view of the implication $\Rightarrow$ in (h), the uniqueness of an S-successor, and Lemma 1.4.1.

Let us formalize relation (i) and prove two parts:

$$D(x), \ H(x, y) \ \& \ D(y) \ \vdash \ x \sim y \qquad\qquad (\Rightarrow)$$
$$H(x, y) \ \& \ D(y), \ x \sim y \ \vdash \ D(x) \qquad\qquad (\Leftarrow)$$

$$(\Rightarrow) \quad {}_6\dfrac{D(x), \ D(y), \ H(x, y)}{{}_4\dfrac{x \approx y}{x \sim y}} \qquad (\Leftarrow) \quad {}_4\dfrac{H(x, y), \ x \sim y}{{}_{27}\dfrac{x \approx y, \ D(y)}{D(x)}}$$

By (1.4.5) and (1.4.6), to prove (j) and (k) it suffices to establish the joined relation $U(x) \vee V(x) \Leftrightarrow r_1(x) = r_2(x)$ which is easily derived from Axiom

B.18 and relation (i). To prove the remaining relations (l) and (m), we note that from Axiom B.22 and relations (a)–(k) we can obtain the relation

$$\Pi(x, y) \vee \Sigma(x, y) \Leftrightarrow r_1(x) \neq r_1(y) \ \& \ r_2(x) \neq r_2(y) \ \&$$
$$r_1(x) = r_2(y) \ \& \ r_2(x) = r_1(y)$$

With the help of (1.4.7) and (1.4.8), it is not hard to divide this relation into the parts corresponding to (l) and (m).   □

1.6. Main Properties of the Quasisuccessor Theory

We pass to the investigation of the most important properties of the quasisuccessor theory.

Lemma 1.6.1. *Let $\mathfrak{M}$ be an arbitrary model of the ordinary successor theory without endpoints and cycles of the signature $\sigma = \{\lhd^2\}$. There exists a model $\mathfrak{N}$ of the theory QS such that $\mathfrak{M}$ is isomorphic to the quotient model $\mathfrak{N}/\sim$ in the signature σ.*

PROOF. In the model $\mathfrak{M}$, the successor predicate will be written in the form $y = x + 1$. Based on the relations in Lemma 1.5.1, we suggest the following method of defining $\mathfrak{N}$. In the model $\mathfrak{M}$, we fix a discrete linear order $<$ such that $\lhd$ is the successor relation. Starting from $\mathfrak{M}$ and $<$, we construct the model $\mathfrak{N}$ of the theory QS as follows. We set $|\mathfrak{N}| = |\mathfrak{M}|^2 \times \{-1, 1\}$. For arbitrary elements $x = \langle \alpha_1, \alpha_2, \sigma \rangle$ and $y = \langle \beta_1, \beta_2, \tau \rangle$ of $|\mathfrak{N}|$, we set

$$x \lhd y \Leftrightarrow \beta_1 = \alpha_1 + 1$$
$$x \sim y \Leftrightarrow \beta_1 = \alpha_1$$
$$M(x, y) \Leftrightarrow \beta_2 = \alpha_2 + 1$$
$$H(x, y) \Leftrightarrow \beta_2 = \alpha_2$$
$$x \approx y \Leftrightarrow \beta_1 = \alpha_1 \ \& \ \beta_2 = \alpha_2$$
$$Q(x, y) \Leftrightarrow \beta_1 = \alpha_1 \ \& \ \beta_2 = \alpha_2 \ \& \ \tau \neq \sigma$$
$$R(x, y) \Leftrightarrow \beta_1 = \alpha_1 + 1 \ \& \ \beta_2 = \alpha_2 \ \& \ \tau = \sigma$$
$$S(x, y) \Leftrightarrow \beta_1 = \alpha_1 \ \& \ \beta_2 = \alpha_2 + 1 \ \& \ \tau = \sigma$$
$$D(x) \Leftrightarrow \alpha_1 = \alpha_2$$
$$U(x) \Leftrightarrow \alpha_1 = \alpha_2 \ \& \ \sigma = 1$$
$$V(x) \Leftrightarrow \alpha_1 = \alpha_2 \ \& \ \sigma = -1$$

$$\Pi(x, y) \Leftrightarrow \beta_1 \neq \alpha_1 \ \& \ \beta_1 = \alpha_2 \ \& \ \beta_2 = \alpha_1 \ \&$$
$$(\tau = \sigma \Leftrightarrow \alpha_1 < \beta_1)$$
$$\Sigma(x, y) \Leftrightarrow \beta_1 \neq \alpha_1 \ \& \ \beta_1 = \alpha_2 \ \& \ \beta_2 = \alpha_1 \ \&$$
$$(\tau = \sigma \Leftrightarrow \beta_1 < \alpha_1)$$

A routine verification shows that all the axioms of the theory QS hold in $\mathfrak{N}$. It is not hard to show that the natural mapping $\langle \alpha_1, \alpha_2, \sigma \rangle \mapsto \alpha_1$ defines a $\lhd$-isomorphism between $\mathfrak{N}/\!\!\sim$ and $\mathfrak{M}$. □

Lemma 1.6.2. *Let $\mathfrak{N}'$ and $\mathfrak{N}''$ be models of the theory QS with the same number of $\lhd$-chains. Then an arbitrary $\lhd$-isomorphism $\mu \colon \mathfrak{N}'/\!\!\sim \ \to \mathfrak{N}''/\!\!\sim$ between the quotient models can be extended to the σ_{QS}-isomorphism $\mu^* \colon \mathfrak{N}' \to \mathfrak{N}''$. For a given μ, the number of such extensions μ^* is equal to the number of methods of defining the sign coordinate in the model $\mathfrak{N}''$ (for fixed $<''$ in $\mathfrak{N}''/\!\!\sim$).*

PROOF. Let $\mu \colon \mathfrak{N}'/\!\!\sim \ \to \mathfrak{N}''/\!\!\sim$ be a given isomorphism between quotient models. We fix isomorphic discrete linear orders $<'$ and $<''$ that extend the corresponding successor relations in the quotient models $\mathfrak{N}'/\!\!\sim$ and $\mathfrak{N}''/\!\!\sim$ and fix a sign coordinate s' in the model $\mathfrak{N}'$. We define arbitrarily the sign coordinate s'' in the model $\mathfrak{N}''$ and construct the natural mapping $\mu^* \colon |\mathfrak{N}'| \to |\mathfrak{N}''|$ in accordance with the following rule: for $a \in |\mathfrak{N}'|$ and $b \in |\mathfrak{N}''|$, we set $\mu^*(a) = b \Leftrightarrow \underset{i=1}{\overset{2}{\&}} \big(r_i(b) = \mu(r_i(a)) \big) \ \& \ s''(b) = s'(a)$. By Lemmas 1.4.1 and 1.4.2, μ^* is a bijection. By Lemma 1.5.1, all predicates of the theory QS can be expressed in terms of the coordinates of its arguments; moreover, for every predicate these expressions are the same in $\mathfrak{N}'$ and $\mathfrak{N}''$. Consequently, μ^* is an isomorphism. It is easy to verify that $s'' \mapsto \mu^*$ is a bijection between all methods of defining the sign coordinate s'' in the model $\mathfrak{N}''$ and all isomorphisms μ^* that coinside with μ after passage to quotient models. □

Theorem 1.6.1. *The theory QS is consistent, ω_1-categorical, complete, and model-complete.*

PROOF. The consistency of the theory QS follows from Lemma 1.6.1. We prove that the theory QS is ω_1-categorical. Let $\mathfrak{N}'$ and $\mathfrak{N}''$ be models of the theory QS of cardinality $\alpha \geqslant \omega_1$. By Corollary 1.4.1, the quotient models $\mathfrak{N}'/\!\!\sim$ and $\mathfrak{N}''/\!\!\sim$ have cardinalities α and, consequently, are isomorphic with respect to the signature $\sigma' = \{\lhd, \sim\}$. By Lemma 1.6.2, the models $\mathfrak{N}'$ and $\mathfrak{N}''$ are isomorphic. The completeness of the theory QS follows from the Vaught theorem, since QS is ω_1-categorical and finite models are absent. By the Lindström theorem, the theory QS is model-complete because it is $\forall\exists$-axiomatizable and ω_1-categorical.   □

Theorem 1.6.2. *The theory QS is almost strongly minimal, and the Morley rank of the theory QS is equal to 3. The formula $U(x)$ is strongly minimal and $|\mathfrak{N}| = \text{acl}\big(U(\mathfrak{N})\big)$ for any model $\mathfrak{N} \in \text{Mod}(QS)$.*

PROOF. To prove the strong minimality of the formula $U(x)$, we assume the contrary. There is a model $\mathfrak{N} \in \text{Mod}(QS)$ and a formula $\varphi(x, c_0, \ldots, c_{n-1})$ with constants in $\mathfrak{N}$ which divides $U(\mathfrak{N})$ into two infinite parts. In the signature, we introduce the symbols of the constants c_i, $i < n$, and consider the following collection of formulas in variables x and y:

$$\Sigma^*(x, y) = \{U(x),\, \varphi(x, c_0, \ldots, c_{n-1})\} \cup \Sigma'(x) \cup$$
$$\{U(y),\, \neg\varphi(y, c_0, \ldots, c_{n-1})\} \cup \Sigma''(y)$$

where $\Sigma'(x)$ means that, in the $\lhd$-successor, the class $[x]_\sim$ is at a distance of more than k from the values of the coordinates $r_i(c_j)$ for all $i \in \{1, 2\}$, $j < n$, $k \in \mathbb{N}$; and $\Sigma''(y)$ means the same for y. This assumption guarantees the local consistency of the set $\Sigma^*(x, y)$. Consider a model $\mathfrak{N}^*$ of the theory QS with the constants $c_0, \ldots, c_{n-1}$ realizing the set $\Sigma^*(x, y)$ (e.g., for $x = a$, $y = b$, $a, b \in |\mathfrak{N}^*|$). By Lemma 1.6.2 and the relations $r_i(x) = [x]_\sim$ and $s(x) = 1$, under the condition that x is a U-element, it is possible to construct an automorphism μ of the model $\mathfrak{N}^*$ such that $\mu(c_i) = c_i$ for all $i < n$ and $\mu(a) = b$. However, this contradicts the existence of $\varphi(x, \bar{c})$ and $\neg\varphi(y, \bar{c})$ in $\Sigma^*(x, y)$, which proves the strong minimality of the formula $U(x)$. In view of the properties of coordinates, the formula $\theta(x, y_1, y_2) = \underset{i=1}{\overset{2}{\&}} \big(\Pi(y_i) \,\&\, r_i(x) = [y_i]_\sim\big)$ provides the equality $|\mathfrak{N}| = \text{acl}\big(\Pi(\mathfrak{N})\big)$. Hence the theory QS is almost strongly minimal and the rank of its formula is equal to 2. Therefore, the Morley rank of the theory QS is equal to 3.   □

1.7. Exterior Properties of the Basic Theory

In this section, we give some properties of the theory QS which follow from the description of QS. Models of the theory QS are considered from two points of view: the exterior form and the internal structure. The exterior form concerns those properties of QS that will be necessary to use QS as a basic theory in the construction from Chapter 3.

Let $\mathfrak{N}$ and $\mathfrak{N}'$ be models of the theory QS. If $\mu : \mathfrak{N} \to \mathfrak{N}'$ is an isomorphic embedding and $\mu^* : \mathfrak{N}/_\sim \to \mathfrak{N}'/_\sim$ is the isomorphic embedding induced by μ, then we say that μ is a *realization* of μ^*. From the description of the theory QS we obtain the following lemma.

Lemma 1.7.1. *The following assertions hold:*

(a) *if $\mathfrak{N}'$ and $\mathfrak{N}''$ are models of the theory QS, then an arbitrary isomorphic embedding $\mu^* : \mathfrak{N}'/_\sim \to \mathfrak{N}''/_\sim$ is realized by the isomorphic embedding $\mu : \mathfrak{N}' \to \mathfrak{N}''$; moreover, if μ^* is a surjection, then any realization μ of μ^* is a surjection,*

(b) *if $\mathfrak{N}$, $\mathfrak{N}'$, and $\mathfrak{N}''$ are models of the theory QS such that $\mathfrak{N} \subseteq \mathfrak{N}'$, $\mathfrak{N} \subseteq \mathfrak{N}''$, and $\mu^* : \mathfrak{N}'/_\sim \to \mathfrak{N}''/_\sim$ is an isomorphic embedding with the consistency condition $\mu^*([a]_\sim^{\mathfrak{N}'}) = [a]_\sim^{\mathfrak{N}''}$ for all $a \in |\mathfrak{N}|$, then there exists a realization $\mu : \mathfrak{N}' \to \mathfrak{N}''$ of μ^* which is identical on $\mathfrak{N}$.*

Exercises

1. We make the following modification QS^* of the theory QS. We translate the predicates Π and Σ into unary ones. In a model of one diagonal block (cf. Fig. 1.3.1), Π is true on the front plane below the diagonal and on the back plane above the diagonal. Respectively, Σ is true on the two remaining half-planes. Axioms B.1–B.20 remain unchanged, and the remaining axioms are changed in accordance with the new form of the predicates Π and Σ. Show that the theory QS^* obtained is not uncountably categorical and complete.

2. Show that, in a model of the theory QS of length 3, there are 8 methods of defining the sign coordinate and 6 methods of defining the order $<$. Therefore, the following modification of Lemma 1.6.2 is false: "... the number of such extensions μ^* is equal to the number of methods of defining $<$ in $\mathfrak{N}''$ for a fixed sign coordinate in $\mathfrak{N}''$."

3. Show that the modification mentioned in Exercise 2 is true if for the order $<$ we take any relation that is asymmetric on $\lhd$-chains; moreover, all the assertions in this chapter remain valid.

Chapter 2

Binary Trees

The notion of a tree is fundamental in mathematics. The author [51] introduced a class of trees, which turned out to be useful in the study of constructive models of decidable theories. In particular, it was applied in [10, 16, 17, 22, 46, 52]. This class plays an important role in the intermediate construction of a finitely axiomatizable theory. Describing an algorithmic procedure of constructing a tree, we obtain, on the basis of the intermediate construction, a finitely axiomatizable theory whose properties depend on the properties of the tree constructed.

2.1. Basic Definitions

By a *complete binary tree* we mean the partially ordered set $\mathcal{D}_0 = \langle \mathbb{N}, \preccurlyeq \rangle$ of the form shown in Fig. 2.1.1. In particular, $1 \preccurlyeq 8$, $\neg(6 \preccurlyeq 9)$, $\neg(9 \preccurlyeq 6)$.

We introduce two natural operations:

$L(n) = 2n + 1$ is the left successor of n,

$R(n) = 2n + 2$ is the right successor of n.

31

By a *tree* we mean a set $\mathcal{D} \subseteq \mathbb{N}$ such that

(D1) $m \preccurlyeq n \ \& \ n \in \mathcal{D} \Rightarrow m \in \mathcal{D}$ for all $m, n \in \mathbb{N}$,

(D2) $L(n) \in \mathcal{D} \Leftrightarrow R(n) \in \mathcal{D}$ for all $n \in \mathbb{N}$.

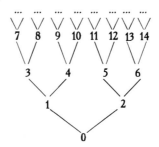

Fig. 2.1.1. Example of a complete binary tree.

In Fig. 2.1.2, three examples of trees are presented. Trees (a) and (b) are finite, whereas tree (c) is infinite.

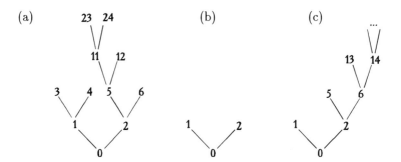

Fig. 2.1.2. Examples of trees.

An element n of a tree $\mathcal{D}$ such that $L(n) \notin \mathcal{D}$ is called a *dead end* of the tree $\mathcal{D}$. The set of all dead ends of a tree $\mathcal{D}$ is denoted by $\text{dend}\,(\mathcal{D})$. A tree is called *atomic* if there is at least one dead end over each of its elements. All the trees depicted in Fig. 2.1.2 are atomic.

By a *chain* we mean a set $\pi \subseteq \mathbb{N}$ such that

(C1) $m, n \in \pi \Rightarrow m \preccurlyeq n \vee n \preccurlyeq m$ for all $m, n \in \mathbb{N}$,

(C2) $m \preccurlyeq n \ \& \ n \in \pi \Rightarrow m \in \pi$ for all $m, n \in \mathbb{N}$.

For a tree $\mathcal{D}$ the set of all maximal chains is denoted by $\Pi(\mathcal{D})$, and the set of all finite maximal chains is denoted by $\Pi^{\text{fin}}(\mathcal{D})$.

We define the general recursive function $f_n(x)$ by the recursive scheme $f_n(0) = n$, $f_n(L(x)) = L(f_n(x))$, $f_n(R(x)) = R(f_n(x))$. It is easy to see that f_n is an isomorphism between a complete tree and the domain $\{x \mid n \preccurlyeq x\}$. The set $\mathcal{D}_1 \oplus \mathcal{D}_2 = \{0\} \cup f_1(\mathcal{D}_1) \cup f_2(\mathcal{D}_2)$ is called the *direct sum* of trees $\mathcal{D}_1$ and $\mathcal{D}_2$. It is easy to check that $\mathcal{D}_1 \oplus \mathcal{D}_2$ is a tree. We note that the tree $\mathcal{D}_1 \oplus \mathcal{D}_2$ can be obtained by "attaching" isomorphic copies of $\mathcal{D}_1$ and $\mathcal{D}_2$ to two dead ends of the three-element tree depicted in Fig 2.1.2(b). The *direct sum of a sequence of trees* $\mathcal{D}_m$, $m \in \mathbb{N}$, is the tree $\oplus \langle \mathcal{D}_m \mid m \in \mathbb{N} \rangle = \{0\} \cup f_1(\mathcal{D}_0) \cup f_5(\mathcal{D}_1) \cup \ldots \cup f_{k_s}(\mathcal{D}_s) \cup \ldots$, where $k_s = 2^{s+2} - 3$. This tree can be obtained by "attaching" copies of $\mathcal{D}_s$ to the sequential dead ends of the tree depicted in Fig. 2.1.2(c). We note that the direct sum is an atomic tree if and only if each summand is an atomic tree.

We study the family of maximal chains of an arbitrary tree $\mathcal{D}$. Let G be a subset of $\Pi(\mathcal{D})$. A chain $\pi \in G$ is called *isolated* in G if there exists an element $t \in \pi$ such that π is a unique chain of G passing through t. Let G' stand for the set of all nonisolated chains in G. By induction on ordinals, we define the subset $\Pi_\alpha(\mathcal{D}) \subseteq \Pi(\mathcal{D})$ as follows:

$$\Pi_0(\mathcal{D}) = \Pi(\mathcal{D}), \quad \Pi_{\alpha+1}(\mathcal{D}) = (\Pi_\alpha(\mathcal{D}))'$$
$$\Pi_\gamma(\mathcal{D}) = \cap\{\Pi_\beta(\mathcal{D}) \mid \beta < \gamma\} \text{ if } \gamma \text{ is a limit ordinal}$$

The least α such that $\Pi_{\alpha+1}(\mathcal{D}) = \Pi_\alpha(\mathcal{D})$ is called the *rank* of the tree $\mathcal{D}$ (denoted by rank($\mathcal{D}$)). The *rank* of a chain $\pi \in \Pi(\mathcal{D})$ is an ordinal α (denoted by rank(π)) such that $\pi \in \Pi_\alpha(\mathcal{D}) \setminus \Pi_{\alpha+1}(\mathcal{D})$. In general, the rank function is partially defined on the set $\Pi(\mathcal{D})$. A tree $\mathcal{D}$ is called *superatomic* if the rank function is defined everywhere on $\Pi(\mathcal{D})$. In other words, a tree $\mathcal{D}$ is superatomic if $\Pi_\alpha(\mathcal{D}) = \varnothing$ for some α.

From definitions it follows that

$$(\forall \alpha < \text{rank}(\mathcal{D}))(\exists \pi \in \Pi(\mathcal{D})) \, \text{rank}(\pi) = \alpha \qquad (2.1.1)$$

$$\text{rank}(\mathcal{D}) = \sup\{\text{rank}(\pi) \mid \pi \in \Pi^R(\mathcal{D})\} \qquad (2.1.2)$$

where $\Pi^R(\mathcal{D})$ denotes the set of all those chains of $\Pi(\mathcal{D})$ that have rank. We indicate some properties of the rank function on chains.

Lemma 2.1.1. *For any tree $\mathcal{D}$ the set of chains $\pi \in \Pi(\mathcal{D})$ having rank is at most countable.*

PROOF. Let $\pi \in \Pi_\alpha(\mathcal{D}) \setminus \Pi_{\alpha+1}(\mathcal{D})$. Denote by $I(\pi)$ the set of elements isolating the chain π in $\Pi_\alpha(\mathcal{D})$. It is easy to check that if chains π_1 and π_2 have rank, then $I(\pi_1) \cap I(\pi_2) = \varnothing$ for $\pi_1 \neq \pi_2$. Since the set $\mathbb{N}$ is countable, the lemma is proved. $\square$

Lemma 2.1.2. *The rank of any tree is a countable ordinal. If $\mathcal{D}$ is a superatomic tree, then* $\mathrm{rank}(\mathcal{D})$ *is not a limit ordinal.*

PROOF. The first assertion follows from Lemma 2.1.1 and formulas (2.1.1) and (2.1.2). To prove the second assertion, we suppose that $\mathcal{D}$ is a superatomic tree. In view of (2.1.2), it suffices to show that the fact that γ is a limit countable ordinal such that for any $\beta < \gamma$ in $\Pi(\mathcal{D})$ there is a chain of rank β implies that, in $\Pi(\mathcal{D})$, there is a chain of rank at least γ. Indeed, let us choose a sequence of chains π_m, $m \in \mathbb{N}$, in $\Pi(\mathcal{D})$ such that $\sup\{\mathrm{rank}(\pi_m) \mid m \in \mathbb{N}\} = \gamma$. There exists a chain $\pi^* \in \Pi(\mathcal{D})$ such that infinitely many chains π_m pass through each of its elements. Since the tree $\mathcal{D}$ is superatomic, the rank of the chain π^* is defined. Hence $\mathrm{rank}(\pi^*) \geqslant \gamma$ by definition. $\qquad\square$

The following theorem characterizes superatomic trees.

Theorem 2.1.1. *For any tree $\mathcal{D}$ the following conditions are equivalent:*

(a) *a tree $\mathcal{D}$ is superatomic;*

(b) *the set $\Pi(\mathcal{D})$ is at most countable;*

(c) *the cardinality of the set $\Pi(\mathcal{D})$ is less than 2^ω;*

(d) *every countable set $G \subseteq \Pi(\mathcal{D})$ has isolated chains;*

(e) *every set $G \subseteq \Pi(\mathcal{D})$ has isolated chains.*

PROOF. We proceed in accordance with the following scheme: (a) $\Rightarrow$ (b) $\Rightarrow$ (c) $\Rightarrow$ (d) $\Rightarrow$ (e) $\Rightarrow$ (a).

The implication (a)$\Rightarrow$(b) follows from Lemma 2.1.1. The implication (b)$\Rightarrow$(c) is obvious. To prove (c)$\Rightarrow$(d) we assume the contrary: there is a countable set $G \subseteq \Pi(\mathcal{D})$ without isolated chains. Then $V = \cup\{\pi \mid \pi \in G\}$ is a subset of $\mathcal{D}$ and satisfies the conditions $x \preccurlyeq y$ & $y \in V \to x \in V$ and $(\forall x \in V)(\exists yz \in V)[x \preccurlyeq y$ & $x \preccurlyeq z$ & $\neg(y \preccurlyeq z)$ & $\neg(z \preccurlyeq y)]$. Therefore, among subsets of V, there are 2^ω different infinite chains; moreover, all of them belong to $\Pi(\mathcal{D})$, which contradicts (c). The implication (d)$\Rightarrow$(e) holds because, from any nonempty set $G \subseteq \Pi(\mathcal{D})$ without isolated chains, it is possible to choose a countable subset $G_0 \subseteq G$ that has no isolated chain. The implication (e)$\Rightarrow$(a) follows from the above definitions. $\qquad\square$

We denote by λ the first nonconstructive ordinal.

Lemma 2.1.3. *Let $\mathcal{D}$ be an arbitrary recursively enumerable tree. The following assertions hold:*

(a) $\mathrm{rank}(\mathcal{D}) \leqslant \lambda$;

(b) $\mathrm{rank}(\mathcal{D}) < \lambda$ *if the tree $\mathcal{D}$ is superatomic.*

PROOF. (a). In $\mathcal{D}$, we take a chain π that has rank. For definiteness, let $\pi \in \Pi_\alpha(\mathcal{D}) \setminus \Pi_{\alpha+1}(\mathcal{D})$. By (2.1.2), it suffices to show that the ordinal α is constructive. We choose the element t that isolates the chain π in the family $\Pi_\alpha(\mathcal{D})$. By definition, every chain of the tree $\mathcal{D}$ passing through t has rank at most α. We consider the set

$$E = \{x \in \mathcal{D} \mid x \preccurlyeq t \lor t \preccurlyeq x\}$$

and construct a monotonically decreasing sequence of sets

$$E = E_0 \supseteq E_1 \supseteq \ldots \supseteq E_\beta \supseteq \ldots \supseteq E_\alpha \supseteq E_{\alpha+1} = \varnothing$$

where

$$E_\beta = \cup\{\pi \in \Pi_\beta(\mathcal{D}) \mid t \in \pi\}, \quad 0 \leqslant \beta \leqslant \alpha + 1$$

by definition. We now introduce binary relations R, $\sim$, and $\lhd$ as follows:

$$R(x,y) \Leftrightarrow (\exists \beta, \gamma)\big(0 \leqslant \beta \leqslant \gamma < \alpha \ \& \ x \in E_\beta \setminus E_{\beta+1} \ \& \ y \in E_\gamma \setminus E_{\gamma+1}\big)$$

$$x \sim y \Leftrightarrow R(x,y) \ \& \ R(y,x) \tag{2.1.3}$$

$$x \lhd y \Leftrightarrow R(x,y) \ \& \ (\forall z)\big(R(x,z) \ \& \ R(z,y) \to x \sim z \lor z \sim y\big)$$

By construction, R is a linear preorder on E, $\sim$ is the corresponding equivalence relation, and $\lhd$ is the successor relation with respect to this preorder. In accordance with (2.1.3), the quotient $E/{\sim}$ with respect to the relation R is an order of ordinal type α, in which the iteration of eliminating isolated chains of the tree $\mathcal{D}$ is presented in the set E. Each step of the iteration of eliminating isolated chains presented by the relation $\lhd$ can be formalized with the help of formulas with R, $\lhd$, $\sim$; moreover, this preorder is well ordered. Therefore, R is implicitly definable in the elementary arithmetic. Consequently, it is hyperarithmetic [56]. But, in this case, the linear order

$$x \preccurlyeq^* y \Leftrightarrow (x, y \in E) \ \& \ \big[(R(x,y) \ \& \ \neg(x \sim y)) \lor (x \sim y \ \& \ x < y)\big]$$

which is expressed in terms of R, is also hyperarithmetic and has the ordinal type $\alpha_1 = \alpha \cdot \omega$. In accordance with Spector's result (cf. [56]), the ordinal α_1 (and the ordinal α along with it) is constructive.

(b). Assertion follows from assertion (a) and Lemma 2.1.2.   □

REMARK 2.1.1. The proof of Lemma 2.1.3 remains the same if, instead of the condition of the recursive enumerability of $\mathcal{D}$, we require the weaker condition that it is arithmetic. Furthermore, this condition can be weakened up to the condition of hyperarithmetic; however, the proof of the lemma becomes more involved.

2.2. Construction of Trees

In this section, we describe a standard procedure for translating constructive linear orders in recursive trees. This procedure allows us to construct superatomic trees of any constructive rank. It will also be used to estimate the complexity of sets. Starting from a finite sequence of natural numbers

$$\varkappa = \langle n_0, n_1, \dots, n_{s-1} \rangle \tag{2.2.1}$$

we construct an infinite chain $\pi[\varkappa]$ as follows: beginning from 0, we make $n_0 + 1$ steps to the right, $n_1 + 1$ steps to the left, then $n_2 + 1$ steps to the right and so on, until we reach the last element of the sequence. After that we make the next turn (the direction is changed) and make ω steps. A step to the left from t means passage to $L(t)$, a step to the right means passage to $R(t)$, where L and R are defined in Sec. 2.1. By definition, the empty sequence corresponds to the last right chain.

Let $\langle \mathcal{L}, \nu \rangle$ be a constructive linear order. We use it to define the following set of sequences of natural numbers:

$$K(\mathcal{L}, \nu) = \{\langle n_0, n_1, \dots, n_{s-1} \rangle \mid \nu(n_0) > \nu(n_1) > \dots > \nu(n_{s-1})\}$$

We denote by $\mathcal{D}(\mathcal{L}, \nu)$ the least tree containing the set

$$\cup\{\pi[\varkappa] \mid \varkappa \in K(\mathcal{L}, \nu)\} \tag{2.2.2}$$

By definition, we obtain the tree $\mathcal{D}(\mathcal{L}, \nu)$ from the set (2.2.2) by adding the L-successors and the R-successors of those elements that are dead ends of this tree. On the other hand, the tree $\mathcal{D}(\mathcal{L}, \nu)$ has no other dead ends. Therefore,

$$\mathcal{D}(\mathcal{L}, \nu) = \cup\{\pi[\varkappa] \mid \varkappa \in K(\mathcal{L}, \nu)\} \cup \operatorname{dend}(\mathcal{D}(\mathcal{L}, \nu)) \tag{2.2.3}$$

We are now in a position to describe the correlation between the properties of the initial linear order and those of the constructed tree.

Lemma 2.2.1. *Let $(\mathcal{L}, \nu)$ be an arbitrary constructive linear order. The following assertions hold.*

(a) *The tree $\mathcal{D}(\mathcal{L}, \nu)$ is superatomic if and only if $\mathcal{L}$ is well ordered.*

(b) If $\mathcal{L}$ is well ordered, then the family of chains $\Pi(\mathcal{D}(\mathcal{L}, \nu))$ is computable.

(c) $\mathrm{rank}(\mathcal{D}(\mathcal{L}, \nu)) = 1 + \alpha + 1$ if $\mathcal{L}$ is well ordered by ordinal type α and the numbering ν satisfies the relation

$$(\forall x \in |\mathcal{L}|)(\nu^{-1}(x) \text{ is infinite}) \qquad (2.2.4)$$

PROOF. (a). Assume that $\mathcal{L}$ is not well ordered. Then there is a decreasing sequence of elements

$$\nu(n_0) > \nu(n_1) > \ldots > \nu(n_k) > \ldots \qquad (2.2.5)$$

We consider its arbitrary subsequence

$$\nu(n_{i_0}) > \nu(n_{i_1}) > \ldots > \nu(n_{i_s}) > \ldots \qquad (2.2.6)$$

From the infinite sequence of natural numbers $\varkappa = \langle n_{i_s} \mid s < \omega \rangle$ we construct a chain π with infinitely many turns, as was defined for the finite sequence (2.2.1). Since every finite part of the chain π is covered by at least one of the chains $\pi[\varkappa]$, $\varkappa \in K(\mathcal{L}, \nu)$, we obtain $\pi \subseteq \mathcal{D}(\mathcal{L}, \nu)$. It is obvious that different subsequences (2.2.6) yield different chains π. Consequently, in the family $\Pi(\mathcal{D}(\mathcal{L}, \nu))$, there is the continuum of different infinite chains. By Theorem 2.1.1, the tree $\mathcal{D}(\mathcal{L}, \nu)$ cannot be superatomic.

To prove the converse assertion, we assume that the tree $\mathcal{D}(\mathcal{L}, \nu)$ is not superatomic. By Theorem 2.1.1, the family $\Pi(\mathcal{D}(\mathcal{L}, \nu))$ is uncountable. Therefore, there is an infinite chain $\pi \subseteq \mathcal{D}(\mathcal{L}, \nu)$ with an infinite number of turns. In view of (2.2.3), every finite part of the chain π must be covered by at least one of the chains $\pi[\varkappa]$, $\varkappa \in K(\mathcal{L}, \nu)$. Therefore, the lengths of intervals situated between the turns of this chain define an infinite sequence $\varkappa = \langle n_i \mid i < \omega \rangle$ subject to the condition (2.2.5). Consequently, $\mathcal{L}$ cannot be well ordered.

(b). The aforesaid shows that, if $\mathcal{L}$ is well ordered, none of the infinite chains $\pi \subseteq \mathcal{D}(\mathcal{L}, \nu)$ has infinitely many turns. Therefore, the chains $\pi[\varkappa]$, $\varkappa \in K(\mathcal{L}, \nu)$, exhaust all infinite chains of the above tree. By construction, $\mathcal{D}(\mathcal{L}, \nu)$ is a recursive tree, from which follows the computability of the family of its finite chains. As a result, we obtain the computability of the family of all chains of this tree.

(c). For a sequence $\varkappa \in K(\mathcal{L}, \nu)$, $\varkappa = \langle n_0, n_1, \ldots, n_{s-1} \rangle$, we denote by $\delta(\varkappa)$ the ordinal type of the set $\mathcal{L} \restriction \{y \mid \nu(n_{s-1}) > y\}$. By definition, $\delta(\varkappa)$ is equal to the ordinal type α of the set $\mathcal{L}$ for the empty sequence $\varkappa = \langle \, \rangle$.

Using (2.2.4) and proceeding by induction on β, we can show that the following general relation holds:

$$\Pi_\beta(\mathcal{D}(\mathcal{L}, \nu)) = \begin{cases} \Pi(\mathcal{D}(\mathcal{L}, \nu)) & \text{for } \beta = 0 \\ \{\pi_\varkappa \mid \varkappa \in K(\mathcal{L}, \nu) \,\&\, 1 + \delta(\beta) \geqslant \beta\} & \text{for } \beta > 0 \end{cases}$$

In particular, $\Pi_{1+\alpha}(\mathcal{D}(\mathcal{L}, \nu)) = \{\pi_{(\,)}\}$, $\Pi_{1+\alpha+1}(\mathcal{D}(\mathcal{L}, \nu)) = \varnothing$. By definition, $\mathrm{rank}(\mathcal{D}(\mathcal{L}, \nu)) = 1 + \alpha + 1$. □

2.3. Standard Sets

The results of this section will be used to estimate the complexity for different classes of sentences. Let $[W]_\mathcal{D}$ denote the closure of the set $W \subseteq \mathbb{N}$ up to a tree. It is easy to see that, in general, the set can be constructed from W in two steps as follows:

$$W \mapsto W' \mapsto W'' = [W]_\mathcal{D}$$

where

$$W' = \{y \mid (\exists x \in W)\, y \preccurlyeq x\}$$
$$W'' = W' \cup \{L(x) \mid x \in W'\} \cup \{R(x) \mid x \in W'\}$$

Let W_n, $n \in \mathbb{N}$, be a standard Post numbering of a family of recursively enumerable sets and let W_n^A, $n \in \mathbb{N}$, be a standard numbering of sets enumerated with oracle $A \subseteq \mathbb{N}$. We introduce the sets

$$\mathcal{D}_n = [W_n]_\mathcal{D}, \quad n \in \mathbb{N}$$
$$\mathcal{D}_n^A = [W_n^A]_\mathcal{D}, \quad n \in \mathbb{N}, \quad A \subseteq \mathbb{N}$$

These sets play the role of universal computable numberings for families of recursively enumerable trees and recursively enumerable trees with oracle A. The constructed numberings of trees possess all required good properties. For example, one can prove that there exist general recursive functions $f(x)$ and $g(x)$ that reduce new numberings to old ones with the help of the relations $\mathcal{D}_n = W_{f(n)}$, $\mathcal{D}_n^A = W_{g(n)}^A$, $n \in \mathbb{N}$, $A \subseteq \mathbb{N}$. If X is a set and Σ is a class of arithmetic or analytic hierarchy, then the notation $X \approx \Sigma$ means that the set X belongs to the class Σ and is m-universal for this class. As is known, for a fixed Σ such conditions define X uniquely up to a recursive isomorphism.

We now proceed to exact estimates of the algorithmic complexity for some natural sets of indices in the above numbering of trees.

Lemma 2.3.1. *The following estimates of the algorithmic complexity hold*:

(a) $A = \{n \mid \mathcal{D}_n \ is \ atomic\} \approx \Pi_3^0$;

(b) $A' = \{n \mid \mathcal{D}_n \ is \ atomic \ \& \ \Pi^{\mathrm{fin}}(\mathcal{D}) \ is \ computable\} \approx \Sigma_4^0$;

(c) $A'' = \{n \mid \mathcal{D}_n \ is \ atomic \ \& \ \Pi^{\mathrm{fin}}(\mathcal{D}) \ is \ not \ computable\} \approx \Pi_4^0$;

(d) $A^* = \{n \mid (\exists A \subseteq \mathbb{N}) \, \mathcal{D}_n^A \ is \ atomic\} \approx \Sigma_1^1$;

(e) $B_0 = \{n \mid (\exists A \subseteq \mathbb{N}) \, |A| \leqslant 1 \ \& \ \mathcal{D}_n^A \ is \ atomic\} \approx \Sigma_4^0$;

(f) $B_1 = \{n \mid (\exists A \subseteq \mathbb{N}) \, |A| = 1 \ \& \ \mathcal{D}_n^A \ is \ atomic\} \approx \Sigma_4^0$;

(g) $H = \{n \mid \mathcal{D}_n \ is \ superatomic\} \approx \Pi_1^1$;

(h) $H' = \{n \mid \Pi(\mathcal{D}_n) \ is \ computable\} \approx \Pi_1^1$;

(i) $H^* = \{n \mid (\exists A \subseteq \mathbb{N}) \, \mathcal{D}_n^A \ is \ superatomic\} \approx \Sigma_2^1$.

PROOF. The estimates from above can be established immediately. In cases (g) and (i), it is necessary to use the characterization of superatomic trees from Theorem 2.1.1. We proceed to estimates from below.

(a). Consider the standard Π_3^0-set (cf. [56])

$$A_3 = \{n \mid (\forall k \in W_n) \, W_k \ is \ finite\} \tag{2.3.1}$$

and fix some value of the parameter n. We construct a computable sequence of trees $\mathcal{D}_n^{(k)}$, $k \in \mathbb{N}$, depending on n in the following way:

$$\mathcal{D}_n^{(k)} = \begin{cases} \{0\} & \text{if } k \notin W_n \\ \{t \mid (\exists y) \, y \in W_k \ \& \ t \leqslant 2y\} & \text{if } k \in W_n \end{cases}$$

If $k \in W_n$ and W_k is infinite, then $\mathcal{D}_n^{(k)} = \mathbb{N}$; otherwise, $\mathcal{D}_n^{(k)}$ is a finite tree. By the s–m–n-theorem, there exists a general recursive function $f(n)$ such that $\mathcal{D}_{f(n)} = \oplus \langle \mathcal{D}_n^{(k)} \mid k \in \mathbb{N} \rangle$. By construction,

$$n \in A_3 \Leftrightarrow \mathcal{D}_{f(n)} \ is \ an \ atomic \ tree \tag{2.3.2}$$

which yields the necessary estimates from below.

(b), (c). To obtain estimates from below, we use the standard set defined in [56]:

$$A_4 = \{n \mid (\exists^\omega k) \, k \in W_n \ \& \ W_k \ is \ infinite\}$$

We describe an effective process of constructing a recursively enumerable tree $\mathcal{D}^{(n)}$, which depends on the natural parameter n so that

$$\mathcal{D}^{(n)} \text{ is an atomic tree for any } n \in \mathbb{N} \qquad (2.3.3)$$

$$n \in A_4 \Leftrightarrow \text{ the family } \Pi^{\text{fin}}(\mathcal{D}^{(n)}) \text{ is computable} \qquad (2.3.4)$$

We denote by W_n^t a finite part of the set W_n that can be computed in t steps. We define a general recursive function

$$h_n(k, t) = \begin{cases} 0 & \text{if } k \notin W_n^t \\ |W_k^t| & \text{if } k \in W_n^t \end{cases}$$

It is obvious that for any k the function $h_n(k, t)$ is monotone in t. By the definition of the set A_4, we have

$$n \in A_4 \Leftrightarrow (\exists^\omega k) \lim_{t \to \infty} h_n(k, t) = \infty \qquad (2.3.5)$$

There exists a strong sequence of finite sets $\pi_{m,n}^t$ such that

(1) $\pi_{m,n}^t$ is a chain for any m, n, t,

(2) $\pi_{m,n}^t \subseteq \pi_{m,n}^{t+1}$ for any m, n, t,

(3) any computable family of chains coincides with one of the families of the form $\Pi_m = \{\pi_{m,n} \mid n \in \mathbb{N}\}$, $m \in \mathbb{N}$, $\pi_{m,n} = \cup\{\pi_{m,n}^t \mid t \in \mathbb{N}\}$.

We now construct the tree $\mathcal{D}^{(n)}$. For the sake of brevity we denote it by $\mathcal{D}$. Let $\mathcal{D}^t$ denote the part of $\mathcal{D}$ which is already constructed at the moment t. We will use the marks $\square_m^s$, $m < s$. At some moment, the mark $\square_m^s$ may be placed on a chain Π_m passing through the element

$$q_s = 2^{s+2} - 3 \qquad (2.3.6)$$

After that the mark is not moved. Some marks can be unused. At each step, we consider a pair (m, s), $m < s$; moreover, such a pair must be considered infinitely many times.

Construction

STEP $t = 0$. Let

$$\mathcal{D}^0 = \{0\} \cup \{1, 2, 5, 6, \dots, 2^{n+1} - 3, 2^{n+1} - 2, \dots ; n \in \mathbb{N}\}$$

The set $\mathcal{D}^0$ is an infinite tree having the last right chain as in Fig. 2.1.2(c). None of the marks is used at the initial moment.

STEP $t > 0$. We consider a pair (m, s).

CASE 1: the mark $\square_m^s$ is not yet placed. In this case, we verify if there exists $k < t$ such that $q_s \in \pi_{m,k}^t$. In the case of a positive answer, we put the mark $\square_m^s$ on one of the chains $\pi_{m,k}$ subject to this condition. In addition, we set $\mathcal{D}^t = \mathcal{D}^{t-1}$.

CASE 2: the mark $\square_m^s$ is already placed, and it marks the chain $\pi_{m,k}$. Let e be a maximal element of the set $\pi_{m,k}^t \cap \mathcal{D}^{t-1}$. We put

$$
\mathcal{D}^t = \begin{cases} \mathcal{D}^{t-1} \cup \{L(e), R(e)\} & \text{if } e < h_n(s,t) \\ \mathcal{D}^{t-1} & \text{otherwise} \end{cases}
$$

The step t is completed.

We assume that π is an infinite chain of the constructed tree $\mathcal{D} = \cup\{\mathcal{D}^t \mid t \in \mathbb{N}\}$. We also assume that π is not the last right chain. Then π passes through one of the elements (2.3.6). By construction, π must coincide with some chain $\pi_{m,k}$ marked by one of the marks $\square_m^s$. Moreover, it is necessary that the condition $\lim_{t \to \infty} h_n(s,t) = \infty$ hold. The above reasoning implies that at most s different infinite chains can pass through the element q_s, but only in the case of growth of the function $h_n(s,t)$ as t increases; moreover, all these chains are recursively enumerable and, consequently, they are recursive. Thus, in any case, the tree $\mathcal{D}$ is superatomic, which implies (2.3.3).

We now prove (2.3.4). Let $n \notin A_4$. By (2.3.5),

$$
(\exists s_0)(\forall s > s_0) \lim_{t \to \infty} h_n(s,t) < \infty
$$

In this case, the tree $\mathcal{D}$ can contain only a finite set of infinite chains, and each of these sets is recursive. Hence the family $\Pi^{\text{fin}}(\mathcal{D})$ is computable.

Let $n \in A_4$. Consider a family Π_m containing only finite chains. It is required to prove that $\Pi_m \neq \Pi^{\text{fin}}(\mathcal{D})$. To this end, we take $s > m$ such that $\lim_{t \to \infty} h_n(s,t) = \infty$. The fact that the mark $\square_m^s$ is not used means that none of the chains of the family Π_m passes through q_s; therefore, Π_m cannot coincide with the family of all finite chains of the tree $\mathcal{D}$.

Let the mark $\square_m^s$ be placed on $\pi_{m,k} \in \Pi_m$ at some moment. Let e be its maximal element. Since the function $(\lambda t)h_n(s,t)$ increases unboundedly, the elements $L(e)$ and $R(e)$ are included in $\mathcal{D}^t$. As a result, the chain $\pi_{m,k}$ is not a maximal chain of the tree $\mathcal{D}$. In this case, we also obtain $\Pi_m \neq \Pi^{\text{fin}}(\mathcal{D})$. Thus, (2.3.4) is proved.

By the s–m–n-theorem, there exists a general recursive function $f(n)$ such that the constructed tree $\mathcal{D} = \mathcal{D}^{(n)}$ coincides with $\mathcal{D}_{f(n)}$. The above process provides the relations

$n \in A_4 \to \mathcal{D}_{f(n)}$ is atomic & $\Pi^{\mathrm{fin}}(\mathcal{D}_{f(n)})$ is computable,

$n \notin A_4 \to \mathcal{D}_{f(n)}$ is atomic & $\Pi^{\mathrm{fin}}(\mathcal{D}_{f(n)})$ is not computable.

These relations yield the necessary estimates from below.

(d). To derive an estimate from below, we use the standard Σ_1^1-set defined in [56]:

$$E = \{n \mid (\exists f)(\forall w)\, S(\overline{f}(w), n)\} \tag{2.3.7}$$

where S is a suitable recursive relation and $\overline{f}$ is defined by the function f as follows: $\overline{f}(w) = \mathrm{Num}\langle f(0), f(1), \dots, f(w-1)\rangle$, where Num denotes the standard effective enumeration of a list.

We say that a set $A \subseteq \mathbb{N}$ *codes* a function $f \colon \mathbb{N} \to \mathbb{N}$ if the following conditions are satisfied:

$A = \{k_0, k_1, \dots, k_s, \dots\}, \quad k_0 < k_1 < \dots < k_s < \dots,$

$f(s) = k_{s+1} - k_s - 1, \quad s \in \mathbb{N}.$

A set A coding a function f is denoted by A_f. The coding defined in such a way is a one-to-one correspondence between the set of functions $f \colon \mathbb{N} \to \mathbb{N}$ and the set of infinite sets $A \subseteq \mathbb{N}$.

Given S, it is easy to construct a new recursive relation R such that

$$(\forall n)(\forall f)\big[(\forall w)\, S(\overline{f}(w), n) \Leftrightarrow (\forall t)\, R\big(\mathrm{Num}(A_f \cap \{0, 1, \dots, t-1\}), n\big)\big]$$

Therefore, the set (2.3.7) can be represented in the form

$$E = \big\{n \mid \text{there exists an infinite set } A \subseteq \mathbb{N}$$
$$\text{such that } (\forall t)\, R\big(\mathrm{Num}(A \cap \{0, 1, \dots, t-1\}), n\big)\big\}$$

From $s \in \mathbb{N}$ and $A \subseteq \mathbb{N}$ we construct the tree $\mathcal{D}(s, A) = \oplus\langle \mathcal{D}^{(k)} \mid k \in \mathbb{N}\rangle$, where

$$\mathcal{D}^{(0)} = \begin{cases} \{0\} & \text{if } (\forall t)\, R\big(\mathrm{Num}(A \cap \{0, 1, \dots, t-1\}), n\big) \\ \mathbb{N} & \text{if } \neg(\forall t)\, R\big(\mathrm{Num}(A \cap \{0, 1, \dots, t-1\}), n\big) \end{cases}$$

$$\mathcal{D}^{(k)} = \begin{cases} \{0, 1, \dots, 2t\} & \text{if } t = (\mu x)[x \in A \ \& \ x > k] \\ \mathbb{N} & \text{if the set } \{x \in A \mid x > k\} \text{ is empty} \end{cases}$$

for $k > 0$. It is obvious that the tree $\mathcal{D}(s, A)$ can be effectively constructed from s and A. By the s–m–n-theorem, there exists a general recursive function $f(s)$ such that $\mathcal{D}(s, A) = \mathcal{D}_{f(s)}^A$.

If A is finite, then for some $k > 0$ the tree $\mathcal{D}^{(k)}$ is complete; therefore, $\mathcal{D}(s, A)$ cannot be atomic. If A is infinite, then all the trees $\mathcal{D}^{(k)}$, $k > 0$, are finite. Consequently, the fact that the tree $\mathcal{D}(s, A)$ is atomic depends on $\mathcal{D}^{(0)}$. As a result, for any natural s we can conclude that $s \in A$ if and only if there exists $A \subseteq \mathbb{N}$ such that $\mathcal{D}^A_{f(s)}$ is atomic. This gives us the required estimate from below.

(e), (f). To derive estimates from below, we use the standard set A_4 introduced in the proof of (b) and (c). We describe the process of constructing the tree $\mathcal{D}$, which depends on two parameters

$$s \in \mathbb{N}, \quad A \subseteq \mathbb{N} \qquad (2.3.8)$$

We construct step-by-step a sequence of finite sets $\mathcal{D}^t_m$ as follows.

STEP $t = 0$. We set $\mathcal{D}^t_m = \{0\}$ for all $m \in \mathbb{N}$.

STEP $t > 0$.

CASE 1: $\{0, 1, \dots, t\} \cap A = \varnothing$. We set

$$\mathcal{D}^t_m = \begin{cases} \{0, 1, \dots, 2t\} & \text{if } m = 0 \\ \mathcal{D}^{t-1}_m & \text{otherwise} \end{cases}$$

CASE 2: $\{0, 1, \dots, t\} \cap A \neq \varnothing$. Let k be the least element of this set. We put

$$\mathcal{D}^t_m = \begin{cases} \{0, 1, \dots, 2t\} & \text{if } k < m < t \ \& \ m \in W^t_s \ \& \ W^t_m \neq W^{t-1}_m \\ \mathcal{D}^{t-1}_m & \text{otherwise} \end{cases}$$

The step t is completed.

Let $\mathcal{D}^\omega_m = \cup\{\mathcal{D}^t_m \mid t \in \mathbb{N}\}$, $\mathcal{D} = \oplus\langle \mathcal{D}_m \mid m \in \mathbb{N} \rangle$. By construction, the tree $\mathcal{D}$ is effectively constructed from the given parameters (2.3.8). Therefore, there exists a general recursive function $f(s)$ such that $\mathcal{D}$ coincides with $\mathcal{D}^A_{f(s)}$ for all values of the parameters (2.3.8).

For $A = \varnothing$, the tree $\mathcal{D}$ cannot be atomic since the tree $\mathcal{D}^\omega_0$ is complete. For $A = \{k\}$, the tree $\mathcal{D}$ is atomic if and only if the set $W_s \cap \{k+1, k+2, \dots\}$ does not contain recursively enumerable indices of infinite sets. As a result, we obtain the following relations:

$$s \in A_4 \Leftrightarrow (\exists A \subseteq \mathbb{N}) \, |A| \leqslant 1 \ \& \ \mathcal{D}^A_s \text{ is atomic}$$

$$s \in A_4 \Leftrightarrow (\exists A \subseteq \mathbb{N}) \, |A| = 1 \ \& \ \mathcal{D}^A_n \text{ is atomic}$$

which yield the necessary estimates from below.

(g), (h). To obtain estimates from below, we use the standard Π^1_1-set defined in [56]:

$$W = \{s \mid \varphi^{(2)}_s \quad \text{is the characteristic function of some well ordering of some set of natural numbers}\}$$

Now, we construct the constructive linear order $(\mathcal{L}_s, \nu_s)$ uniformly with respect to a given parameter s so that the following relation is satisfied:

$$s \in A \Leftrightarrow \mathcal{L}_s \text{ is well ordered} \qquad (2.3.9)$$

We fix s. The function $\varphi_s^{(2)}$ is denoted by $\varphi(x, y)$ and the order $(\mathcal{L}_s, \nu_s)$ is denoted by $(\mathcal{L}, \nu)$. A given partial recursive function $\varphi(x, y)$ is the characteristic function of some ordering of a set of natural numbers if and only if the following conditions are satisfied for all natural m, n, t:

$$(m, n) \in \mathrm{dom}(\varphi) \Leftrightarrow (m, m) \in \mathrm{dom}(\varphi) \ \& \ (n, n) \in \mathrm{dom}(\varphi)$$
$$(m, n) \in \mathrm{dom}(\varphi) \ \& \ m \neq n \to \varphi(m, n) + \varphi(n, m) = 1$$
$$(m, m) \in \mathrm{dom}(\varphi) \to \varphi(m, m) = 1$$
$$\varphi(m, n) = 1 \ \& \ \varphi(n, t) = 1 \to \varphi(m, t) = 1$$

Let C_k, $k \in \mathbb{N}$, be an effective enumeration of all possible conditions of the form just indicated and let C_k^t denote the condition obtained from C_k by replacing φ by φ^t, where φ^t is a finite part of φ that is computed in t steps. For the steps $t = 0, 1, 2, \ldots$, we construct the constructive linear order $(\mathcal{L}, \nu)$ as the sum

$$\alpha + \gamma_0 + \gamma_1 + \ldots + \gamma_k + \ldots \qquad (2.3.10)$$

At the step $t = 0$, we set $\alpha = 0$ and $\gamma_k = 1$ for all k. In the sequel, the summand α will be constructed in accordance with what the function φ gives and where it defines a linear order. Furthermore, at the step t, to the beginning of the order γ_k, $k < t$, we annex one more element if, at this moment, the condition C_k^t is false. As a result, the summand γ_k is finite provided that the condition C_k is true at the limit. If C_k is false, then, beginning from some moment, new elements are constantly added to the beginning of γ_k and, finally, the order type of this summand will be equal to ω^*.

If $\varphi(x, y)$ defines a linear order of type β, then all the conditions C_k are true. Therefore, the order (2.3.10) has type $\beta + \omega$. Otherwise, this order is not well ordered due to some summand γ_k. In fact, the above construction provides the validity of the condition (2.3.9).

To derive estimates from below, we use the translation of linear orders in trees, which was described in Sec. 2.2. By the s–m–n-theorem, we find a general recursive function $f(s)$ such that $\mathcal{D}(\mathcal{L}_s, \nu_s) = \mathcal{D}_{f(s)}$ for all $s \in \mathbb{N}$. By Lemma 2.2.1, we have

$s \in W$ if and only if $\mathcal{D}_{f(s)}$ is a superatomic tree,

$s \in W$ if and only if the family $\Pi(\mathcal{D}_{f(s)})$ is computable.

These assertions yield the necessary estimates from below.

(i). To obtain an estimate from below, we take the standard Σ_2^1-set defined in [56]:

$$T^2 = \left\{ s \mid (\exists A \subseteq \mathbb{N}) \, \varphi_s^A \text{ is the characteristic function defining} \right.$$
$$\left. \text{a functional tree with finite paths} \right\}.$$

Translating the characteristic function in the Kleene–Brouwer ordering for the corresponding functional trees [56], passing to functional trees, and using the procedure of translating the linear order $(\mathcal{L}, \nu)$ in the tree $\mathcal{D}(\mathcal{L}, \nu)$, in view of the s–m–n-theorem, we obtain the general recursive function $f(x)$ such that for all $s \in \mathbb{N}$ and $A \subseteq \mathbb{N}$ the function φ_s^A is the characteristic function defining a functional tree with finite paths if and only if $\mathcal{D}_{f(s)}^A$ is a superatomic tree. Finally, we arrive at the following estimate from below:

$$s \in T^2 \Leftrightarrow (\exists A \subseteq \mathbb{N}) \, \mathcal{D}_{f(s)}^A \text{ is a superatomic tree.} \qquad \Box$$

2.4. Axiomatizable Construction

In [17], a general method of constructing decidable theories of the signature with unary predicates was described. From an arbitrary constructive Boolean algebra $(\mathcal{B}, \nu)$, a complete theory $\mathrm{AT}(\mathcal{B}, \nu)$ is constructed such that the embeddedness of unary predicates corresponds to the structure of the algebra $(\mathcal{B}, \nu)$ and all nonempty sets that are distinguished by predicates are assumed to be infinite. To construct constructive Boolean algebras, which is required in this construction, binary recursively enumerable trees were used. They were especially designed for constructing Boolean algebras. In this section, we sketch a modification of such a construction. Starting from an arbitrary recursively enumerable tree $\mathcal{D}$, we construct an axiomatizable complete theory $\mathbb{T}(\mathcal{D})$ of the signature with unary predicates while an intermediate factor (i.e., Boolean algebra) is absent. There is a simple table of translating the properties of a tree into those of the theory obtained. Thus, the problem of construction of decidable theories with prescribed properties is reduced to the problem of construction of binary recursively enumerable trees. The last problem is usually simpler in comparison with the initial one.

We proceed to the construction. Let $\mathcal{D}$ be a recursively enumerable tree and let $f \colon \mathbb{N} \to \mathcal{D}$ be a general recursive function enumerating the tree $\mathcal{D}$. The latter contains 0; therefore, it cannot be empty. Our goal is to construct a recursively axiomatizable theory $\mathbb{T}(\mathcal{D}, f)$ of the signature with countable number of unary predicates:

$$\sigma = \{=, U_0^1, U_1^1, \ldots, U_s^1, \ldots; s \in \mathbb{N}\}$$

We say that a *predicate U_s is ascribed to an element d of a tree $\mathcal{D}$* if $f(s) = d$. Generally speaking, several signature predicates can be ascribed to the same element.

The axioms of the theory $\mathbb{T}(\mathcal{D}, f)$ are the following:

A1. $(\forall x)\, U_s(x)$ if $f(s) = 0$.

A2. $(\forall x)[\neg U_i(x)] \vee [\neg U_j(x)]$ if $f(i) = L(k)$ and $f(j) = R(k)$.

A3. $(\forall x)\big[U_k(x) \Leftrightarrow U_i(x) \vee U_j(x)\big]$ if $f(k) = t$, $f(i) = L(t)$, and $f(j) = R(t)$.

A4. $(\exists^{\geqslant n} n) U_i(x)$ for all $n, i \in \mathbb{N}$.

The structure of the set of axioms is very simple. First, each predicate ascribed to a root of the tree $\mathcal{D}$ is assumed to be true on the universe. Second, if the triple of predicates U_k, U_i, and U_j is ascribed to the same branching point t, $L(t)$, $R(t)$, then the following claim is adopted as a postulate: the predicates U_i and U_j are disjoint partitions of the predicate U_k. Finally, the last axiom requires that all the predicates distinguish infinite sets. From the axioms we obtain the following relations:

$$f(i) \preccurlyeq f(j) \Leftrightarrow \mathbb{T}(\mathcal{D}, f) \vdash (\forall x)\, (U_i(x) \to U_j(x))$$
$$f(i) = f(j) \Leftrightarrow \mathbb{T}(\mathcal{D}, f) \vdash (\forall x)\, (U_i(x) \leftrightarrow U_j(x))$$

The following lemma can be proved by constructing models.

Lemma 2.4.1. *For any $\mathcal{D}$ and f, the theory $\mathbb{T}(\mathcal{D}, f)$ is consistent and has no prime models.*

The following obvious lemma shows that the properties of the theory just described are determined by the form of the tree $\mathcal{D}$, whereas the role of the numbering function f is unessential.

Lemma 2.4.2. *Given a tree $\mathcal{D}$, let $f'\colon \mathbb{N} \to \mathcal{D}$ and $f''\colon \mathbb{N} \to \mathcal{D}$ be two enumerating general recursive functions. Then, starting from the Kleene numbers of these functions, it is possible to construct effectively a recursive mapping $\mu\colon FL(\sigma) \to FL(\sigma)$ which is an isomorphism between these theories.*

In view of Lemma 2.4.2, we may write $\mathbb{T}(\mathcal{D})$ omitting f, since the theory is defined in a unique way up to an isomorphism and is independent of the choice of the numbering function.

We study how properties of the theory $\mathbb{T}(\mathcal{D})$ depend on the properties of a tree $\mathcal{D}$. From the construction we obtain the following theorem.

Theorem 2.4.1. *The theory $\mathbb{T}(\mathcal{D})$ is countably categorical if and only if the tree $\mathcal{D}$ is finite.*

Theorem 2.4.2. *For any recursively enumerated tree $\mathcal{D}$, the theory $\mathbb{T}(\mathcal{D})$ is complete.*

PROOF. Let φ be a sentence of the signature of the theory $\mathbb{T}(\mathcal{D})$. Since φ contains only a finite number of different unary predicates, it is possible to indicate a natural number e such that all the predicates in φ are relative to those levels of the tree $\mathcal{D}$ that do not exceed e. Then φ or $\neg\varphi$ is provable in the subtheory $\mathbb{T}(\mathcal{D}')$ that is defined by the finite tree $\mathcal{D}'$. This subtheory is complete in view of the Vaught theorem, since it is countably categorical by Theorem 2.4.1. $\qquad\square$

Developing the method of proof of Theorem 2.4.2, one can obtain a detailed characterization of the first-order definable relations of the theories $\mathbb{T}(\mathcal{D})$. We denote by E the class of formulas of the signature σ that will be written as $\varepsilon(x_1, x_2, \ldots, x_n)$ and are conjunctions of formulas of the form $x_i = x_j$ or $\neg(x_i = x_j)$. In addition, for every pair of indices i, j only one of the above-mentioned formulas occurs in ε, and ε in total must be consistent. In other words, formulas ε of the class E are atomic diagrams over different numbers of variables such that they are constructed from the single equality predicate.

A quantifier-free formula of the signature σ of the form

$$\theta = \varepsilon(x_1, x_2, \ldots, x_s)\ \&\ U_{k_1}(x_1)\ \&\ U_{k_2}(x_2)\ \&\ \ldots\ \&\ U_{k_s}(x_s)$$

is called a *primitive formula* of the theory $\mathbb{T}(\mathcal{D}, f)$ if $f(k_i) = f(k_j)$ provided that $x_i = x_j$ occurs in ε.

Acting as in the proof of Theorem 2.4.2, we can prove the following assertion.

Lemma 2.4.3. *In the theory $\mathbb{T}(\mathcal{D}, f)$, any formula is equivalent to a disjunction of primitive formulas of this theory (the empty disjunction, which presents the identically false formula, is also admitted).*

Eliminating quantifiers, we obtain the following assertion.

Lemma 2.4.4. *The theory* $\mathbb{T}(\mathcal{D}, f)$ *is model-complete.*

To study the properties of the theory $\mathbb{T}(\mathcal{D})$, it is useful to have a description up to an isomorphism of all models of this theory. Let $\mathfrak{M}$ be a model of the theory $\mathbb{T}(\mathcal{D})$ and let π be a maximal chain of the tree $\mathcal{D}$. We denote by $\ker_\pi(\mathfrak{M})$ the set of all elements $\alpha \in |\mathfrak{M}|$ on which unary predicates related to elements of the chain π are true. This set is called the *kernel of the chain* π in $\mathfrak{M}$. It is obvious that the kernels of different chains in the same model are mutually disjoint and the union of kernels over all chains of the tree $\mathcal{D}$ forms the universe. In the case of a finite chain $\pi \in \Pi(\mathcal{D})$, the kernel $\ker_\pi(\mathfrak{M})$ is exactly the set of the truth of the predicate $P_i(x)$ related to the last element of this chain. In the case of an infinite chain $\pi \in \Pi(\mathcal{D})$, eliminating any number of elements of the kernel $\ker_\pi(\mathfrak{M})$ or adding new elements, we obtain a model of the same theory $\mathbb{T}(\mathcal{D}, f)$ as above. Thus, the kernels of infinite chains can have any number of elements. The axiom A4 remains true because any element of an infinite chain in a tree is also covered by other chains.

Now, we are in a position to describe all the isomorphism types of models.

Lemma 2.4.5. *The following assertions hold*:

(a) *Let* λ *be an arbitrary mapping from* $\Pi(\mathcal{D})$ *into the cardinals. Then there exists a model* $\mathfrak{M}$ *of the theory* $\mathbb{T}(\mathcal{D})$ *such that* $(\forall \pi \in \Pi(\mathcal{D})) \, | \ker_\pi(\mathfrak{M})| = \lambda(\pi)$ *if and only if the sum of cardinals* $\lambda(\pi)$ *over all chains* $\pi \in \Pi(\mathcal{D})$ *is infinite and* $\lambda(\pi)$ *is infinite for every finite chain* $\pi \in \Pi(\mathcal{D})$.

(b) *Models* $\mathfrak{M}$ *and* $\mathfrak{N}$ *of the theory* $\mathbb{T}(\mathcal{D})$ *are isomorphic if and only if for any chain* $\pi \in \Pi(\mathcal{D})$ *the cardinalities of the kernels* $\ker_\pi(\mathfrak{M})$ *and* $\ker_\pi(\mathfrak{N})$ *are equal.*

(c) *Let* $\mathfrak{M}$ *and* $\mathfrak{N}$ *be models of the theory* $\mathbb{T}(\mathcal{D})$. *There exists an isomorphic (thereby, elementary) embedding of the model* $\mathfrak{M}$ *into the model* $\mathfrak{N}$ *if and only if* $|\ker_\pi(\mathfrak{M})| \leqslant |\ker_\pi(\mathfrak{N})|$ *for any chain* $\pi \in \Pi(\mathcal{D})$.

PROOF. The assertions can be established by direct construction of models and embeddings. Any isomorphic embedding is elementary in view of the model completeness. $\square$

We now characterize prime and countable saturated models. From the description of models we obtain the following lemma.

Lemma 2.4.6. *The following assertions hold*:

(a) *A model $\mathfrak{M}$ of the theory $\mathbb{T}(\mathcal{D})$ is prime if and only if the kernels $\ker_\pi(\mathfrak{M})$ are empty for any infinite chain $\pi \in \Pi(\mathcal{D})$.*

(b) *A model $\mathfrak{M}$ of the theory $\mathbb{T}(\mathcal{D})$ is countable saturated if and only if the kernels $\ker_\pi(\mathfrak{M})$ are infinite for any infinite chain $\pi \in \Pi(\mathcal{D})$.*

Now, we are ready to state the main assertion about the dependence of the properties of the theory $\mathbb{T}(\mathcal{D})$ on the properties of a tree $\mathcal{D}$. Its proof can be obtained by direct construction of the corresponding models and their constructivizations with the help of the description of the class of models and first-order definable relations in the theory $\mathbb{T}(\mathcal{D})$.

Theorem 2.4.3. *Let $\mathcal{D}$ be a recursively enumerable tree. Then the following relations hold*:

(a) *The theory $\mathbb{T}(\mathcal{D})$ has a prime model if and only if the tree $\mathcal{D}$ is atomic.*

(b) *A prime model of the theory $\mathbb{T}(\mathcal{D})$ (if it exists) is strongly constructivizable if and only if the family of chains $\Pi^{\mathrm{fin}}(\mathcal{D})$ is computable.*

(c) *A prime model of the theory $\mathbb{T}(\mathcal{D})$ (if it exists and is strongly constructivizable) is autostable with respect to strong constructivizations if and only if the tree $\mathcal{D}$ is recursive.*

(d) *The theory $\mathbb{T}(\mathcal{D})$ has a countable saturated model if and only if the tree $\mathcal{D}$ is superatomic.*

(e) *A countable saturated model of the theory $\mathbb{T}(\mathcal{D})$ is strongly constructivizable if and only if the family of chains $\Pi(\mathcal{D})$ is computable.*

(f) *The theory $\mathbb{T}(\mathcal{D})$ is totally transcendental if and only if the tree $\mathcal{D}$ is superatomic.*

(g) *The Morley rank of the theory $\mathbb{T}(\mathcal{D})$ is equal to $1 + \mathrm{rank}(\mathcal{D})$.*

To conclude the chapter, we note that the construction of the axiomatizable theory $\mathbb{T}(\mathcal{D})$ described is, in a sense, the preimage of the intermediate construction of finitely axiomatizable theories which is described in Chapter 3.

Exercises

1. Simplify the proof of assertion (b) in Lemma 2.3.1 to obtain a direct proof of the following assertion: "there exists a superatomic recursively enumerable tree $\mathcal{D}$ such that $\Pi^{\text{fin}}(\mathcal{D})$ is not computable."

2. Prove that if $\mathcal{D}$ is a superatomic tree and $\Pi(\mathcal{D})$ is computable, then $\Pi^{\text{fin}}(\mathcal{D})$ is also computable [21].

3. Prove that for the tree $\mathcal{D}$ from Exercise 1 $\Pi(\mathcal{D})$, is not computable.

4. Construct an atomic but not superatomic recursively enumerable tree $\mathcal{D}$ such that $\Pi^{\text{fin}}(\mathcal{D})$ is computable.

5. Use the trees from the above exercises to construct complete decidable theories with different properties of prime models and saturated models.

6. Using the result from Secs. 2.2 and 2.4, construct a complete decidable ω-stable theory of Morley rank α for a constructive ordinal α as large as desired.

7. Let π be an arbitrary chain. Prove that the following assertions are equivalent: "π is recursive," "π is recursively enumerable," and "$\mathbb{N}\backslash\pi$ is recursively enumerable."

8. Prove that any tree $\mathcal{D}$ is recursive if and only if dend $(\mathcal{D})$ is recursive.

9. To a root of the tree $\mathcal{D}$ we assign the unit of a countable atomless Boolean algebra. At a branching point, we divide the corresponding element into two nonzero elements. Denote by $\mathcal{B}_{\mathcal{D}}$ the Boolean subalgebra generated by the set obtained. Prove the following assertions.

    (a) If $\mathcal{D}$ is recursively enumerable, then $\mathcal{B}_{\mathcal{D}}$ is constructivizable.

    (b) Every constructive Boolean algebra is defined from some recursively enumerable tree in the above method.

    (c) A tree $\mathcal{D}$ is atomic if and only if the algebra $\mathcal{B}_{\mathcal{D}}$ is atomic.

    (d) A tree $\mathcal{D}$ is superatomic if and only if the algebra $\mathcal{B}_{\mathcal{D}}$ is superatomic.

Chapter 3

The Construction over a Unary List

In this chapter, we study the intermediate construction $\mathbb{FM}$. An understanding of its principal components facilitates the study of the universal
construction. The intermediate construction uses the tool of binary trees,
which provides a unique method in applications of the construction. Most
of the results presented in the book were obtained on the basis of the intermediate construction. Furthermore, some of the results cannot be proved
by means of the universal construction. Describing the intermediate construction, we present notions, methods, and ideas that will be applied in
the construction $\mathbb{FU}$.

3.1. The Main Theorem for the Intermediate Construction

Let ε_k, $k \in \mathbb{N}$, denote the condition that "a set contains an element
k." By a *truth-table condition* (*tt-condition*) we mean a propositional
formula constructed from elementary propositions ε_k. The assertion that
a truth-table condition τ is true in $A \subseteq \mathbb{N}$ is symbolically written in the
form $A \vDash \tau$. We introduce the Gödel numbering of truth-table conditions.

As is known, the Scheffer stroke $\xi_1 \mid \xi_2 \Leftrightarrow \neg(\xi_1 \ \& \ \xi_2)$ forms a function-
ally complete system for the two-valued logic. Therefore, for truth-table
conditions we can use propositional formulas constructed from elementary
truth-table conditions by using only the Scheffer stroke. For every truth-
table condition τ we inductively define $\mathrm{Nom}(\tau)$ as follows:

(a) $\mathrm{Nom}(\varepsilon_k) = 2k$

(b) $\mathrm{Nom}(\tau' \mid \tau'') = 2c(\mathrm{Nom}(\tau'), \mathrm{Nom}(\tau'')) + 1$

where $c(x, y)$ denotes the standard function of numbering of pairs of natural numbers. We denote by τ_k the truth-table condition with number k.
We fix the described numbering and we introduce the notation

$$\mathcal{R}_m = \{A \subseteq \mathbb{N} \mid (\forall k \in W_m) \, A \vDash \tau_k\}, \quad m \in \mathbb{N}$$

Let $[W]_{\mathcal{D}}$ denote the closure of a set $W \subseteq \mathbb{N}$ to a tree (cf. Sec. 2.3). We
also introduce the notation

$$\mathcal{D}_n = [W_n]_{\mathcal{D}}, \ n \in \mathbb{N}, \quad \mathcal{D}_n^A = [W_n^A]_{\mathcal{D}}, \ n \in \mathbb{N}, \ A \subseteq \mathbb{N}$$

where W_n is a recursively enumerable set with the Post number n and
$\mathcal{D}_n^A$ is the nth recursively enumerable set with respect to some standard
numbering with respect to the oracle $A \subseteq \mathbb{N}$. We will write $\mathbb{F}(m, s)$ instead
of $\mathbb{FM}(m, s)$ for brevity.

Theorem 3.1.1 [intermediate construction]. *For a finite rich sig-
nature σ and a pair of natural numbers $\langle m, s \rangle$ it is possible to construct
effectively a finitely axiomatizable model-complete theory $F = \mathbb{F}(m, s)$
of the signature σ and a recursive sequence of sentences Ψ_n, $n \in \mathbb{N}$, of
the same signature such that*

(1) *sentences Ψ_n, $n \in \mathbb{N}$, generate the Lindenbaum algebra of the
theory $\mathbb{F}(m, s)$,*

(2) *the theory $\mathbb{F}(m, s)[A] = \mathbb{F}(m, s) \cup \{\Psi_i \mid i \in A\} \cup \{\neg \Psi_j \mid j \in \mathbb{N} \setminus A\}$,
where $A \subseteq \mathbb{N}$, is consistent if and only if $A \in \mathcal{R}_m$,*

(3) *for any $A \in \mathcal{R}_m$ the following relations hold:*

(a) *the theory $\mathbb{F}(m, s)[A]$ has a prime model if and only if the tree
$\mathcal{D}_s^A$ is atomic,*

(b) *a prime model of the theory $\mathbb{F}(m, s)[A]$ (if it exists) is strongly
constructivizable if and only if the set A is recursive and the
family of chains $\Pi^{\mathrm{fin}}(\mathcal{D}_s^A)$ is computable,*

(c) *a prime model of the theory* $\mathbb{F}(m, s)[A]$ *(if it exists and is strongly constructivizable) is autostable with respect to strong constructivizations if and only if the tree* $\mathcal{D}_s^A$ *is recursive,*

(d) *the theory* $\mathbb{F}(m, s)[A]$ *has a countable saturated model if and only if the tree* $\mathcal{D}_s^A$ *is superatomic,*

(e) *a countable saturated model of the theory* $\mathbb{F}(m, s)[A]$ *is strongly constructivizable if and only if the set* A *is recursive and the family of chains* $\Pi(\mathcal{D}_s^A)$ *is computable,*

(f) *the theory* $\mathbb{F}(m, s)[A]$ *is transcendental if and only if the tree* $\mathcal{D}_s^A$ *is superatomic,*

(g) *the Morley rank of the theory* $\mathbb{F}(m, s)[A]$ *is* $\max\{15, 1+\mathrm{rank}(\mathcal{D}_s^A)+\gamma\}$, *where* $\gamma = 2$ *if the tree* $\mathcal{D}_s^A$ *is superatomic and* $\gamma = 0$ *otherwise.*

The proof of Theorem 3.1.1 occupies all of this chapter. The plan of the proof is as follows. Starting from an arbitrary Turing machine $\mathcal{M}$, some class indicated below, and arbitrary parameters $m, s \in \mathbb{N}$, we construct the finitely axiomatizable theory

$$F(\mathcal{M}, m, s) = \mathbf{FRM} + \mathbf{NET} + \mathbf{MT}(\mathcal{M}, m, s) + \mathbf{TRN} + \mathbf{NRM}(\mathcal{M}, m, s)$$
$$(3.1.1)$$

such that the program of the Turing machine $\mathcal{M}$ is contained in the axiomatics and the natural numbers m and s are regarded as input parameters. Then we construct a specific Turing machine $\mathcal{M}^*$ for which the required finitely axiomatizable theory is represented in the form

$$\mathbb{F}(m, s) = F(\mathcal{M}^*, m, s) \qquad (3.1.2)$$

3.2. Turing Machines with Division of Cells

We describe a Turing machine that will be used in the intermediate construction. We consider the machine $\mathcal{M}$ with a finite collection a_i, $i < d$, of tape symbols and a finite collection q_j, $j < e$, of symbols of states of $\mathcal{M}$. The program of $\mathcal{M}$ consists of instructions of the form

$$a_i q_j \longrightarrow a_m q_t L \qquad (3.2.1)$$
$$a_i q_j \longrightarrow a_m q_t R \qquad (3.2.2)$$
$$a_i q_j \longrightarrow a_m a_n q_t L \qquad (3.2.3)$$
$$a_i q_j \longrightarrow a_m a_n q_t R \qquad (3.2.4)$$

The tape of the machine is infinite in both directions and is marked off into cells. At every instant, the head of the machine observes one of the cells. Every cell contains one of the symbols a_i, $i < d$. At every instant, the machine is in one of the states q_j, $j < e$. The machine starts from the state q_0. The command (3.2.1) is the instruction that if the machine is in the state q_j at some time and the tape cell being scanned contains a_i, then the machine writes down the symbol a_m in the tape cell, takes the state q_t, and moves the head to the left by a cell. The command (3.2.2) is similar to (3.2.1), but it moves the head to the right. We say that these commands are *applicable* to the state $a_i q_j$. The command (3.2.3) is the instruction that, in the state $a_i q_j$, the machine must *divide* the tape cell being scanned into two cells, write a_m in the left cell, and write a_n in the right cell. After that the machine takes the state q_t and moves the head to the left from the cell that is neighboring to the pair of new cells. The command (3.2.4) is the same, but the tape is moved to the right.

It is required that at most one command can be applied in any situation. The machine halts if none of its commands can be applied. Thus, the work of the machine is uniquely determined by its program and the initial information located on the tape. We do not specify what we mean by a result because only the work of a Turing machine is of interest for our purposes.

The set of machines satisfying the above description will be called the *class of Turing machines with division of cells*. Such machines will appear in axioms of the theory F in Sec. 3.3. The choice of the machine and the input parameters will be made in Sec. 3.15.

3.3. Signature

First of all, we construct the finitely axiomatizable theory F satisfying all the assumptions of Theorem 3.1.1 except the condition on a signature. The reduction to a given finite rich signature σ, as is required in Theorem 3.1.1, can be made by standard methods presented below.

The signature σ of the theory F contains 16 binary predicates and $18 + d + e$ unary predicates, where d, e denote parameters of the Turing machine $\mathcal{M}$ occurring in the description of the theory F. In fact, we use the machine with three $(d = 3)$ tape symbols a_0, a_1, and a_2, which will be denoted by 0, 1, and B for illustration. The superscript "2" indicates binary predicates. The remaining symbols are unary predicates.

Table 3.3.1. The list of predicates of the intermediate construction

Groups of predicates	Step
Predicates of the skeleton mechanism $\triangleleft^2$, $\sim^2$, ... quasisuccession (crossing)	**QS**
Basis sets X basis set of crossing of higher level Y basis set of horizontal crossing W basis set of branching structure	**FRM**
Predicates of the structure of a higher-level frame D^2 succession along the first dimension (time) H^2 instants of time, B the upper part of a model C the lower part of a model, J the initial instant of time	
Predicates of the internal structure of the frame P^2 a point at successive instants of time S^2 the neighborhood relation of points E^2 a point of the frame, O the origin of coordinates	**NET**
Predicates of the marking-out mechanism G the defining marking-out line M the marking-out line	
Predicates of the work of the Turing machine A tape cells, Q the position of the head on the tape L, R the indicators of the direction of motion of the head Z the saturation line A_0, A_1, A_2 the tape symbols (respectively, $0, 1, B$) Q_0, Q_1, ... , Q_{e-1} the symbols of the internal states	**MT**
Predicates of division of the interpretation domain U the basic interpretation domain V^2 the branching structure of interpretations	**TRN**

The predicates are divided into seven groups. Table 3.3.1 briefly characterizes the role of each predicate in the theory F. Dots indicate the omitted predicate symbols of the language of the theory QS so that σ is an extension of σ_{QS}.

For the language of the theory QS we adopt the following convention: the predicate symbols $\lhd^2$ and $\sim^2$ (cf. Table 3.3.1) are related to the theory QS and the latter is included in the frame with the help of these predicates. The rest of the predicate symbols are assumed to be free.

We begin with the characterization of the theory as a unit. The structure of F is of a multistep character. We distinguish five steps.

STEP 1 (the quasisuccessor theory QS) [Chapter 1]. This step provides the strictly plane form of the frame without cycles.

STEP 2 (the frame **FRM**) [Sec. 3.4]. Axioms of this step describe the general structure of the frame of models that has the form of a plane two-dimensional net on which the remaining steps act.

STEP 3 (the net structure **NET**) [Sec. 3.5]. By means of local combinatorial conditions, axioms of this step form those types of net structure in the frame that will be required for the work of the Turing machine.

STEP 4 (the interpretation of the Turing machine **MT**) [Sec. 3.7]. This step defines the action of the given Turing machine $\mathcal{M}$. With the help of special unary predicates, the states of the tape and the internal states of the Turing machine are presented at each instant of time. Axioms of this step describe relationships between the head and the tape in accordance with the program of the machine $\mathcal{M}$.

STEP 5 (the translator **TRN**) [Sec. 3.8]. The goal of this step is to translate the results of the work of the Turing machine into the structure of first-order definable subsets of the special unary predicate U. By means of the translator, the Turing machine $\mathcal{M}$ affects the model-theoretic properties of each completion of the theory F in accordance with Theorem 3.1.1.

3.4. Axiomatics. The Frame (FRM)

We describe axioms of the theory F, combining them in groups according to their purposes. Every group of axioms is relative to one of the terms on the right-hand side of (3.1.1) (this partition of axioms into groups is adopted in Table 3.3.1). We use continuous enumeration of all the axioms and a separate enumeration in every group of axioms.

In this section, we begin to describe the theory F, starting from the *frame axioms*.

(FRM.1) Foundation of superstructure

$1°$. X, Y, W is a partition of the universe; moreover, each of these sets is nonempty.

(FRM.2) The structure of the highest level

$2°$. H is an equivalence relation and D is a successor relation on H-classes.

$3°$. The predicate J distinguishes a unique H-class.

$4°$. B, C is a partition of the universe.

$5°$. The predicates $B(x)$ and $C(x)$ are well defined on H-classes.

$6°$. $D(x, y)$ and $\neg J(y)$ imply $B(x) \leftrightarrow B(y)$.

$7°$. $D(x, y)$ and $J(y)$ imply $C(x)$ & $B(y)$.

$8°$. In every H-class there exist Y-elements and X-elements.

(FRM.3) The internal structure

$9°$. E is an equivalence relation on $X \cup W$ and is false outside $X \cup W$.

$10°$. $E(x, y) \longrightarrow H(x, y)$.

$11°$. In every E-class there exist X-elements.

(FRM.4) The skeleton mechanism

$12°$. On the set Y, the predicates $\lhd$, $\sim$, and other predicates of the signature σ_{QS} define a model of the quasisuccessor theory such that $\sim$-classes are sets of the form $[a]_H \cap Y$ for all H-classes $[a]_H$.

$13°$. $Y(u)$ & $Y(v)$ & $u \lhd v \to D(u, v)$.

$14°$. On the set of X-elements of each H-class, $\lhd$, $\sim$, and other predicates of the signature σ_{QS} define a model of the quasisuccessor theory such that $\sim$-classes are sets of the form $[a]_E \cap X$ for all E-classes $[a]_E$ from the H-class.

$15°$. In the rest of the cases, the predicates of σ_{QS} are false.

(FRM.5) The basis of the geometry of the internal structure

16°. S is a successor relation on E-classes and the predicate S is false outside $X \cup W$.

17°. $S(x, y) \rightarrow H(x, y)$.

18°. $S(x, y) \leftrightarrow (\exists u, v)\big[E(x, u) \ \& \ E(y, v) \ \& \ X(u)\&X(v) \ \& \ u \lhd v\big]$.

19°. P is a successor relation on E-classes, and initial classes (without P-predecessors) are admissible. The predicate P is false outside $X \cup W$.

20°. $P(x, y) \rightarrow D(x, y)$.

21°. If $S(x, y) \ \& \ P(x, x') \ \& \ P(y, y')$, then y' is either the first S-predecessor or the second S-predecessor of the element x'. In the last case, the first S-predecessor of the element x' has no P-predecessor.

22°. If $S(x, y)$, then at least one of the elements x and y has the P-predecessor.

23°. The predicate O distinguishes a unique E-class in J.

The axioms of the group **FRM** describe a general form of the frame, include the frame in the quasisuccessor theory (in order to exclude any cyclic formations), and describe a general structure of S–P-connections in the form of a two-dimensional plane. The general form of the model of the above axioms is presented in Fig. 3.4.1(a). The circles correspond to $\sim$-classes of the quasisuccessor theory, and the arrows between them indicate $\lhd$-connections. H-classes and E-classes are represented by ovals.

A set distinguished by the predicate C (respectively B) is called the *lower* (*upper*) part of the model. The predicate J distinguishes the H-class located at the interface between the upper part and the lower part. The predicate O marks a unique E-class located in J. It plays the role of the "origin" in the net coordinate plane formed by the frame.

In the further description of axioms, we consider P-connections and S-connections between E-classes and unary predicates that are well defined on E-classes. Omitting details, we will represent E-classes by points, as in Fig. 3.4.1(b). E-classes are also called *points*.

By a *block* we mean a minimal nonempty set closed under P-connections and S-connections. A block containing the point O is called *standard*. This block will be used for the computation controller; therefore, it is also referred to as the *computing block*. A collection of blocks intersected by an H-class is called a *zone*. In other words, a zone consists of blocks in the framework of one D-chain.

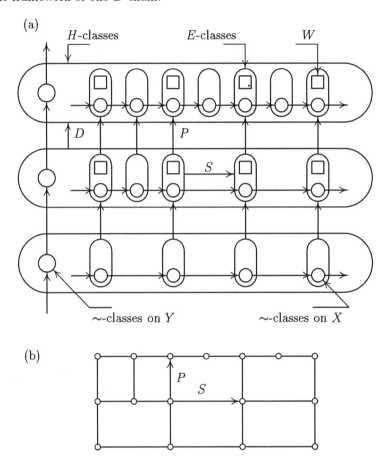

Fig. 3.4.1. The scheme of the frame of the intermediate construction.

The general scheme of the block-zone structure of models subject to the axioms of the group **FRM** is shown in Fig. 3.4.2. In the general case, a model may consist of many zones and each zone may consist of many blocks. The structure of blocks and zones will be specified by further axioms.

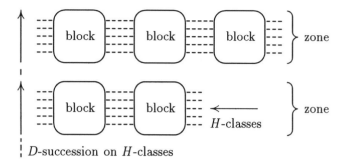

Fig. 3.4.2. Global block-zone form of a model.

3.5. Axiomatics. The Net Structure (NET)

In the formation of the local structure of blocks, the key role belongs to P–S-connections between points as well as unary predicates from the lists $\mathfrak{X} = \{O, G, M, A, Q, L, R, Z\}$ and $\mathfrak{X}' = \{B, C, J\}$. Let $\mathfrak{N}$ be a model subject to all the above-mentioned axioms and let a_i, $i \leqslant k$, and b_j, $j \leqslant s$, be finite sequences of elements of $\mathfrak{N}$ such that $S(a_i, a_{i+1})$, $i < k$, $S(b_j, b_{j+1})$, $j < s$, $P(a_i, b_{f(i)})$, $i \leqslant k$, $f(0) = 0$, $f(k) = s$, $1 \leqslant f(i+1) - f(i) \leqslant 2$, $i < k$, where $f(x)$ is some integer-valued function. In the case described, if the index j differs from $f(0), \ldots, f(k)$, then the elements b_j cannot have P-predecessors by Axiom 21°. The above collection of elements, together with the structure of P–S-connections and the indication of the truth of predicates from the lists $\mathfrak{X} \cup \mathfrak{X}'$, is called a *chain*.

We represent chains in a geometrical way by placing their S-connections along the horizontal line and P-connections along the vertical line. Elements of chains are indicated by the notation of the true predicates from $\mathfrak{X}$ or a white square if all these predicates are false. Double lines of S-connections mean that the elements belong to the upper part; bold lines, to the class J; and simple lines, to the lower part of the model. In the catalogue Catal 3.1, a number of chains is listed. These chains, called *elementary chains*, are used in the definition of the combinatorial form of blocks. The chain in Catal 3.1(i) is denoted by ε_i, $1 \leqslant i \leqslant 23$.

With the help of the catalogue of chains, we can formulate axioms concerning the form of nets of the frame.

(NET.1) Geometry of the internal structure of blocks

24°. The unary predicates O, G, M, A, Q, L, R, and Z distinguish subsets in $X \cup W$. All these predicates are well defined on E-classes.

25°. In a neighborhood of an arbitrary pair of points $[x]_E$ and $[y]_E$, under the condition $S(x, y)$ the structure of the model must be such that the neighborhood is covered by two elementary chains ε_i, ε_j, $1 \leqslant i, j \leqslant 23$, with correct transmission of the configuration of P–S-connections and all the predicates from $\mathfrak{X} \cup \mathfrak{X}'$. The covering must be such that the pair of points x, y is located on the upper line of the image of the chain ε_i and on the lower line of the image of the chain ε_j.

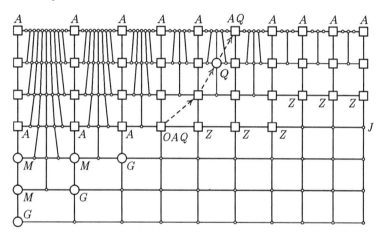

Fig. 3.5.1. A standard block of the intermediate construction.

REMARK 3.5.1. Axiom 25° holds if for every pair of S-connected elements x, y there exists a covering of only one of the indicated forms, e.g., a covering by the upper line of some elementary chain.

26°. In every H-class, in the lower part of the model there is a unique point satisfying G.

27°. In every H-class, in the upper part of the model there is a unique point satisfying Q.

28°. In every H-class, in the upper part of the model there are exactly three points satisfying Z.

A general form of P–S-connections in a neighborhood of the point O is presented in Figs. 3.5.1 and 3.5.2. We give some comments below.

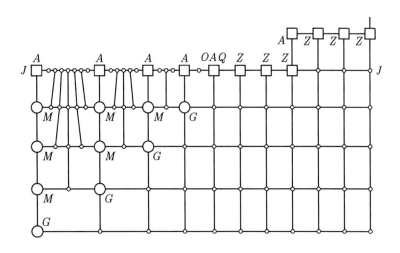

Fig. 3.5.2. The basis of a standard block.

3.6. Description of Blocks

Out goal is to describe all possible forms of blocks defined by the axioms from Secs. 3.4 and 3.5.

The catalogue Catal 3.2 contains brief descriptions, names, and notation for 11 types of blocks. As is shown below, Catal 3.2 is a complete catalogue, i.e., it contains descriptions of all types of blocks admitted by the above axioms.

Table 3.6.1 demonstrates the role of parameters used in the description of nonstandard blocks in Catal 3.2. There are several manners of handling the catalogue which differ by abstraction level.

At the abstraction level **H**, we indicate only a name, e.g., A, AQ, and ignore the remaining details of the block. At the abstraction level **X**, we additionally indicate the parameter ν, provided that it appears in the description of the block, e.g., $AQ(L)$ and so on. At the next two abstraction levels, we also indicate new parameters the meaning of which will be explained below.

In this section, we establish that Catal 3.2 is a complete catalogue at the abstraction level **X**. We begin with the study of the structure of the standard block (the computing block) generated by the point O. We consider an arbitrary infinite sequence

$$\lambda = \langle \lambda_i, \ i < \omega \rangle, \ \lambda_i \in \{L1, R1, L2, R2\} \tag{3.6.1}$$

where λ_i are indicators of the command types in the block, $L1$ and $R1$ denote the type of command without division of cells, and $L2$ and $R2$ denote the type of command with division of cells. The symbols $L1$ and $L2$ mean that the head moves to the left, whereas the symbols $R1$ and $R2$ mean that the head moves to the right after realization of the command. The symbols $L1$, $R1$, $L2$, and $R2$ are relative to the points of AQ in realizations of elementary chains ε_{12}, ε_{13}, ε_{14}, and ε_{15}. We say that a standard block $\mathcal{B}$ *has type* λ if along the Q-line of the block, starting from the point O, the points of execution of commands of type λ_i, $i = 0, 1, 2, \ldots$, are successively located. A standard block of type λ is denoted by $O(\lambda)$.

Table 3.6.1. Abstraction levels and parameters

Abstraction levels	Characterization	Parameters
H	a characteristic fragment of a block	Name
X	P–S-connections and predicates in $\mathfrak{X} \cup \mathfrak{X}'$	ν
I	the informational predicates A_i, Q_j	i, j
W	the number of components inside A-paths	τ

Lemma 3.6.1. *For any sequence* λ *such that* $\lambda_0 = \lambda_1 = R1$*, the block* $O(\lambda)$ *exists and is unique up to an isomorphism at the abstraction level* **X**.

PROOF. We describe how to line the block $O(\lambda)$ with elementary chains. We first gather, independently of the parameter λ, the lower half-block, called the *basis* (cf. Fig. 3.5.2). Then we construct the upper half-block for the sequence λ (cf. Fig. 3.5.1). We first place the elementary chain ε_{23}. By Axiom 25°, the upper part of this chain must cover the point O since there are no other suitable chains. Then we place the elementary chain ε_{22}. Its lower part covers the same point O. We note that the upper part of the chain ε_6 contains a point marked by the predicate G and there are no other chains possessing the same property. Hence the point G defined from the covering by the chain ε_{23} must be covered by the upper part of the chain ε_6 in accordance with Axiom 25°. The chain ε_6 defines a new point G by its lower part. This point can be covered by the upper part of the chain ε_6 and so on. Continuing the process, in the block B, we construct an unbounded line of points G directed toward the lower left from the initial point O.

We note that two extreme elements on the top of the second chain ε_{22} are marked by the predicates Z and $\varnothing$ (the symbol $\varnothing$ means that all predicates in the list $\mathcal{X}$ are false). However, ε_{20} is a single chain that can cover the pair Z and $\varnothing$ by its lower part. But the upper part of the chain ε_{20} defines a new pair of similar points. Therefore, we have to repeat the covering by the chain ε_{20} many times. As a result, we obtain a slanting steplike Z-line directed toward the upper right from the initial point O (cf. Figs. 3.5.1 and 3.5.2).

Using the chains ε_1 and ε_2, we line a row at the bottom along the Z-line by extending its lower steps. We can realize it by placing ε_2 and then ε_1 at successive corners under the Z-line. The chain ε_{17} is suitable for such a corner, but it is inadmissible because the use of this chain leads to a contradiction. The procedure of lining rows can be repeated so as to raise the width of the lined strip under the Z-line. Finally, the region between the lines Z and J will be lined by the chains ε_1 and ε_2.

Then a unique possible chain ε_3 is used to line the horizontal row under the line J to the right of the G-line. The same rows below are formed by the chain ε_4. Thus, the part of the basis located toward the lower right from the G-line and Z-line has the form of a simple predicate-free net.

The domain of the basis to the left of the G-line can be lined by the chains ε_3, ε_4, ε_7, ε_8 in a unique way. The next M-line is formed by the chain ε_8 from bottom to top, starting from the M-point near the G-line. Simultaneously, the region between the previous M-line and the present one is filled by the chain ε_4. In the last row going to J, the chains ε_7 and ε_3 are used. Completing the M-line, we pass to the left neighboring one and so on. As a result, all the domain of the basis to the left of the G-line will be filled. Thus, the basis of a standard block is as is shown in Fig. 3.5.2.

The upper half-block must be lined by horizontal rows from right to left, starting from the Z-line. It is easy to see that the process is strictly determinate. Lack of uniqueness occurs only if, in the next row, a point of type AQ appears (in this case, any of the chains ε_{12}, ε_{13}, ε_{14}, ε_{15} can be taken for the cover in the next row). If this happens, the next component λ_i of the sequence λ is used to choose a variant.

Completing the construction, we obtain the block $O(\lambda)$ satisfying the requirements of the lemma. The condition $\lambda_0 = \lambda_1 = R1$ is necessary because the point O itself is the first point at which the command is realized, and a unique possible covering ε_{22} of this point defines a command of type $R1$. The second point of the command AQ is covered by only the chain ε_{21}, which also defines a command of type $R1$. The inclusion of other variants requires an unjustified extension of the catalogue of elementary chains. The determination of lining the standard block $O(\lambda)$ for a given sequence λ guarantees the uniqueness up to an isomorphism at the abstraction level $\mathbf{X}$ for the fixed parameter λ. □

The ideas presented in the proof of the lemma lead to the notions of a singular line and a singular point. By a *singular line* we mean a row of touching realizations by an elementary chain. A point at which some elementary chain is realized is called *singular* if several singular lines start from this point. We define one more type of line. A row of touching realizations of a chain is called a *generating line* if it generates a fan of singular lines of a certain type. Singular points of type O and AQ, singular lines of types A, M, and Q, and generating lines of type G and Z appear in the process of lining a standard block.

Lemma 3.6.2. *If a block $\mathcal{B}$ contains a point satisfying $J(x)$ and a point satisfying $Q(x) \vee Z(x) \vee G(x)$, then it is standard.*

PROOF. By condition, in $\mathcal{B}$ there is a J-line. Therefore, the Q-line necessarily is brought down, and this happens at O. In the same way, if in the block with the J-line there is a Z-line, then its extension to the lower left must reach J, which defines the point O. Similarly, the G-line must reach J, which defines the point O. □

Lemma 3.6.3. *If a block $\mathcal{B}$ contains a point satisfying $Q(x)$ and a point satisfying $Z(x)$, then it is standard.*

PROOF. We note that the Q-line cannot be lower than the Z-line since the "slope coefficient" of the Q-line with respect to the surrounding net structure is equal to $1/2$ or $-1/2$ and the same coefficient of the Z-line with respect to the domain at the bottom is equal to $1/3$; otherwise, they must touch, which contradicts Catal 3.1. Thus, the Q-line must be located higher than the Z-line; therefore, the Q-line necessarily is brought down, which defines the point O. □

Now, we describe a series of blocks by a certain unique method. We consider the following classes of elementary chains:

$$K_1 = \{\varepsilon_1\}, \quad K_2 = \{\varepsilon_2, \varepsilon_1, \varepsilon_3, \varepsilon_4\}, \quad K_3 = \{\varepsilon_4\},$$
$$K_4 = \{\varepsilon_8, \varepsilon_4\}, \quad K_5 = \{\varepsilon_9, \varepsilon_8, \varepsilon_4\}, \quad K_6 = \{\varepsilon_5, \varepsilon_1\},$$
$$K_7 = \{\varepsilon_7, \varepsilon_1, \varepsilon_2, \varepsilon_3, \varepsilon_4, \varepsilon_5, \varepsilon_6, \varepsilon_8\}, \quad K_8 = \{\varepsilon_{20}, \varepsilon_1, \varepsilon_5, \varepsilon_6, \varepsilon_{18}, \varepsilon_{19}\},$$
$$K_9^1 = \{\varepsilon_9, \varepsilon_1\}, \quad K_9^2 = \{\varepsilon_{10}, \varepsilon_{14}, \varepsilon_7, \varepsilon_8, \varepsilon_1\},$$
$$K_{10}^1 = \{\varepsilon_{12}, \varepsilon_{16}, \varepsilon_1, \varepsilon_5, \varepsilon_{10}\}, \quad K_{10}^2 = \{\varepsilon_{12}, \varepsilon_{17}, \varepsilon_1, \varepsilon_5, \varepsilon_{10}, \varepsilon_{11}\},$$
$$K_{10}^3 = \{\varepsilon_{13}, \varepsilon_{16}, \varepsilon_1, \varepsilon_5, \varepsilon_{10}, \varepsilon_{11}\}, \quad K_{10}^4 = \{\varepsilon_{13}, \varepsilon_{17}, \varepsilon_1, \varepsilon_5, \varepsilon_{11}\},$$
$$K_{10}^5 = \{\varepsilon_{14}, \varepsilon_{16}, \varepsilon_1, \varepsilon_5, \varepsilon_{10}\}, \quad K_{10}^6 = \{\varepsilon_{14}, \varepsilon_{17}, \varepsilon_1, \varepsilon_5, \varepsilon_{10}, \varepsilon_{11}\},$$
$$K_{10}^7 = \{\varepsilon_{15}, \varepsilon_{16}, \varepsilon_1, \varepsilon_5, \varepsilon_{10}, \varepsilon_{11}\}, \quad K_{10}^8 = \{\varepsilon_{15}, \varepsilon_{17}, \varepsilon_1, \varepsilon_5, \varepsilon_{11}\}.$$

We use the notation K_n^i for all the above classes, assuming that the superscript i is absent for $n < 9$. The first chain in K_n^i is called *principal*.

Lemma 3.6.4. *The following assertions hold:*

(a) *For every class K_n^i, $1 \leqslant n \leqslant 10$, there exists a block $\mathcal{B} = \mathcal{B}_n^i$ such that only chains of the class K_n^i are realized in $\mathcal{B}$. Moreover, the principal chain is also realized in $\mathcal{B}$, the block $\mathcal{B}$ is unique up to an isomorphism at the abstraction level $\mathbf{X}$, and $\mathcal{B}$ is a block of the fragment n from* Catal 3.2 *(one version of such a block if there are several in the catalogue).*

(b) *If all chains realized in the block $\mathcal{B}$ are contained in the same class K_n^i and the principal chain of K_n^i is not realized in $\mathcal{B}$, then the block $\mathcal{B}$ has type $\mathcal{B}_m^k$ in accordance with the description (a) for one of the classes K_m^k, $m < n$.*

PROOF. (a). For a given class, the block is lined by elementary chains in accordance with the following scheme. We first place the principal chain. Then, analyzing the possible cases, we establish that the entire block can be uniquely lined by available chains.

(b). We analyze all possible cases. ☐

Lemma 3.6.5. *If in a block $\mathcal{B}$ there is exactly one point of execution of commands, then the isomorphism type of the block $\mathcal{B}$ at the abstraction level $\mathbf{X}$ is uniquely defined by the type of the point of execution of commands; moreover, it is defined by some class K_n^i.*

PROOF. We first place a coupled pair of elementary chains $\varepsilon_i \in \{\varepsilon_{12}, \varepsilon_{13}, \varepsilon_{14}, \varepsilon_{15}\}$ and $\varepsilon_j \in \{\varepsilon_{16}, \varepsilon_{17}\}$, which defines the point AQ of execution of commands of a given type. Further, the block is lined by any elementary chains, so that no other points of execution of commands appear. In particular, none of the chains $\varepsilon_{12} - \varepsilon_{17}$ may be used once more;

otherwise, we obtain one more point of execution of commands. We exclude the chains $\varepsilon_{21}-\varepsilon_{23}$, which yield a standard block with many points of execution of commands, the chains $\varepsilon_2-\varepsilon_4$ and $\varepsilon_6-\varepsilon_9$ containing points from J and the lower part of the model, and the chains $\varepsilon_{18}-\varepsilon_{20}$ defining the stepwise Z-line; otherwise, we obtain a standard block in view of Lemma 3.6.3.

Having the first two chains, we further must use only the chains ε_1, ε_5, ε_{10}, and ε_{11}, which represent a pure net, as well as pieces of A-lines and Q-lines. Having analyzed some variants, we can show that the result is the block defined by Lemma 3.6.4(a) with a suitable class K_n^i. □

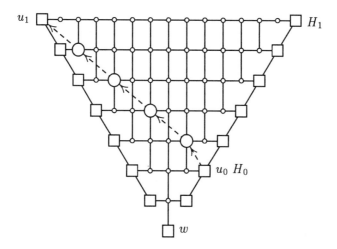

Fig. 3.6.1. The downward-directed crowd principle.

Lemma 3.6.6. *If in a block B there are at least two points of execution of commands, then the block B is standard.*

PROOF. In the block B, we take two H-classes, H_0 and H_1, containing points of execution of commands u_0 and u_1 such that the distance t between H_0 and H_1 is minimal for B. By Axiom $27°$, in every H-class of the part of the block B considered, there is exactly one Q-point. By the above choice, in the region between H_0 and H_1 there are no other points of execution of commands. Therefore, the Q-line must join the points u_0 and u_1 (cf. Fig. 3.6.1). Consequently, A-lines passing through the points u_0 and u_1 are different and cannot be unboundedly extended downward.

In accordance with Catal 3.1, the A-line can be brought down in one of the following situations.

CASE 1: on the J-line.

CASE 2: on the Z-line.

CASE 3: near a point of execution of commands with division of cells.

In Cases 1 and 2, the block is standard by Lemmas 3.6.2 and 3.6.3. Case 3 contradicts the fact that t is minimal (cf. Fig. 3.6.1). Other types of commands at the points u_0, u_1 change the configuration of the picture, but those parts are inessential in our consideration. ☐

Now, we are in a position to proceed to the proof of the key assertion about the completeness of the catalogue Catal 3.2. The above-mentioned collections of elementary chains completely define all the types of nonstandard blocks in Catal 3.2. Therefore, it remains to prove that other blocks are impossible. It suffices to establish that if there is a collection of chains that are realized in some block and are covered by none of the classes from the above list, then the block contains the point O.

Lemma 3.6.7 [the main combinatorial lemma]. *Let a model $\mathfrak{N}$ satisfy Axioms $1°$–$23°$ and let $\mathcal{B}$ be one of its blocks. In the block $\mathcal{B}$, Axioms $24°$–$28°$ hold if and only if the description of $\mathcal{B}$ at the abstraction level* **X** *is contained in* Catal 3.2.

PROOF. If the description of the block $\mathcal{B}$ is contained in Catal 3.2, the truth of Axioms $24°$–$28°$ in $\mathcal{B}$ can be directly established.

To prove the converse assertion, we consider an arbitrary block $\mathcal{B}$ in some model satisfying Axioms $24°$–$28°$. We show that the isomorphism type of the block $\mathcal{B}$ at the abstraction level **X** is contained in Catal 3.2. We consider three possible cases.

CASE 1: the block $\mathcal{B}$ is located in the lower part. Only ε_4, ε_8, and ε_9, which constitute exactly the class K_5, can be realized in $\mathcal{B}$. The block $\mathcal{B}$ is in Catal 3.2 by Lemma 3.6.4.

CASE 2: the block $\mathcal{B}$ is located on the interface between the upper and lower parts, i.e., in $\mathcal{B}$ there exist J-points. A block $\mathcal{B}$ is standard by definition if the chain ε_{22} or the chain ε_{23} is realized in $\mathcal{B}$. The chain ε_{21} is inadmissible because it requires realization of the chain ε_{22}. If only one of the chains ε_9–ε_{20} is realized in $\mathcal{B}$, then the block $\mathcal{B}$ is standard by Lemma 3.6.2. Thus, it remains to consider the chains ε_1–ε_8, which constitute exactly the class K_7. In this case, the block $\mathcal{B}$ is contained in Catal 3.2 by Lemma 3.6.4.

CASE 3: the block is located in the upper part. We immediately can exclude the chains ε_2–ε_4, ε_6–ε_9, and ε_{21}–ε_{23} from our analysis. Any of the chains ε_{12}–ε_{17} defines the points of execution of commands. If one of these chains is realized, then the block is contained in the catalogue by Lemmas 3.6.4–3.6.6. Therefore, these chains can be eliminated from consideration. It remains to consider seven chains:

$$\varepsilon_1, \ \varepsilon_5, \ \varepsilon_{10}, \ \varepsilon_{11}, \ \varepsilon_{18}, \ \varepsilon_{19}, \ \varepsilon_{20} \qquad (3.6.2)$$

the last three chains are equivalent in the sense that they are realized or not simultaneously in any block. Analyzing variants, it is possible to check the following assertion: a subset M of the set (3.6.2) is included in none of the classes K_n^i if and only if M contains one of the following sets:

$$\begin{aligned} &\{\varepsilon_{10},\varepsilon_{11}\}, \ \{\varepsilon_{10},\varepsilon_{18}\}, \ \{\varepsilon_{10},\varepsilon_{19}\}, \ \{\varepsilon_{10},\varepsilon_{20}\}, \\ &\{\varepsilon_{11},\varepsilon_{18}\}, \ \{\varepsilon_{11},\varepsilon_{19}\}, \ \{\varepsilon_{11},\varepsilon_{20}\} \end{aligned} \qquad (3.6.3)$$

Let X denote the set of elementary chains realized in the block $\mathcal{B}$. We can assume that X is included in (3.6.2) because all other variants were considered above. If X is included in one of the classes K_n^i, then the block $\mathcal{B}$ is contained in the catalogue by Lemma 3.6.4; otherwise, X contains one of the pairs (3.6.3). If X contains $\{\varepsilon_{10},\varepsilon_{11}\}$, then the block $\mathcal{B}$ must contain points of execution of commands, but this case was studied above. If X contains any of the rest of the sets (3.6.3), then the block is standard by Lemma 3.6.3. $\qquad\square$

3.7. Axiomatics. The Turing Machine (MT)

(MT.1) General form of informational lines

29°. $A(x) \leftrightarrow A_0(x) \vee A_1(x) \vee A_2(x)$; moreover, the predicates $A_0(x)$, $A_1(x)$, and $A_2(x)$ are well defined on E-classes and distinguish mutually disjoint domains.

30°. $Q(x) \leftrightarrow Q_0(x) \vee Q_1(x) \vee \ldots \vee Q_{e-1}(x)$; moreover, the predicates $Q_0(x), Q_1(x), \ldots, Q_{e-1}(x)$ are well defined on E-classes and distinguish mutually disjoint domains.

(MT.2) The initial information of the Turing machine

31°. $S(x,y) \ \& \ S(y,z) \ \& \ O(z) \rightarrow A_2(x) \ \& \ A_1(z) \ \& \ Q_0(z)$.

$32°$. Let k be the least integer such that $m + s + 5 \leqslant 3k$, where m, s are the parameters used in the theory F in accordance with (3.1.1). Let $x_1, x_2, \ldots, x_{3k}$ be representatives of E-classes that are successively located on steps over Z-elements to the right of the point O. More exactly, the following conditions must be satisfied:

(a) $(\exists u)(\exists v)\, O(u)\ \&\ S(u, v)\ \&\ P(v, x_1)$,

(b) $S(x_{i+1}, x_{i+2})\ \&\ S(x_{i+2}, x_{i+3})$, $i = 0, 3, \ldots, 3k - 3$,

(c) $(\exists z)\, S(x_i, z)\ \&\ P(z, x_{i+1})$, $i = 3, 6, \ldots, 3k - 3$.

Then the above elements must be subject to the conditions $A_2(x_1)$, $A_1(x_2), \ldots, A_1(x_{i-1}), A_0(x_i), A_1(x_{i+1}), \ldots, A_1(x_{j-1}), A_0(x_j), \ldots,$ $A_0(x_{3k})$, where $i = m + 3$, $j = m + s + 5$.

$33°$. Let $Z(u)$, $Z(v)$, $Z(w)$, $S(u, v)$, $S(v, w)$, $P(u, x)$, $P(v, y)$, and $P(w, z)$; moreover, x, y, z are at a distance of more than k from the H-class J along the D-successor, where k is defined in Axiom $32°$. Then $A_2(x)\ \&\ A_0(y)\ \&\ A_1(z)$.

$34°$. $J(x)\ \&\ (\exists z)\big[P(z, x)\ \&\ M(z)\big] \rightarrow A_0(x) \vee A_1(x)$.

(MT.3) Messages of actions

$35°$. $P(x, y)\ \&\ A(x)\ \&\ A(y)\ \&\ \neg Q(x) \rightarrow \big(A_i(x) \leftrightarrow A_i(y)\big)$, $i = 0, 1, 2$.

$36°$. $D(x, y)\ \&\ Q(x)\ \&\ Q(y)\ \&\ \neg A(x) \rightarrow \big(Q_j(x) \leftrightarrow Q_j(y)\big)$, $j = 0, 1, \ldots,$ $e - 1$.

$37°$. Let $\varphi = P(x, y_0)\ \&\ S(y_{-3}, y_{-2})\ \&\ S(y_{-2}, y_{-1})\ \&\ S(y_{-1}, y_0)\ \&\ S(y_0, y_1)\ \&\ S(y_1, y_2)\ \&\ S(y_2, y_3)$. It is required that the following sentences be true:

(a) $\varphi\ \&\ A_i(x)\ \&\ Q_j(x) \rightarrow A_m(y_0)\ \&\ Q_t(y_{-2})$ for any command of the Turing machine $\mathcal{M}$ of the form $a_i q_j \rightarrow a_m q_t L$;

(b) $\varphi\ \&\ A_i(x)\ \&\ Q_j(x) \rightarrow A_m(y_0)\ \&\ Q_t(y_2)$ for any command of the Turing machine $\mathcal{M}$ of the form $a_i q_j \rightarrow a_m q_t R$;

(c) $\varphi\ \&\ A_i(x)\ \&\ Q_j(x) \rightarrow A_m(y_{-1})\ \&\ A_n(y_1)\ \&\ Q_t(y_{-3})$ for any command of the Turing machine $\mathcal{M}$ of the form $a_i q_j \rightarrow a_m a_n q_t L$;

(d) $\varphi\ \&\ A_i(x)\ \&\ Q_j(x) \rightarrow A_m(y_{-1})\ \&\ A_n(y_1)\ \&\ Q_t(y_3)$ for any command of the Turing machine $\mathcal{M}$ of the form $a_i q_j \rightarrow a_m a_n q_t R$.

38°. $\neg(\exists x)\,A_i(x)\ \&\ Q_j(x)$ for any pair of indices $\langle i, j\rangle$, $i < 3$, $j < e$, such that the situation $a_i q_j$ does not occur in commands of the Turing machine $\mathcal{M}$.

H-classes correspond to *instants of time* of work of the Turing machine and a displacement along the D-chain corresponds to the process of the job with time. The distinguished H-class J is the *initial instant of time* of operation of the Turing machine. The set of A-points of an H-class presents *tape cells* at one point in time. The segment of the P-chain consisting of A-points is called an *A-route*. Each A-route represents a tape cell at sequential instants of time. In the upper part, a Q-point which is unique in every H-class indicates the position of the *head* of the machine with respect to the tape. Furthermore, E-classes such that $A(x)\ \&\ Q(x)$ correspond to a *point of execution of commands* of the Turing machine. In regions between cells of the tape, there are predicate-free points so that the head is shifted to a neighboring cell in several instants of time.

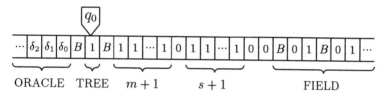

Fig. 3.7.1. The start of working of the Turing machine.

Axioms 31°–34° describe the state of the tape and the head at the initial moment when the Turing machine starts (cf. Fig. 3.7.1). At that instant the head is fixed on the field TREE of length 1. The role of the field TREE is explained below. To the right of the field the above-mentioned parameters m and s are represented by two groups of units (cf. Sec. 3.1). To the right of them the free field FIELD of repeating symbols a_2, a_0, and a_1 is located. This field has a certain meaning for model-theoretic properties and is used by the Turing machine. To the left of the field TREE there is the field ORACLE containing some arbitrary messages given by 0's and 1's. The designation of this field will be discussed below. The next states of the tape and the head are defined by the initial information and commands of the Turing machine $\mathcal{M}$ during the process of the job.

We note that the parameters of the abstraction level **I** in Catal 3.2 indicate the state of the head and the tape symbol for the informational lines A and Q.

3.8. Axiomatics. The Translator (TRN)

We pass to the axioms of the last step connecting the operation of the Turing machine and the model-theoretic properties of the theory.

(TRN.1) The branching structure of the translator

$39°$. The class $[x]_E$ contains W-elements if and only if $A(x)$.

$40°$. $V(x,y) \to W(x) \,\&\, W(y)$.

$41°$. For every x from W, there exists a unique element y such that $V(x,y)$.

$42°$. If $(\exists z)\, V(z,y)$, then for the element y there exist exactly two elements z such that $V(z,y)$.

$43°$. $A(x)$, $P(x,y_0)$, $P(y_{-1},y_0) \,\&\, P(y_0,y_1)$ imply $V(x,y) \to y \in [y_{-1}]_E \cup [y_0]_E \cup [y_1]_E$.

$44°$. $U(x) \leftrightarrow O(x) \,\&\, W(x)$.

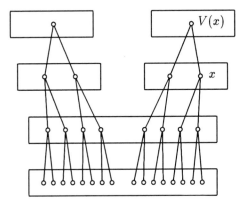

Fig. 3.8.1. Branching components.

The axioms of this group describe a special branching-downward structure of V-connections on elements of the basis W included in the A-points which represent the tape cells of the Turing machine (cf. Fig. 3.8.1). By the above axioms, the predicate V can be regarded as a one-place operation on the set W. Therefore, we write $y = V(x)$ instead of $V(x,y)$.

By a *component* we mean a minimal nonempty subset $\varkappa \subseteq W$ that is closed under V-images and V-preimages. By the axioms, any component is located in the upper part of the model within a block along A-lines. Passing through a point of execution of commands with division of cells, the component must "choose" one of two possible extensions of the path along the A-line (cf. Fig. 3.8.1), where rectangles represent the sets of W-elements in E-classes in a neighborhood of the point of execution of commands with division of cells.

The boundedness from below of a component depends on the presence of points at which the A-line is brought down. Any component is unbounded from above; therefore, the intersection with an E-class (if it is nonempty) contains infinitely many elements.

To characterize a location of a component in a block, we introduce the notion of the *A-path* of the block $\mathcal{B}$ as an arbitrary maximal linear chain of its A-routes passing through points of execution of commands with division of cells. By definition, any A-path is unboundedly extended to the top. An A-path is unbounded from below or breaks at a J-point or Z-point. Every component is located at one A-path; moreover, many components can go along the same A-path. It is possible that none of components pass along some A-path (for example, if a standard block contains the continuum of different A-paths and the model is countable).

An A-path is called *finite-linked* if it consists of a finite number of A-lines, and *infinite-linked* otherwise. At least one component is in every finite-linked A-path. On the contrary, an infinite-linked A-path can contain none of the components because each of its A-lines is covered by components of other A-paths. The parameters of the abstraction level $\mathbf{W}$ in Catal 3.2 (also, cf. Table 3.6.1) indicate the number of components in every A-path of the block under consideration.

The last axiom of the theory $F = F(\mathcal{M}, m, s)$ is stated in Sec. 3.11. We denote by $F'(\mathcal{M}, m, s)$ the subtheory defined by Axioms $1°$–$44°$.

3.9. One-Block Models

In this section, we deduce the consistency condition for the theory $F'(\mathcal{M}, m, s)$. For this purpose, we study models (called one-block models) consisting of a single standard block. The axioms of the group **(MT.2)** admit that, at the initial instant of time of operation of the Turing machine, in the field ORACLE located on the left-side of the tape there is an arbitrary sequence $\{\delta_i \mid i < \omega\}$, where $\delta_i \in \{0, 1\}$, $i < \omega$ (cf. Fig. 3.7.1).

Sequential cells of the field ORACLE are written as $\Delta_0, \Delta_1, \Delta_2, \ldots$, and are referred to as *storages* with the oracle information. In the general case, by *storage* we mean a tape cell or a group of tape cells with some special information.

Because of the form of the standard block, at the initial instant of time there are only three cells to the right of the head on the tape of the Turing machine. However, the saturation line Z is constructed independently of the operation of the Turing machine, advancing the speed of the movement of the head. If the head moves to the right, the tape does not break and all the required cells are found. Therefore, we can assume that the tape is unbounded in both direction beginning from the initial moment, and the right-hand side of the tape contains the parameters m, s and the periodically repeating sequence of symbols which is defined by Axiom 32° (cf. Fig. 3.7.1). Let $\mathfrak{M}$ be an arbitrary model of the theory $F'(\mathcal{M}, m, s)$. We consider the initial instant of time of the Turing machine $\mathcal{M}$ corresponding to the standard block of this model. Let

$$\Delta = \{i \mid \text{in the storage } \Delta_i \text{ the symbol 1 is written}\}$$

The set Δ defined by a standard block of a model $\mathfrak{M}$ is denoted by Oracle $(\mathfrak{M})$.

We fix a Turing machine $\mathcal{M}$ with three tape symbols and some values of the parameters $m, s \in \mathbb{N}$, $\Delta \subseteq \mathbb{N}$. By the (Δ, m, s)-*computation*, we mean the job of the Turing machine $\mathcal{M}$ starting from the state shown in Fig. 3.7.1, where information on the left-hand side of the tape is defined by the set Δ as follows:

$$\delta_i = \begin{cases} 1 & \text{if } i \in \Delta \\ 0 & \text{if } i \notin \Delta \end{cases}$$

Let $\text{Nonstop}(\mathcal{M}, m, s)$ denote the set of all Δ such that the (Δ, m, s)-computation of the Turing machine $\mathcal{M}$ does not lead to a halt.

The main dependence between the notions introduced is given by the following lemma.

Lemma 3.9.1. *Let $\mathcal{M}$ be a Turing machine with three tape symbols and number parameters $m, s \in \mathbb{N}$. We suppose that for any $\Delta \subseteq \mathbb{N}$ the first two realized commands in the (Δ, m, s)-computation of the Turing machine $\mathcal{M}$ have type R1 ("move the head to the right without division of cells"). Then the following assertions hold:*

(a) Oracle $(\mathfrak{M}) \in \text{Nonstop}(\mathcal{M}, m, s)$ *for any model $\mathfrak{M}$ of the theory* $F'(\mathcal{M}, m, s)$,

(b) *for any set* $\Delta \in \mathrm{Nonstop}(\mathcal{M}, m, s)$, *there exists a model* $\mathfrak{M} \in$ $\mathrm{Mod}(F'(\mathcal{M}, m, s))$ *consisting of one standard block and satisfying the condition* $\mathrm{Oracle}\,(\mathfrak{M}) = \Delta$.

PROOF. (a). We consider a model $\mathfrak{M}$ of the theory $F'(\mathcal{M}, m, s)$. Let $\Delta = \mathrm{Oracle}\,(\mathfrak{M})$. By the above axioms, the standard block of the model $\mathfrak{M}$ is the (Δ, m, s)-computation of the Turing machine $\mathcal{M}$ considered in time. Since Axiom $38°$ holds in $\mathfrak{M}$, we have $\Delta \in \mathrm{Nonstop}(\mathfrak{M})$.

(b). Let $\Delta \in \mathrm{Nonstop}(\mathcal{M}, m, s)$. Considering the $(\mathcal{M}, m, s)$-computation of the Turing machine $\mathcal{M}$ in time, we can build on the upper half of the standard block which is defined by this computation. Then we put the obtained block on the frame and insert a component in each of its K-paths. We obtain the model for which all Axioms $1°$–$44°$ hold.   □

Lemma 3.9.1 implies the following lemma.

Lemma 3.9.2. *For any Turing machine $\mathcal{M}$ with three tape symbols such that, in any (Δ, m, s)-computation of the Turing machine, the first two realized commands have type R1, the following assertions hold*:

$$\mathrm{Nonstop}(\mathcal{M}, m, s) = \{\mathrm{Oracle}\,(\mathfrak{M}) \mid \mathfrak{M} \in \mathrm{Mod}\,(F'(\mathcal{M}, m, s))\},$$

$F'(\mathcal{M}, m, s)$ *is consistent if and only if* $\mathrm{Nonstop}(\mathcal{M}, m, s) \neq \varnothing$.

3.10. Global Structure of Models

Based of the above description of blocks, we specify the block-zone form of models subject to Axioms $1°$–$44°$ (cf. Fig. 3.4.2). The structure of zones of different types is described in Table 3.10.1. Each zone can contain a collection of blocks of certain types. In the column named "essential," the types of blocks are indicated such that a single copy of the block occurs in the zone. The existence of such blocks is guaranteed by Axioms $23°$, $26°$–$28°$. In the column "inessential," the types of blocks are indicated such that any number of such blocks or none of the blocks can occur in the zone.

Let Z_1 and Z_2 be zones of some models. We call them *isomorphic at the abstraction level* **X** *or* **I** if there is a one-to-one correspondence between points (i.e., between E-classes) of these zones that preserves all predicates relative to this level. It is obvious that two zones are isomorphic if they have the same type and contain the same collections of blocks of inessential types. This is true because only essential blocks O and

AQ contain singular points, but the rest of the blocks are homogeneous with respect to shifts along the P-chain. We note that models having identical descriptions in the language of types of blocks and zones (even at the abstraction level **I**) can turn out to be nonisomorphic in view of the difference in the number of components along the A-paths.

Table 3.10.1

Notation for type of zone	Types of blocks in zones		Zone
	Essential	Inessential	
$\mathbf{J}(\Delta)$	$O(\Delta)$	BH, MA	Standard zone
$\mathbf{B}''(\nu, j)$	$Q(\nu, j), Z$	B, A	Zone in the upper part with point of execution of commands
$\mathbf{B}'(\nu, i, j)$	$AQ(\nu, i, j), Z$	B, A	Zone in the upper part with the line of moving the head
$\mathbf{C}$	G	$C, \mathbb{M}$	Zone in the lower part

In any model $\mathfrak{M}$ of the theory $F'(\mathcal{M}, m, s)$, there is exactly one standard zone, whereas other zones may occur in all proportions or be absent altogether.

3.11. Admissible Turing Machines and the Normalization Axiom

In this section, we indicate the class of Turing machines that will be considered in what follows.

A Turing machine $\mathcal{M}$ with division of cells is called *admissible* if it has three tape symbols and satisfies the following conditions:

(a) for any initial parameters $\Delta \subseteq \mathbb{N}$, $m, s \in \mathbb{N}$, the first two commands in the (Δ, m, s)-computation of the Turing machine yield the motion of the head to the right; moreover, these commands do not divide cells,

(b) $\Delta \in \mathrm{Nonstop}(\mathcal{M}, m, s) \Rightarrow (\Delta$ and $\mathbb{N} \setminus \Delta$ are infinite) for any parameters $\Delta \subseteq \mathbb{N}$, $m, s \in \mathbb{N}$,

(c) at the beginning of the (Δ, m, s)-computation of the Turing machine $\mathcal{M}$, commands with division of cells can occur only in the cell of the field TREE as well as the cells obtained as the result of division. Thereby, the field TREE is extended; moreover, none of the tape cells located outside the field TREE can be divided,

(d) for any $\Delta \subseteq \mathbb{N}$, $m, s \in \mathbb{N}$, in the case $\Delta \in \mathrm{Nonstop}(\mathcal{M}, m, s)$, in the (Δ, m, s)-computation of the Turing machine $\mathcal{M}$ there are infinitely many points of execution of commands with division of cells.

We pass to the description of the last axiom of the theory $F(\mathcal{M}, m, s)$, which provides the model completeness of the theory. We first introduce some auxiliary notions. Let $\nu \in \{L, R\}$, $i < 3$, $j < e$, $\mathfrak{M} \in \mathrm{Mod}\left(F'(\mathcal{M}, m, s)\right)$. A point of the model $\mathfrak{M}$ is called

— a $\langle \nu, j \rangle$-*point* if it is situated in the part of the Q-route with slope ν and is marked by the predicate Q_j,

— a $\langle \nu, i, j \rangle$-*point* if it is a point of execution of commands at the state $a_i q_j$ and, before the execution of a command, the head moves in the direction ν.

We adopt the convention that every $\langle \nu, i, j \rangle$-point is simultaneously a $\langle \nu, j \rangle$-point.

Let J_k, $k \in \mathbb{N}$, denote an H-class which is the kth D-successor of the class J. By the *weight* of the triple $\langle \nu, i, j \rangle$ in the model $\mathfrak{M}$ we mean the number

$$\|\langle \nu, i, j \rangle\|_{\mathfrak{M}} = \begin{cases} m & \text{if the set of } \langle \nu, i, j \rangle\text{-points in the standard block of the model } \mathfrak{M} \text{ is finite and the latter, counting along the } D\text{-chain, is in the class } J_m \\ \omega & \text{if the set of } \langle \nu, i, j \rangle\text{-points in the standard block of the model } \mathfrak{M} \text{ is infinite} \end{cases}$$

Similarly, we can define the number $\|\langle \nu, j \rangle\|_{\mathfrak{M}}$, which is called the *weight of the pair* $\langle \nu, j \rangle$ in the model $\mathfrak{M}$.

A Turing machine $\mathcal{M}$ is called *normal* if there exists a natural number $n_{\mathcal{M}}$ (called the *boundary of realizations* for $\mathcal{M}$) such that for any triple

$\langle \nu, i, j \rangle$ one of the following conditions holds:

$$(\forall \mathfrak{M} \in \mathrm{Mod}\,(F'(\mathcal{M}, m, s)))\ \|\langle \nu, i, j \rangle\|_{\mathfrak{M}} < n_{\mathcal{M}} \qquad (3.11.1)$$

$$(\forall \mathfrak{M} \in \mathrm{Mod}\,(F'(\mathcal{M}, m, s)))\ \|\langle \nu, i, j \rangle\|_{\mathfrak{M}} = \omega \qquad (3.11.2)$$

A triple $\langle \nu, i, j \rangle$ is said to be *finitely realized* or *infinitely realized* if (3.11.1) or (3.11.2) holds respectively. Similar notions can be introduced for pairs $\langle \nu, j \rangle$.

The following lemma is obvious.

Lemma 3.11.1. *Let* $j < e$, $\nu \in \{L, R\}$ *and let* $\mathfrak{M}$ *be a model of the theory* $F'(\mathcal{M}, m, s)$. *Then* $\|\langle \nu, j \rangle\|_{\mathfrak{M}} = \max\{\|\langle \nu, i, j \rangle\|_{\mathfrak{M}} : 0 \leqslant i < 3\}$.

By Lemma 3.11.1, if all triples $\langle \nu, i, j \rangle$, satisfy the normality condition, then a similar condition is valid for all pairs $\langle \nu, j \rangle$.

Regarding a normal Turing machine $\mathcal{M}$, we introduce the following axiom.

(NRM) The normalization axiom

$45°$. Outside the classes $J_0, J_1, \ldots, J_{n(\mathcal{M})}$ there are no $\langle \nu, i, j \rangle$-points for every finitely realized triple $\langle \nu, i, j \rangle$ and there are no $\langle \nu, j \rangle$-points for every finitely realized pair $\langle \nu, j \rangle$.

Adding the normalization axiom to the theory $F'(\mathcal{M}, m, s)$, we obtain the theory $F(\mathcal{M}, m, s)$, which is the final purpose of the construction in accordance with the scheme from Sec. 3.1. In the remaining part of the chapter, we study model-theoretic properties of the theory $F(\mathcal{M}, m, s)$, construct a suitable Turing machine, and choose parameters so as to obtain, as a result, the theory possessing the properties indicated in Theorem 3.1.1.

By Lemma 3.11.1, for the effective construction of the normalization axiom, it suffices to know the list of all finitely realized triples and the boundary of their realizations $n_{\mathcal{M}}$. The normalization axiom is always true in a standard block and, consequently, in any model of the theory $F'(\mathcal{M}, m, s)$ consisting of a single standard zone. However, the normalization axiom affects nonstandard zones of models. In particular, it inhibits nonstandard zones of types $\mathbf{B}'(\nu, i, j)$ and $\mathbf{B}''(\nu, j)$ for all finitely realized triples $\langle \nu, i, j \rangle$ and pairs $\langle \nu, j \rangle$, which prevents the indetermination (nonprovability and irrefutability) of some propositions in the theory $F(\mathcal{M}, m, s)$.

The aforesaid allows us to establish some properties of the theory $F(\mathcal{M}, m, s)$ which are similar to those of the subtheory that does not contain the normalization axiom.

Lemma 3.11.2. *Let $\mathcal{M}$ be an admissible normal Turing machine. Then* $\mathrm{Nonstop}(\mathcal{M}, m, s) = \{\mathrm{Oracle}(\mathfrak{M}) \mid \mathfrak{M} \in \mathrm{Mod}(F(\mathcal{M}, m, s))\}$.

PROOF. The arguments are similar to those in the proof of the corresponding assertion in Sec. 3.9, but it is necessary to take into account that the normalization axiom is always true in the standard block.   □

3.12. Perfect Models and the Model Completeness

In this section, we develop a special technique to study model-theoretic properties of the theory $F(\mathcal{M}, m, s)$. Let $\mathfrak{M}$ be a model of the theory $F(\mathcal{M}, m, s)$. Denote by $\mathrm{TREE}(\mathfrak{M})$ a minimal set of A-points of the standard block of the model $\mathfrak{M}$ such that $\mathrm{TREE}(\mathfrak{M})$ contains the point O and is closed under P-successors in A and under newly generated A-points after points of execution of commands with division of cells. This definition depends only on the standard block; therefore, we will use the notation $\mathrm{TREE}(\mathcal{B})$, where $\mathcal{B}$ is the standard block of the model $\mathfrak{M}$.

Points of execution of commands are binary branchings. Hence the set $\mathrm{TREE}(\mathfrak{M})$ is, in a sense, a tree such that, between branchings, the length of the lines may be as large as desired. Abstracting from lengths of A-routes in $\mathrm{TREE}(\mathfrak{M})$, we can regard the indivisible parts as an element. Then we obtain an ordinary binary tree (cf. the definition in Chapter 2), which will be denoted by $\mathrm{Tree}(\mathfrak{M})$ or $\mathrm{Tree}(\mathcal{B})$, where $\mathcal{B}$ is a standard block of the model $\mathfrak{M}$.

It is obvious that $\mathrm{Tree}(\mathfrak{M})$ is uniquely determined by the process of the job of the Turing machine $\mathcal{M}$ which, in turn, is uniquely defined by the initial information on the tape of the Turing machine. Therefore, the tree $\mathrm{Tree}(\mathfrak{M})$ is also denoted by $\mathrm{Tree}(\mathcal{M}, \Delta, m, s)$, where $\Delta = \mathrm{Oracle}(\mathfrak{M})$ and $\mathcal{M}, m, s$ are parameters appearing in the theory considered. We note that $\mathrm{Tree}(\mathcal{M}, \Delta, m, s)$ is constructed as an enumerable set over the oracle Δ with parameters m and s. By condition (d) from the definition of an admissible Turing machine, this tree is infinite for an admissible machine provided that the parameters do not lead to a halt.

We pass to model-theoretic properties. Let $\mathfrak{M}$ be a model of the theory $F(\mathcal{M}, m, s)$ of cardinality α.

The model $\mathfrak{M}$ is called *perfect over the set* $K \subseteq |\mathfrak{M}|$ if the following conditions hold:

(a) $\mathfrak{M}$ has α zones of type $\mathbf{B}'(\nu, i, j)$ which do not intersect K for every infinitely realized triple $\langle \nu, i, j \rangle$,

(b) $\mathfrak{M}$ has α zones of type $\mathbf{B}''(\nu, j)$ which do not intersect K for every infinitely realized pair $\langle \nu, j \rangle$,

(c) $\mathfrak{M}$ has α zones of type $\mathbf{C}$ which do not intersect K,

(d) in every zone, the model $\mathfrak{M}$ has α blocks which do not intersect K of every inessential type for this zone,

(e) every A-path of the model $\mathfrak{M}$ contains α components which do not intersect K.

The following lemma is a consequence of definitions.

Lemma 3.12.1. *If* $\mathfrak{M} \in \mathrm{Mod}\,(F(\mathcal{M}, m, s))$, $N \subseteq K \subseteq |\mathfrak{M}|$, *and the model* $\mathfrak{M}$ *is perfect over* K, *then* $\mathfrak{M}$ *is perfect over* N.

We simply say a perfect model if the latter is perfect over the empty set. If $\mathfrak{M} \subseteq \mathfrak{N}$ and a model $\mathfrak{N}$ is perfect over $|\mathfrak{M}|$, then $\mathfrak{N}$ is called a *perfect extension* of the model $\mathfrak{M}$.

Let $t_0(\mathfrak{M})$ be the cardinality of the set $\Pi\,(\mathrm{Tree}\,(\mathfrak{M}))$ and let $t(\mathfrak{M})$ be the number of all A-paths of the model $\mathfrak{M}$.

Lemma 3.12.2. *Let* $\mathcal{M}$ *be a normal admissible Turing machine. For any model* $\mathfrak{M}$ *of the theory* $F(\mathcal{M}, m, s)$, *the following relations hold:*

$$\omega \leqslant t_0(\mathfrak{M}) \leqslant 2^\omega \tag{3.12.1}$$

$$t_0(\mathfrak{M}) \leqslant t(\mathfrak{M}) \leqslant \max\{t_0(\mathfrak{M}), \|\mathfrak{M}\|\} \tag{3.12.2}$$

$$t(\mathfrak{M}) = t_0(\mathfrak{M}) \text{ if } \|\mathfrak{M}\| = \omega \tag{3.12.3}$$

$$t(\mathfrak{M}) \leqslant \|\mathfrak{M}\| \text{ if } \|\mathfrak{M}\| \geqslant 2^\omega \tag{3.12.4}$$

PROOF. The inequalities (3.12.1) and the first inequality in (3.12.2) are obvious. The second inequality in (3.12.2) is valid since there are at most a countable number of A-paths in every nonstandard block. The relations (3.12.3) and (3.12.4) follow from the previous relations. □

Lemmas 3.12.3 and 3.12.4 establish important properties of perfect extensions.

Lemma 3.12.3. *Let $\mathcal{M}$ be a normal admissible Turing machine and let m, s be arbitrary parameters. Each model $\mathfrak{M} \in \mathrm{Mod}\,(F(\mathcal{M}, m, s))$ has an elementary perfect extension of any cardinality*

$$\alpha \geqslant \max\{t_0(\mathfrak{M}), \|\mathfrak{M}\|\}$$

PROOF. We consider an arbitrary elementary extension $\mathfrak{N}$ of the model $\mathfrak{M}$ of cardinality α. By Lemma 3.12.2, $t(\mathfrak{N}) \leqslant \alpha$. We consider the signature $\sigma^* = \sigma \cup \{c_\alpha \mid \alpha \in |\mathfrak{N}|\}$. Let $\mathfrak{N}^*$ denote the natural enrichment of the model $\mathfrak{N}$ to a model of the signature σ^*. We describe various methods of construction of different types of elementary extensions of the model $\mathfrak{N}$ which are necessary for constructing a perfect model.

To construct an elementary extension $\mathfrak{N}'$, $\mathfrak{N} \preccurlyeq \mathfrak{N}'$, of cardinality α in which there is at least one new zone of type $\mathbf{B}'(\nu, i, j)$, we consider the set of formulas $\Sigma(x)$ containing the following assertions:

x is a $\langle \nu, i, j \rangle$-point,

$\neg H(x, c_\alpha)$ for all $\alpha \in |\mathfrak{N}|$.

Since the triple $\langle \nu, i, j \rangle$ is infinitely realized, the set $\Sigma(x)$ is locally consistent with $\mathrm{Th}(\mathfrak{N}^*)$. The elementary extension $\mathfrak{N}'$ of the model $\mathfrak{N}^*$ of cardinality α which realizes $\Sigma(x)$ is the required model. The remaining types of zones indicated in conditions (b) and (c) of the definition of a perfect model are considered in a similar way.

Let Z be a zone of a model $\mathfrak{N}$ of type $\mathbf{B}'$. To construct the elementary extension $\mathfrak{N}'$, $\mathfrak{N} \preccurlyeq \mathfrak{N}'$, of cardinality α which has a new block of type $A(i)$ in the zone Z, we consider the set of formulas

$$\Sigma(x) = \{H(x, c_b)\} \cup \{\neg(E(x, c_\alpha) \mid \alpha \in |\mathfrak{N}|\} \cup \{A_i(x)\} \cup p(x)$$

where b is a fixed element of this zone and $p(x)$ is the set of formulas that assert that S–P-connections in a neighborhood of x are of type $A(i)$. In accordance with the description of models, in the zone Z of the model $\mathfrak{N}$, there is the saturation block in which any finite part of the set $\Sigma(x)$ can be realized. For $\mathfrak{N}'$ we can take the elementary extension of the model $\mathfrak{N}^*$ of cardinality α which realizes $\Sigma(x)$. The remaining types of zones and blocks indicated in condition (d) of the definition of a perfect model are considered in a similar way. To study blocks of type $JA(0)$ and $JA(1)$, it is required to use condition (b) from the definition of an admissible Turing machine $\mathcal{M}$.

We consider an arbitrary A-path $\mathcal{A}$ in the model $\mathfrak{N}$ and an element b from one of the E-classes on this A-path. We construct the set of formulas

$$\Sigma(x) = \{E(x, c_b)\} \cup \{V^k(x) \text{ is in } \mathcal{A} \mid k < \omega\}$$
$$\cup \{V^k(x) \neq c_a \mid a \in |\mathfrak{M}|, k < \omega\}$$
$$\cup \{c_a \notin V^{-k}(x) \mid a \in |\mathfrak{M}|, k < \omega\}$$

where $V^0(x) = x$, $V^{k+1}(x) = V(V^k(x))$ and $V^{-k}(x)$ is the kth preimage of x for $k > 0$. Since every finite part of the path $\mathcal{A}$ is covered by some component of the model $\mathfrak{N}$, the set $\Sigma(x)$ is locally consistent with the theory $\mathrm{Th}(\mathfrak{N}^*)$. The elementary extension $\mathfrak{N}'$ of the model $\mathfrak{N}^*$ which realizes the set $\Sigma(x)$ contains a new component located in $\mathcal{A}$.

By a standard method, alternating extensions for different types of zones, blocks, and all possible paths, we construct an elementary chain of length α whose union $\mathfrak{N}^*$ is an elementary extension of the initial model $\mathfrak{M}$ of cardinality α. This model satisfies all the requirements of the definition of a perfect extension of the model $\mathfrak{M}$. $\square$

Lemma 3.12.4. *Let $\mathcal{M}$ be a normal admissible Turing machine and $\mathfrak{M} \in \mathrm{Mod}\,(F(\mathcal{M}, m, s))$, $\alpha \geqslant \max\{t_0(\mathfrak{M}), \|\mathfrak{M}\|\}$. Then the perfect extension of the model $\mathfrak{M}$ of cardinality α is uniquely defined up to an isomorphism.*

PROOF. Let $\mathfrak{N}_1$ and $\mathfrak{N}_2$ be perfect extensions of the model $\mathfrak{M}$ of cardinality α. Using conditions (a)–(c) from the definition of a perfect model, we construct a one-to-one correspondence f_z between zones of the models $\mathfrak{N}_1$ and $\mathfrak{N}_2$ that preserves types of zones and their disposition with respect to the submodel $\mathfrak{M}$. Passing to H-classes and blocks and then to E-classes of these models and using condition (d) from the definition of a perfect model, we construct a bijective correspondence f_e between points (i.e., E-classes) of the models $\mathfrak{N}_1$ and $\mathfrak{N}_2$ which preserves unary predicates correct on E-classes of the abstraction level $\mathbf{I}$ and satisfies the condition $f_e([x]_E^{\mathfrak{N}_1}) = [x]_E^{\mathfrak{N}_2}$ for all x in $|\mathfrak{M}|$.

We proceed by constructing an element-by-element mapping h between the models $\mathfrak{N}_1$ and $\mathfrak{N}_2$; moreover, we independently consider the parts $X \cup Y$ and W of these models.

A MAPPING ON THE SET $X \cup Y$. Let $\mathcal{A}$, $\mathcal{A}_1$, and $\mathcal{A}_2$ be models of the quasisuccessor theory obtained from the models $\mathfrak{M}$, $\mathfrak{N}_1$, and $\mathfrak{N}_2$ by restriction to the set Y or the set X in two corresponding H-classes of the models $\mathfrak{N}_1$ or $\mathfrak{N}_2$. It is easy to see that $\mathcal{A}_1$ and $\mathcal{A}_2$ are extensions of the model $\mathcal{A}$. It is required to construct an isomorphism $h: \mathcal{A}_1 \to \mathcal{A}_2$ that is identical on $\mathcal{A}$ and is coordinated with the mapping f_e between points of the frames of $\mathfrak{N}_1$ and $\mathfrak{N}_2$. It is easy to see that f_e defines an isomorphism between the quotient models $\mathcal{A}_1/\sim$ and $\mathcal{A}_2/\sim$ that is identical on the common submodel $\mathcal{A}/\sim$. Therefore, a complete isomorphism h with the required properties exists in view of Theorem 1.7.1. Gathering all such parts together, we obtain a complete bijective mapping h', which preserves all the signature predicates on the set $X \cup Y$.

A MAPPING ON THE SET W. Using condition (e) from the definition of a perfect model, we construct a bijective mapping between components of the models $\mathfrak{N}_1$ and $\mathfrak{N}_2$ which is identical on elements of their common submodel $\mathfrak{M}$ so that the corresponding components pass along the corresponding A-paths. It is not hard to transform it in an element-by-element mapping h'' that is identical on $\mathfrak{M}$ and preserves V-connections.

Combining h' and h'', we obtain the isomorphism $h\colon \mathfrak{N}_1 \to \mathfrak{N}_2$, which is identical on $\mathfrak{M}$. $\qquad\square$

Lemma 3.12.5. *Let $\mathcal{M}$ be a normal admissible Turing machine. Then any perfect models $\mathfrak{N}_1, \mathfrak{N}_2 \in \mathrm{Mod}\,(F(\mathcal{M}, m, s))$ having the same cardinality and satisfying the equality* $\mathrm{Oracle}\,(\mathfrak{N}_1) = \mathrm{Oracle}\,(\mathfrak{N}_2)$ *are isomorphic.*

PROOF. The reasoning follows the proof of Lemma 3.12.4, but with some simplifications because it is not necessary to take into account h is the construction. The condition $\mathrm{Oracle}\,(\mathfrak{N}_1) = \mathrm{Oracle}\,(\mathfrak{N}_2)$ provides an isomorphism of standard blocks, which is necessary at the stage of constructing the function f_e. $\qquad\square$

Lemma 3.12.6. *Let $\mathcal{M}$ be a normal admissible Turing machine. Then for models of the theory $F(\mathcal{M}, m, s)$ every perfect extension is elementary.*

PROOF. Let $\mathfrak{M}$ be a model of the theory $F(\mathcal{M}, m, s)$ and let $\mathfrak{N}$ be its perfect extension of cardinality α. By Lemma 3.12.6, the model $\mathfrak{N}$ has an elementary perfect extension $\mathfrak{N}'$ of cardinality α and, by Lemma 3.12.4, the models $\mathfrak{N}$ and $\mathfrak{N}'$ are isomorphic over $\mathfrak{M}$. $\qquad\square$

We apply the technique developed to prove the key technical lemma on model completeness.

Lemma 3.12.7. *Let $\mathcal{M}$ be a normal admissible Turing machine. Then the theory $F(\mathcal{M}, m, s)$ is model-complete.*

PROOF. We take arbitrary models $\mathfrak{M}_1$ and $\mathfrak{M}_2$ such that $\mathfrak{M}_1 \subseteq \mathfrak{M}_2$. It suffices to prove that $\mathfrak{M}_1 \preccurlyeq \mathfrak{M}_2$. By Lemma 3.12.3, the model $\mathfrak{M}_2$ has an elementary perfect extension $\mathfrak{N}$, i.e., $\mathfrak{M}_1 \subseteq \mathfrak{M}_2 \preccurlyeq \mathfrak{N}$. By Lemma 3.12.1, $\mathfrak{N}$ is also a perfect extension of the model $\mathfrak{M}_1$. By Lemma 3.12.6, $\mathfrak{M}_1 \preccurlyeq \mathfrak{N}$. As a result, we obtain $\mathfrak{M}_1 \preccurlyeq \mathfrak{M}_2$. $\square$

Lemma 3.12.8. *Let $\mathcal{M}$ be a normal admissible Turing machine and let m, s be arbitrary parameters. Then $\mathfrak{M}_1 \equiv \mathfrak{M}_2 \Leftrightarrow \mathrm{Oracle}\,(\mathfrak{M}_1) = \mathrm{Oracle}\,(\mathfrak{M}_2)$ for any models $\mathfrak{M}_1, \mathfrak{M}_2 \in \mathrm{Mod}\,(F(\mathcal{M}, m, s))$.*

PROOF. The implication $\Rightarrow$ is obvious. To prove the converse implication, we assume that $\mathrm{Oracle}\,(\mathfrak{M}_1) = \mathrm{Oracle}\,(\mathfrak{M}_2)$. We set

$$\alpha = \max\{t_0(\mathfrak{M}_1), t_0(\mathfrak{M}_2), \|\mathfrak{M}_1\|, \|\mathfrak{M}_2\|\}$$

By Lemma 3.12.5, the models $\mathfrak{M}_1$ and $\mathfrak{M}_2$ have elementary perfect extensions $\mathfrak{N}_1$ and $\mathfrak{N}_2$ of cardinality α. By Lemma 3.12.1, $\mathfrak{N}_1$ and $\mathfrak{N}_2$ are perfect models of the same cardinality; moreover, $\mathrm{Oracle}\,(\mathfrak{N}_1) = \mathrm{Oracle}\,(\mathfrak{M}_1) = \mathrm{Oracle}\,(\mathfrak{M}_2) = \mathrm{Oracle}\,(\mathfrak{N}_2)$. By Lemma 3.12.5, $\mathfrak{N}_1 \cong \mathfrak{N}_2$. Consequently, $\mathfrak{M}_1 \cong \mathfrak{M}_2$. $\square$

For $i < \omega$ we denote by Ω_i the sentence asserting that, at the initial instant of time, the cell of oracle Δ_i contains 1. Let Δ be an arbitrary subset of $\mathbb{N}$. We introduce the notation

$$F(\mathcal{M}, m, s)[\Delta] = F(\mathcal{M}, m, s) \cup \{\Omega_i \mid i \in \Delta\} \cup \{\neg\Omega_j \mid j \in \mathbb{N} \setminus \Delta\}$$

The following lemma characterizes the family of all completions of the theory $F(\mathcal{M}, m, s)$ and, thereby, defines the Lindenbaum algebra of this theory.

Lemma 3.12.9. *Let $\mathcal{M}$ be a normal admissible Turing machine. Then*

(a) *if $\Delta \in \mathrm{Nonstop}(\mathcal{M}, m, s)$, then the theory $F(\mathcal{M}, m, s)[\Delta]$ is consistent and is complete,*

(b) *if $\Delta \notin \mathrm{Nonstop}(\mathcal{M}, m, s)$, then the theory $F(\mathcal{M}, m, s)[\Delta]$ is inconsistent,*

(c) *the sentences Ω_i, $i < \omega$, generate the Lindenbaum algebra of the theory $F(\mathcal{M}, m, s)$.*

PROOF. Assertions (a) and (b) follow from the definitions and results of Sec. 3.9 and the present section. Assertion (c) follows from (a) and (b) in view of Lemma 0.3.2. $\square$

3.13. Criteria for Prime and Saturated Models

In this section, we study the existence conditions for prime and saturated models of completions of the theory $F(\mathcal{M}, m, s)$.

Lemma 3.13.1. *Let* $\mathcal{M}$ *be a normal admissible Turing machine and let* $\Delta \in \mathrm{Nonstop}(\mathcal{M}, m, s)$. *A model* $\mathfrak{M}$ *of the theory* $F(\mathcal{M}, m, s)[\Delta]$ *is prime if and only if the following conditions hold:*

(a) $\mathfrak{M}$ *is a one-block model,*

(b) *every finite-linked A-path of the model* $\mathfrak{M}$ *contains exactly one component,*

(c) *every infinite-linked A-path of the model* $\mathfrak{M}$ *contains no component.*

PROOF. Let $\mathfrak{M}$ be a prime model. The necessity of condition (a) is obvious.

We prove (b) by contradiction. We assume that $\mathfrak{M}$ is a one-block model and $\beta \neq 1$ components pass along a finite-linked A-path $\mathcal{A}$ in $\mathfrak{M}$. The case $\beta = 0$ is impossible because the last A-route from $\mathcal{A}$ cannot be covered by components of other lines. It remains to consider the case $\beta \geqslant 2$. We denote by $\mathfrak{M}_1$ the model obtained from $\mathfrak{M}$ by eliminating all components passing along $\mathcal{A}$, except one component. Then the initial model $\mathfrak{M}$ is not embedded in $\mathfrak{M}_1$, which contradicts the assumption that $\mathfrak{M}$ is prime.

We prove (c) by contradiction. We assume that at least one component passes along an infinite-linked A-path $\mathcal{A}$ in the model $\mathfrak{M}$. Since every finite part of this path is covered by components of other A-paths, from $\mathfrak{M}$ we can eliminate all components passing along $\mathcal{A}$. We obtain the model $\mathfrak{M}_1$ in which $\mathfrak{M}$ is not embedded, which contradicts the assumption that the model is prime.

To prove the converse assertion, we assume that a model $\mathfrak{M}$ of the theory $F(\mathcal{M}, m, s)[\Delta]$ satisfies conditions (a)–(c). Let $\mathfrak{M}'$ be an arbitrary model of the theory $F(\mathcal{M}, m, s)[\Delta]$. Standard blocks of the above models are isomorphic. Furthermore, in the model $\mathfrak{M}'$, for any finite-linked A-path at least one component must pass along it. This fact provides the existence of an isomorphic embedding of $\mathfrak{M}$ into $\mathfrak{M}'$ which is elementary by the model completeness. Consequently, the model $\mathfrak{M}$ is prime.   □

Lemma 3.13.1 implies the existence condition for a prime model.

Lemma 3.13.2. *Let* $\mathcal{M}$ *be a normal admissible Turing machine and let* $\Delta \in \mathrm{Nonstop}(\mathcal{M}, m, s)$. *The theory* $F(\mathcal{M}, m, s)[\Delta]$ *has a prime model if and only if* Tree $(\mathcal{M}, \Delta, m, s)$ *is atomic.*

PROOF. If the theory $F(\mathcal{M}, m, s)[\Delta]$ has a prime model, then, by Lemma 3.13.1, its finite-linked paths must cover all A-points, which is equivalent to the fact that Tree $(\mathcal{M}, \Delta, m, s)$ is atomic. Conversely, under the assumption that this tree is atomic, we can construct a one-block model of the theory $F(\mathcal{M}, m, s)$ which satisfies conditions (a)–(c) from Lemma 3.13.1. Therefore, the model is prime.   □

We now study saturated models and the total transcendence.

Lemma 3.13.3. *Let* $\mathcal{M}$ *be a normal admissible Turing machine and let* $\Delta \in \mathrm{Nonstop}(\mathcal{M}, m, s)$. *Then the following assertions hold:*

(a) *if* Tree $(\mathcal{M}, \Delta, m, s)$ *is superatomic, then the theory* $F(\mathcal{M}, m, s)[\Delta]$ *is totally transcendental and has a countable saturated model,*

(b) *if* Tree $(\mathcal{M}, \Delta, m, s)$ *is not superatomic, then* $F(\mathcal{M}, m, s)[\Delta]$ *is not a totally transcendental theory and has no countable saturated model.*

PROOF. (a). Let $\mathfrak{M}$ be an arbitrary countable model of the theory $F(\mathcal{M}, m, s)[\Delta]$. By the superatomicity of Tree $(\mathcal{M}, \Delta, m, s)$ and the conditions (c), (d) from the definition of an admissible Turing machine, we have $t_0(\mathfrak{M}) = \omega$. By Lemma 3.12.5, the model $\mathfrak{M}$ has a countable prefect extension $\mathfrak{N}$. It suffices to show that every type $p(x)$ over $\mathfrak{M}$ is realized in $\mathfrak{N}$. For this purpose we realize the type p in some countable elementary extension $\mathfrak{M}'$ of the model $\mathfrak{M}$. Since $t_0(\mathfrak{M}') = t_0(\mathfrak{M}) = \omega$, the model $\mathfrak{M}'$ has a countable elementary perfect extension $\mathfrak{N}'$ in view of Lemma 3.12.3. Therefore, the type $p(x)$ is realized in $\mathfrak{N}'$, since it is realized in its submodel $\mathfrak{M}'$. By Lemma 3.12.1, the model $\mathfrak{N}'$ is a perfect extension of the model $\mathfrak{M}$. By Lemma 3.12.4, the models $\mathfrak{N}$ and $\mathfrak{N}'$ are isomorphic over $\mathfrak{M}$. Consequently, the type $p(x)$ is realized in $\mathfrak{N}$. Thus, the above theory is totally transcendental. Consequently, it has a countable saturated model.

(b). Let $\mathcal{A}$ be an arbitrary A-path of the standard block of a model of the considered theory and let $\mathcal{A}$ start from the point O. We denote by $\mathcal{A}^*(x)$ the set of formulas without constants which asserts that $U(x)$ and the component generated by the element x passes along $\mathcal{A}$. It easy to see that every such set of form $\mathcal{A}^*(x)$ is consistent with $F(\mathcal{M}, m, s)[\Delta]$, but for $\mathcal{A}_1 \neq \mathcal{A}_2$ the set $\mathcal{A}_1^*(x) \cup \mathcal{A}_2^*(x)$ is inconsistent with $F(\mathcal{M}, m, s)[\Delta]$. Consequently, the theory $F(\mathcal{M}, m, s)[\Delta]$ has an uncountable set of 1-types over the empty set.   □

Lemma 3.13.4. *Let $\mathcal{M}$ be a normal admissible Turing machine and let $\Delta \in$ Nonstop$(\mathcal{M}, m, s)$. A model $\mathfrak{M}$ of the theory $F(\mathcal{M}, m, s)[\Delta]$ is a countable saturated model if and only if $\mathfrak{M}$ is a countable perfect model.*

PROOF. It is obvious that a countable saturated model satisfies the conditions from the definition of a perfect model if α is countable. Conversely, let $\mathfrak{M}$ be a countable perfect model of the theory $F(\mathcal{M}, m, s)[\Delta]$. By condition (e) from the definition of a perfect model, we have $t_0(\mathfrak{M}) = \omega$. By Lemma 3.13.3, the theory $F(\mathcal{M}, m, s)[\Delta]$ has a countable saturated model $\mathfrak{M}'$. Since $\mathfrak{M}'$ satisfies the conditions from the definition of a perfect model, $\mathfrak{M} \cong \mathfrak{M}'$ by Lemma 3.12.5. □

To conclude the section, we establish the superstability of an arbitrary completion of an arbitrary theory of the form $F(\mathcal{M}, m, s)$.

Lemma 3.13.5. *Let $\mathcal{M}$ be a normal admissible Turing machine. If the theory $F(\mathcal{M}, m, s)[\Delta]$ is consistent, then it is superstable.*

PROOF. Let $\mathfrak{M}$ be a model of the theory $F(\mathcal{M}, m, s)[\Delta]$ of cardinality $\alpha \geqslant 2^\omega$. By Lemmas 3.12.2 and 3.12.3, this model has a perfect extension $\mathfrak{N}$ of the same cardinality α. Then every type over $\mathfrak{M}$ is realized in $\mathfrak{N}$, which is proved in the same way as in the proof of Lemma 3.13.3(a). This proves the superstability of the theory considered. □

3.14. Computations of the Turing Machine

In Secs. 3.14–3.16, we explain the choice of the Turing machine $\mathcal{M}^*$ which satisfies the assumptions of Theorem 3.1.1. We use the Gödel numbering of the truth-table conditions τ_n, $n \in \mathbb{N}$, introduced in Sec. 3.1. We begin with the description of those computations that must be carried out by the Turing machine.

As above, Ω_n denotes the formula that asserts that, at the initial instant of time, in the cell of oracle ORC$(n) = \Delta_n$ the predicate A_1 is true, i.e., the cell contains 1. By Axiom 34°, the sentence $\neg\Omega_n$ asserts that the cell contains the symbol 0. We denote by Ψ_n the formula Ω_{2n}. The formulas Ψ_n, $n \in \mathbb{N}$, introduced are those formulas that are required in Theorem 3.1.1. The effectiveness of their construction is obvious. The formula Ω_n and the corresponding cell of the oracle Δ_n are connected with the truth-table condition τ_n. In accordance with the choice of the coding, the formula Ψ_n is related to the elementary truth-table condition ε_n. Let A be an arbitrary subset of $\mathbb{N}$. The set $A^{tt} = \{k \mid A \vDash \tau_k\}$ is called the tt-$degree$ of the set $A \subseteq \mathbb{N}$. Let $\mathcal{M}^*$ be a Turing machine and $\mathfrak{M}$ be a model of the theory $F(\mathcal{M}^*, m, s)$. It is easy to see that the condition

$$(\exists A \subseteq \mathbb{N}) \text{ Oracle}(\mathfrak{M}) = A^{tt} \qquad (3.14.1)$$

holds if and only if all sentences of the form

$$\Omega_n \leftrightarrow (\Omega_i \mid \Omega_j), \ i,j < \omega, \ n = 2c(i,j) + 1 \qquad (3.14.2)$$

are true in $\mathfrak{M}$.

THE FIRST COMPUTATION OF THE TURING MACHINE $\mathcal{M}^*$ is to control the validity of the conditions (3.14.2). The program of the Turing machine must analyze each of these conditions and provide a halt of the program if at least one of them turns out to be false. Then the truth of all the conditions in (3.14.2) is guaranteed by Axiom 38°.

THE SECOND COMPUTATION OF THE TURING MACHINE $\mathcal{M}^*$ provides assertion (2) of Theorem 3.1.1. We denote by R the set of numbers of identically true tt-conditions. It is obvious that R is infinite and recursive. The second computation of the Turing machine $\mathcal{M}^*$ is to verify the truth of the following sequence of sentences:

$$\Omega_i, \ i \in W_m \cup R \qquad (3.14.3)$$

If, in (3.14.3), a false condition is found, the Turing machine must halt. In this case, Axiom 38° provides the validity of the indicated conditions.

THE THIRD COMPUTATION OF THE TURING MACHINE $\mathcal{M}^*$ is connected with the construction of a special tree over the point O by means of commands with division of cells. We fix a recursive superatomic tree of rank 14 and denote it by $\mathcal{D}^*$. The third computation of the Turing machine is to provide that any model $\mathfrak{M} \in F(\mathcal{M}^*, m, s)$ satisfies the relation

$$\text{Oracle}(\mathfrak{M}) = A^{tt} \ \Rightarrow \ \text{Tree}(\mathfrak{M}) = \mathcal{D}_s^A \oplus \mathcal{D}^* \qquad (3.14.4)$$

which is equivalent to the direct relation for the Turing machine $\mathcal{M}^*$

$$\text{Tree}(\mathcal{M}^*, A^{tt}, m, s) = \mathcal{D}_s^A \oplus \mathcal{D}^* \qquad (3.14.5)$$

Let $\varphi_z^A(t)$ stand for a standard numbering of one-place functions that are partially recursive with the oracle A. The tree $\text{Tree}(\mathcal{M}^*, A^{tt}, m, s)$ is given by an explicit procedure of enumeration with respect to the oracle A; therefore, its relative recursively enumerable index is effectively computed from a given s. Consequently, there exists a general recursive function

$g(s)$ such that for all possible values of parameters the tree is constructed by the following sequences of actions presented in terms of elements of models:

$$\mathrm{DIV}(s, i) = [\text{Divide a cell}\,\mathrm{TREE}(\varphi^A_{g(s)}(i))], \quad i = 0, 1, \ldots \quad (3.14.6)$$

where $\mathrm{TREE}\,(k)$ is the kth cell of the field TREE, $0 \leqslant k < \text{length}\,(\mathrm{TREE})$. Furthermore, for any s the function g defines the recursively enumerable index $s' = g(s)$ for the function $\varphi^A_{s'}(i)$ which is recursive with respect to the oracle. This function is everywhere defined and satisfies the natural conditions that are required in constructing the tree

$$\varphi^A_{s'}(i) \leqslant i \quad \text{for all } i < \omega, \ A \subseteq \mathbb{N}$$

The existence of such a function $g(s)$ is proved by applying the s–m–n-theorem in order to pass from one (standard) method of enumerating of trees over an oracle to another (new) method of constructing the same tree by means of (3.14.6).

3.15. Programming Technique

The external alphabet of the Turing machine $\mathcal{M}^*$ contains three symbols a_0, a_1, and a_2, which are identified with the symbols 0, 1, and B. The sequence of $k + 1$ units is called an *integer cell with the value k* if it is bounded from the left and from the right by the symbols 0 or B. The flow diagram of the Turing machine $\mathcal{M}^*$ is presented in Fig. 3.15.1, and the states of the tape at the instant of passage from one block to another are depicted in Fig. 3.15.2. The names of different sections of the tape are also indicated there. We characterize the role of each section.

ORACLE is the left half of the tape. It contains the oracle information. The program preserves this information during the work of the Turing machine, so that it is possible to address it repeatedly.

TREE is the field of constructing the tree (3.14.5). In operation of the Turing machine, this section is extended owing to commands with division of cells which are realized on this section.

M, S are integer cells keeping the parameters m and s (cf. Theorem 3.1.1). These cells, defined by Axiom 32°, are unchained during the entire process of work of the Turing machine $\mathcal{M}^*$, so that it is possible to address them repeatedly.

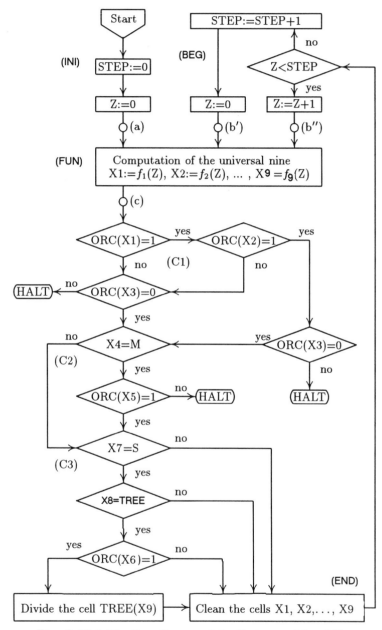

Fig. 3.15.1. The flow diagram of the Turing machine $\mathcal{M}^*$.

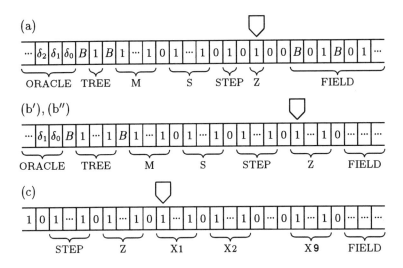

Fig. 3.15.2. Intermediate states of the Turing machine $\mathcal{M}^*$.

STEP is an integer cell which is a basis of the computation process. The initial value is as follows: STEP $= 0$. Then an infinite cycle begins to work with parameter STEP. This cycle is called the *global cycle*.

Z is an integer cell of a *local cycle*. At the next turn of the global cycle STEP $= k$, the local cycle starts with the variable Z, which takes the values $0, 1, 2, \ldots, k$. Then the value of the cell STEP increases by 1 and takes the new value $k + 1$; the local cycle starts for the values of Z from 0 to $k + 1$ and so on.

X1, X2, X3, X4, X5, X6, X7, X8, and X9 are working integer cells. At the first half of each turn of the local cycle, the values of nine special one-place general recursive functions $f_1, f_2, \ldots, f_9$ of Z are computed and are written in the working cells X1, X2, $\ldots$, X9. At the second half of the turn, special executive blocks begin to work. They use the value of the above-mentioned cells in order to realize computations. At the following turn of the local cycle, the same actions are repeated for a new value of the cell Z and so on.

FIELD is a free field on the right which is used as a working field to compute the functions $f_1, f_2, \ldots, f_9$. This field contains a series of zeros and a periodically repeating sequence of the symbols B, 0, and 1 constructed in accordance with Axiom 33°. Having written information in the cells $X1, X2, \ldots, X9$, the Turing machine clears the part of field FIELD that was used in computation by writing zeros.

In accordance with the flow diagram (cf. Fig. 3.15.1), the program of the Turing machine $\mathcal{M}^*$ consists of the initialization block (INI), the block of the changes of cells of local and global cycles (BEG), the block of the computation of nine functions (FUN), three executive blocks (C1), (C2), (C3) for computations, and the block of cleaning fields (END).

We now choose nine functions mentioned above. The functions f_1, f_2, and f_3 are taken so as to satisfy the condition

$$\{\langle f_1(t), f_2(t), f_3(t)\rangle \mid t \in \mathbb{N}\} = \{\langle x, y, 2c(x, y) + 1\rangle \mid x, y \in \mathbb{N}\}$$

i.e., these functions enumerate all triples of numbers on which the truth of (3.14.2) is verified. It is easy to check that part (C1) of the program leads to a halt if one of the conditions (3.14.2) is false.

The functions f_4 and f_5 are taken so as to satisfy the relation

$$\{\langle f_4(t), f_5(t)\rangle \mid t \in \mathbb{N}\} = \{\langle x, y\rangle \mid y \in W_x \cup R\}$$

where R is the set of numbers of identically true truth-table conditions defined above. It is easy to verify that part (C2) of the program is constructed so that it yields a halt if and only if one of the formulas (3.14.3) is false, i.e., this part of the program realizes actions of the second computation of the Turing machine.

We construct the last four functions f_6, f_7, f_8, and f_9. By [56], there exists a general recursive function $\rho(x)$ such that for all $x, y, z \in \mathbb{N}$, $A \subseteq \mathbb{N}$

$$\varphi_x^A(y) = z \iff (\exists u, v)\left[\langle y, z, u, v\rangle \in W_{\rho(x)} \ \& \ D_u \subseteq A \ \& \ D_v \subseteq \mathbb{N} \setminus A\right]$$

where D_n, $n < \omega$, is a standard Gödel numbering of the family of all finite sets of natural numbers.

A quadruple of natural numbers $\langle j, x, y, z\rangle$ is called *elementary* if $j = 49$ and x, y, z are arbitrary or $j \neq 49$ and there exist $u, v \in \mathbb{N}$ such that

(a) $\langle y, z, u, v\rangle \in W_{\rho(g(x))}$,

(b) j is the number of the truth-table condition that asserts that $[D_u \subseteq A \ \& \ D_v \subseteq \mathbb{N} \setminus A]$.

The above-mentioned quadruple $\langle j, x, y, z\rangle$, $j \neq 49$, is called *principal*, and quadruples of the form $\langle 49, x, y, z\rangle$ are referred to as *auxiliary*. It is easy to see that the set E of all elementary quadruples is recursively enumerable. From the definitions we obtain the following main property of the set E:

$$\varphi^A_{g(x)}(y) = z \Leftrightarrow (\exists j) \, [\langle j, x, y, z \rangle \in E \ \& \ A \vDash \tau_j]$$

This property is guaranteed by the principal quadruples, whereas auxiliary quadruples do not prevent it because "49" is the number of the identically false truth-table condition. Thus, elementary quadruples are, in a sense, elements of computation for values of the form $\varphi^A_{g(x)}(y)$. Every value, if it is defined, is computed by a principal elementary quadruple.

We choose general recursive functions f_6, f_7, f_8, f_9 so as to satisfy the condition

$$\{\langle f_6(n), f_7(n), f_8(n), f_9(n) \rangle \mid n \in \mathbb{N}\} = E$$

We consider the process of computing the tree. In accordance with the flow diagram, at each turn of the local cycle, the block (FUN) transmits elementary quadruples to part (C3) of the program via the working cells

$$X6, X7, X8, X9 \qquad (3.15.1)$$

Steps of constructing the tree (3.14.5) are slower than those of the global cycle. This can be explained by the fact that in order to compute the value of $\varphi^A_{g(s)}$ it is necessary to address the oracle many times. For the argument of this function the current length of the field TREE is used; it plays the role of a simple integer cell in this case. For example, the condition "X8 = TREE?" indicated in the flow diagram has the following meaning: Does the number of units of the field X8 coincide with the current length of the field TREE? The next step (3.14.6) adds 1 to the length of the field TREE and, thereby, provides an automatic passage to the next value of the argument.

The computation of the tree is realized as follows. Via cells (3.15.1), the block (FUN) transmits elements of computation for all values of the form $\varphi^B_{g(x)}(y)$ to the block (C3). The third executive part (C3) is waiting for a favorable situation in which the element of computation obtained is a computation for $\varphi^A_{g(s)}(t)$, where t is the length of the field TREE without unit. In this case, the cell X9 containing the value of the indicated function is used for the next action (3.14.6). As a result, we construct the tree (3.14.5).

Thus, the flow diagram of the Turing machine $\mathcal{M}^*$ executes all the required computations. Let us show that, if a program is created in accordance with this flow diagram and follows some requirements, then it is possible to obtain an admissible normal Turing machine and construct effectively the normalization axiom.

ADMISSIBILITY OF THE TURING MACHINE $\mathcal{M}^*$. Let $\mathfrak{M}$ be a model of the theory $F(\mathcal{M}^*, m, s)$. In accordance with the first computation of the Turing machine, the set Oracle $(\mathfrak{M})$ is A^{tt} for some $A \subseteq \mathbb{N}$. The validity of condition (b) from the definition of an admissible Turing machine is obvious. Condition (c) is guaranteed by the flow diagram, where the division commands are used only on cells of the field TREE. Condition (d) holds because, in view of the adopted conventions, the Turing machine $\mathcal{M}^*$ computes the infinite tree (3.14.5); therefore, an infinite number of commands with division of cells will be used in constructing the tree. The validity of (a) is trivially checked.

NORMALITY OF THE TURING MACHINE $\mathcal{M}^*$. This property is guaranteed by the specific form of the flow diagram. Let us explain the main idea.

1. Commands of the section (INI) work one time, and it is easy to count them.

2. The work of the block of the computation of nine functions is independent of the parameters m, s and the content of the oracle. This block repeats the computations for each input value of Z infinitely many times; therefore, each command either is not executed or is executed infinitely many times. We assume that the main program of the block (FUN) works in such a manner if this is the case, the half of the tape to the right of the cell S is completed by zeros, and only symbols 0 and 1 are used in its work. Furthermore, in (FUN) there is an easy subprogram which, meeting the blank B, takes the control, writes zeros in the group of three initial cells, puts the head into next place, and gives back the control to the main program of the block (FUN).

3. The execution blocks (C1), (C2), and (C3) deal with the simplest actions of the type "compare two integer cells," "verify the content of the oracle in the cell address that is written in a given cell," and so on. These actions are easily programmed in the form of the simplest shuttle motions of the head with the auxiliary symbol B for the marker.

Constructing the program that realizes such operations, one can analyze the number of executions of each such command during the work of the Turing machine. It is important that computations along every branch of executing blocks, except blocks that lead to a halt, are repeated infinitely many times. We note that, in the analysis of the condition "X4 = M" two cases appear if the cells do not coincide: X4 < M and X4 > M, which lead to some differences in the execution of commands. However, it is easy to verify that every case is repeated infinitely many times. The same fact is valid for other branches of the flow diagram. To provide the infinite repetition of passages along branches, auxiliary elementary quadruples are required, which cannot induce divisions of cells of the field TREE but yield repeated passages along all branches of the program.

We note that the symbol B plays three roles in the program:

(1) bounds the field TREE from both sides,

(2) serves as an indicator in order to call up a subprogram of (FUN) which clears the next three new cells of the field FIELD,

(3) is used as a symbol-marker in subprograms of shuttle motion of the head in comparing the length of cells and in finding a given position of the field ORACLE to read an information bit.

Following the above flow diagram, one can obtain a normal Turing machine and indicate the list of infinitely realized triples and pairs, as well as the boundary $n_{\mathcal{M}}$ for finitely realized triples and pairs. This allows us to construct effectively the normalization axiom. Then the system of axioms of the theory $F(\mathcal{M}, m, s)$ is effectively constructed.

3.16. The Last Step of the Proof of the Main Theorem

We prove that if $\mathbb{F}(m, s) = F(\mathcal{M}^*, m, s)$ and the sentences Ψ_n, $n \in \mathbb{N}$, are defined by formulas as in Sec. 3.14, then all the conditions, except the condition that the signature of the theory is a given finite rich signature σ, of Theorem 3.1.1 are satisfied. The finite axiomatizability and the effectiveness of the construction of axioms, including the normalization axiom, are direct consequences of the construction. The model completeness of the theory is guaranteed by Lemma 3.12.7.

We verify the assertions of the theorem.

1. By Lemma 0.3.2, it suffices to show that for any $A \subseteq \mathbb{N}$ the theory $\mathbb{F}(m, s)[A]$ is complete provided that it is consistent. We assume that it is consistent. In accordance with the first computation of the Turing machine $\mathcal{M}^*$, for any $A \subseteq \mathbb{N}$

$$\mathbb{F}(m, s)[A] = F(\mathcal{M}^*, m, s)[A^{tt}] \qquad (3.16.1)$$

Indeed, by construction, the left-hand side is a subtheory of the right-hand side. On the other hand, the first computation provides the deduction of all other axioms from the right-hand side in the left-hand side since A^{tt} is uniquely defined by the set A which is presented on even cells of the oracle in the theory. By Lemma 3.12.9 and the relation (3.16.1), the theory $\mathbb{F}(m, s)[A]$ is complete.

2. We assume that the theory $\mathbb{F}(m, s)[A]$ is consistent. Let $\mathfrak{M}$ be a model of the theory. By (3.16.1), Oracle $(\mathfrak{M}) = A^{tt}$. Since the second computation of the Turing machine does not lead to a halt, $W_m \cup R$ is contained in Oracle $(\mathfrak{M})$. In particular, $W_m \subseteq A^{tt}$. This means that, on the set A, all the truth-table conditions τ_k, $k \in W_m$, are true. Consequently, $A \in \mathcal{R}_m$.

Conversely, let $A \in \mathcal{R}_m$. By definition, on the set A, the truth-table conditions τ_k, $k \in W_m$, are true. By the choice of R, on the set A, all the conditions from R are also true. Therefore, $W_m \cup R \subseteq A^{tt}$. The relations obtained mean that $A^{tt} \in \text{Nonstop}(\mathcal{M}^*, m, s)$. By Lemma 3.12.9, the theory $F(\mathcal{M}^*, m, s)[A^{tt}]$ is consistent. By (3.16.1), the theory $\mathbb{F}(m, s)[A]$ is consistent.

3. We consider a set $A \in \mathcal{R}_m$.

(a). The assertion is valid in view of (3.16.1) (cf. Lemma 3.13.2).

(b). Let a prime model $\mathfrak{M}$ of the theory $\mathbb{F}(m, s)[A]$ be strongly constructivizable. By the decidability of the theory, the set A is recursive. Having the constructivization of the model $\mathfrak{M}$, we can enumerate all those components of this model that start from U, which, by Lemma 3.13.1, yields the computability of the family of all finite-linked K-paths that start from U, which can be translated as the computability of the family of chains $\Pi^{\text{fin}}(\mathcal{D}_s^A)$.

We assume that the set A is recursive, the tree $\mathcal{D}_s^A$ is atomic, and the family of its finite paths is computable. Then the family $\Pi^{\text{fin}}(\mathcal{D}_s^A \oplus \mathcal{D}^*)$ is also computable. Let ν be its computable numbering. Since the family consists of maximal chains of the tree, the numbering ν is negative. Hence the family $\Pi^{\text{fin}}(\mathcal{D}_s^A \oplus \mathcal{D}^*)$ has a single computable numbering ν'. Since the set A is recursive and the tree $\mathcal{D}_s^A \oplus \mathcal{D}^*$ is atomic, we can, using the numbering ν', construct a constructive model $(\mathfrak{M}_0, \nu_0)$ of the theory $\mathbb{F}(m, s)[A]$ which satisfies conditions (a)–(c) of Lemma 3.13.1. By the model completeness of the theory, ν_0 is a strong constructivization.

(c). We assume that the set A is recursive and the tree $\mathcal{D}_s^A$ is atomic. Let $\mathfrak{M}$ denote a prime model of the theory $\mathbb{F}(m, s)[A]$. We divide the required assertion into two parts.

We first assume that the tree $\mathcal{D}_s^A$ is not recursive. Then it is recursively enumerable. We denote it $\mathcal{D}$. Let $\text{dend}(\mathcal{D})$ be the set of dead ends of the tree $\mathcal{D}$. We note that $\text{dend}(\mathcal{D})$ cannot be recursively enumerable; otherwise, $\mathcal{D}$ is recursive. From a given $a \in \mathcal{D}$ we effectively construct a formula $\varphi_a(x)$ of the form $W(x)$ & $E(x,c)$, where c is located in the standard block of the model $\mathfrak{M}$ on the A-line which corresponds to a. Furthermore,

$$a \in \mathrm{dend}(\mathcal{D}) \;\Leftrightarrow\; \varphi_a(x) \text{ is an atomic formula in } \mathfrak{M} \qquad (3.16.2)$$

We pass to an arbitrary finite enrichment by constants

$$\mathfrak{M}' = (\mathfrak{M}, d_0, d_1, \dots, d_{k-1}) \qquad (3.16.3)$$

A relation of the form (3.16.2) referred to $\mathfrak{M}'$ remains valid for all dead ends of the tree $\mathcal{D}$, except for a finite number. Taking into account that $\mathrm{dend}(\mathcal{D})$ is not a recursively enumerable set, we conclude that the set of atomic formulas in any model of the form (3.16.3) cannot be recursively enumerable in the corresponding Gödel numbering. By [26, 39], the model is nonautostable with respect to strong constructivizations.

We assume that the tree $\mathcal{D}_s^A$ is recursive. Then the tree $\mathcal{D} = \mathcal{D}_s^A \oplus \mathcal{D}^*$ is also recursive. We consider two strong constructivizations ν_1 and ν_2 of a prime model $\mathfrak{M}$ of the theory considered. It is required to construct a constructive isomorphism

$$h: (\mathfrak{M}, \nu_1) \longrightarrow (\mathfrak{M}, \nu_2)$$

Since $\mathfrak{M}$ is a one-block model, it is easy to construct a correspondence between E-classes that is recursive in given numberings and obtain effectively an element-by-element mapping h' on elements from $X \cup Y$.

We construct an element-by-element mapping h'' on the set W. Since the model $\mathfrak{M}$ is prime, there is a one-to-one correspondence between its components and finite-linked A-paths as well as a correspondence between those finite-linked A-paths of this model that start from U and dead ends of the tree $\mathcal{D}$, which is recursive by assumption. Therefore, it is possible to recognize effectively whether for each of the numberings ν_1 and ν_2 the elements $a, b \in U(\mathfrak{M})$ belong to the same component or to different ones. Then it is easy to construct the required element-by-element mapping h''. Combining the parts h' and h'', we obtain the required recursive isomorphism h.

(d). The assertion is valid by (3.16.1) (cf. Lemma 3.13.3).

(e). We assume that a countable saturated model $\mathfrak{M}$ of the theory $\mathbb{F}(m, s)[A]$ is strongly constructivizable. The decidability of this theory implies that the set A is recursive. By Lemma 3.13.4, the model $\mathfrak{M}$ is perfect. Enumerating those components of this model that start at U and taking into account condition (e) from the definition of a perfect model, we compute the family of all those A-paths that start from the point U.

Passing from A-paths to the corresponding chains of the tree, we obtain the computability of the family $\Pi(\mathcal{D}_s^A \oplus \mathcal{D}^*)$ and, consequently, of the family of all chains $\Pi(\mathcal{D}_s^A)$.

We assume that the set A is recursive and the family $\Pi(\mathcal{D}_s^A)$ is computable. Since the family is computable, it is countable. By Theorem 2.1.1, the tree $\mathcal{D}_s^A$ is superatomic. Consequently, the theory $\mathbb{F}(m, s)[A]$ has a countable saturated model $\mathfrak{M}$. Since the set A is recursive and the family $\Pi(\mathcal{D}_s^A)$ is computable, it is possible to construct a constructive perfect model $(\mathfrak{M}_0, \nu_0)$ of the theory $\mathbb{F}(m, s)[A]$. By the model completeness of the theory, ν_0 is a strong constructivization. By Lemma 3.13.4, the models $\mathfrak{M}$ and $\mathfrak{M}_0$ are isomorphic. Consequently, $\mathfrak{M}$ is a strongly constructivizable model.

(f). The assertion is valid in view of (3.16.1) (cf. Lemma 3.13.3).

(g). For the sake of brevity, we denote $\mathbb{F}(m, s)[A]$ by T and $\mathcal{D}_s^A \oplus \mathcal{D}^*$ by $\mathcal{D}$. We consider a perfect model $\mathfrak{M}$ of the theory T of cardinality $\alpha \leqslant 2^\omega$. Such a model exists in view of Lemmas 3.12.2 and 3.12.3. This model is countable and saturated; therefore, the Morley rank of formulas and types can be computed as the Cantor–Bendikson rank [57]. By Axiom $1°$,

$$\alpha_T = \max\{\operatorname{rank} X(x), \operatorname{rank} Y(x), \sup\{\operatorname{rank} p(x) \mid W(x) \in p(x)\}\}$$

Therefore, the following lemma (assertion (c) is used here) provides the required value of the Morley rank of T.

Lemma 3.16.1 *The following estimates hold*:

(a) $\operatorname{rank} X(x) = 9$,

(b) $\operatorname{rank} Y(x) = 5$,

(c) $\operatorname{rank} \varphi(x, y) = 15$, where $\varphi(x, y) = \neg W(x) \& \neg W(y)$,

(d) $\sup\{\operatorname{rank} p(x) \mid W(x) \in p(x)\} = 1 + \operatorname{rank}(\mathcal{D}) + \gamma$, *where γ is taken from Theorem* 3.1.1(g).

PROOF. (a)–(c). By the model completeness of the theory, any formula $\varphi(x)$ distinguishing a subset of $X(x) \vee Y(x)$ is equivalent to an $\exists$-formula for which quantifiers can be used only inside $X \cup Y$ (by the construction). Therefore, we can perform the computation for (a)–(c), eliminating all W-elements from $\mathfrak{M}$.

Elements from Y belong to a model of the quasisuccessor theory. Therefore, they are defined by two coordinates of the model QS included in the frame. Every coordinate may be in the standard zone (where all

H-classes are first-order definable), in a zone with a singular point, or in a zone with a singular line (homogeneous with respect to shifts along the P-chain). Hence rank $Y(x) = 5$. Two coordinates of elements from X can be in the first-order definable E-classes, at a finite distance from the singular point or at a finite distance from the singular line in a pure block. Moreover, there are 3 additional cases of location in different H-classes. The computation yields the estimate rank $X(x) = 9$. Similarly, we can consider (c).

(d). Let $\pi \in \Pi(\mathcal{D})$ and let $p_\pi(x)$ be the 1-type defined by the condition that $U(x)$ and the component generated by x pass along the A-path corresponding to this chain π. It is easy to show that rank $p_\pi(x) = 1 + \text{rank}(\pi)$. The types defined in other W-points of the standard block have ranks that do not exceed the ranks indicated above. If $\mathcal{D}$ is not superatomic, there are no other types with rank. Therefore, rank $W(x) = 1 + \text{rank}(\mathcal{D})$. If $\mathcal{D}$ is superatomic, then the types in a zone with a singular point and in a zone with a singular line have ranks. In this case, rank $W(x) = 1 + \text{rank}(\mathcal{D}) + 2$.
□

The proof of Theorem 3.1.1 is complete. □

The extension of the result obtained to the case of an arbitrary finite rich signature is studied in Sec. 5.11.

Catalogue **Catal 3.1**

Combinatorial Coding
of the Intermediate Construction

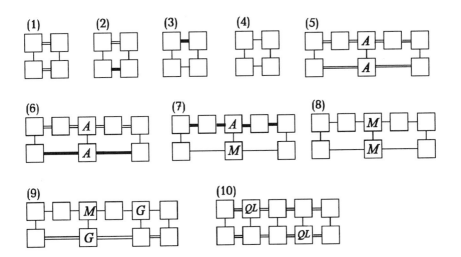

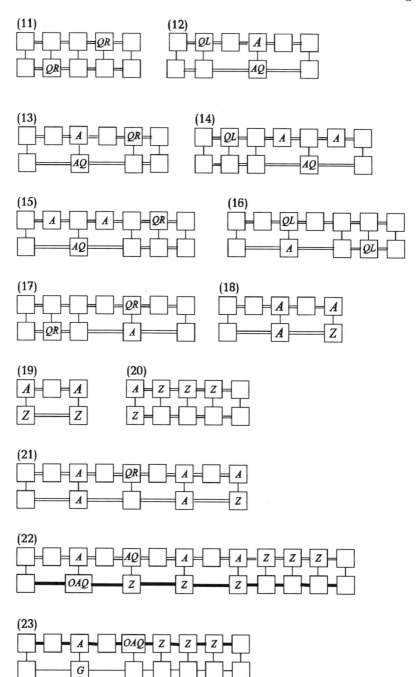

Catalogue Catal 3.2

Blocks
of the Intermediate Construction

Catal 3.2(1)

H is a pure block located in the lower part

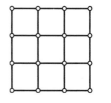

Catal 3.2(2)

BH is a pure block located at the interface between the lower and upper parts

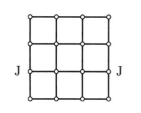

Catal 3.2(3)

B is a pure block located in the upper part
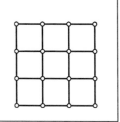

Catal 3.2(4)

M is the marking-out line
The block is located in the lower part

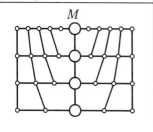

Catal 3.2(5)

G is the generating marking-out line	
The block is located in the lower part	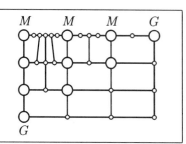

Catal 3.2(6)

A_i, $i < 3$, is the line of a tape cell	
The block is located in the upper part. The predicate A_i is true along the route	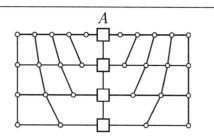

Catal 3.2(7)

$MA(i)$, $i < 3$, is the initial line of a tape cell

The block is located at the interface between the lower and upper parts
The predicate A_i is true along the route

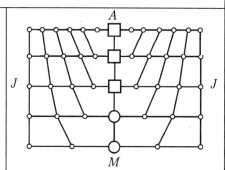

Catal 3.2(8)

Z is the saturation line

The block is located in the upper part
The slanting Z-line generates the fan of A-lines

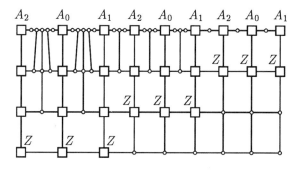

Catal 3.2(9)

$Q(\nu,j)$, $\nu \in \{L,R\}$, $j \in \{0,1,\ldots,e-1\}$, is the line of the motion of the head	Line $Q(L,j)$
	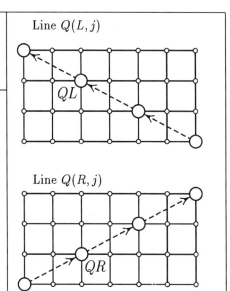
The block is located in the upper part. The line is marked by the predicates $Q\&Q_j$ as well as the direction predicate ν. The head moves toward the direction ν.	Line $Q(R,j)$

Catal 3.2(10)

	Block $AQ(R, i, j)$ command $a_i q_j \to a_m q_t R$ 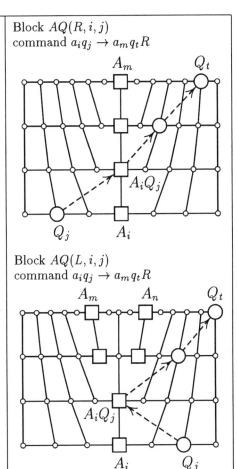
$AQ(\nu, i, j)$, $\nu \in \{L, R\}$, $i < 3$, $j < e$, is a point of execution of a command of the Turing machine	
The block is located in the upper part. The block $AQ(\nu, i, j)$ is defined only if the Turing machine $\mathcal{M}$ has the command $a_i q + j \to a_m q_t D$, $D \in \{L, R\}$. The lines $A(i)$ and $Q(\nu, j)$ meet. In accordance with the command, two or three new lines are generated. There are eight types of forms of a given block (two forms are shown here).	Block $AQ(L, i, j)$ command $a_i q_j \to a_m q_t R$

Catal 3.2(11)

$O(\mathcal{M}, \Delta, m, s)$, $\Delta \subset N$, $m, s \in N$, is a standard block

---

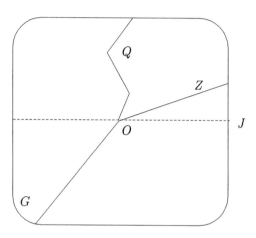

From the neighborhood of the center O, lines of types G, Z, and Q start. The slanting G-line generates the fan of M-lines that are transformed into A-lines after the passage through J. The step-like Z-line generates the fan of A-lines that are periodically marked by the predicates A_2, A_0, and A_1. The block is defined only if the (Δ, m, s)-computation of the Turing machine $\mathcal{M}$ does not lead to a halting. At the level of abstraction of elementary chains, the block is uniquely characterized by the sequence $\lambda = \langle \lambda_i \mid i < \omega \rangle$, $\lambda_i \in \{L1, R1, L2, R2\}$, $\lambda_0 = R1$, which indicates types of commands (with division of cells or not) and the direction of the motion of the head after passage of points of execution of commands.

Chapter 4

Rigid Quasisuccession

In this chapter, we describe a special basic theory for the main construction. This theory is similar to the quasisuccessor theory, but possesses an additional rigidity mechanism. For this reason, we call it the rigid quasisuccessor theory, and denote it by QSR.

First of all, we describe a subtheory QS' of the theory QSR, which is a modification of the first example of an uncountable categorical complete finitely axiomatizable theory of Morley rank 4 [42], i.e., models of this theory are 3-dimensional. Therefore, the theory QS' is called the quasisuccessor theory of rank 3. Then we consider the theory QSR, which is a superstructure over the rigidity mechanism for the theory QS'.

4.1. Axiomatics

In this section, we describe the signature, axioms, and some simple corollaries. The signature of the theories QS' and QSR is presented in Table 4.1.1. The superscript 2 indicates binary predicates. The rest of the symbols are unary predicates.

Table 4.1.1. Signature of the rigid quasisuccession

Predicates and their main destination		Axioms
Predicates of quasisuccession:		
$\lhd^2$, $\sim^2$, $\approx^2$ M^2, H^2	General form of a model	B.1°–B.4°
D	Diagonal	B.5°–B.7°
		B.24°
R^2, S^2	Formation of pairs of nets	B.11°–B.22°
P^2	Connection between neighboring pairs of nets	B.11°–B.22°, B.31°
E^2	Equivalence of pairs of nets	B.8°–B.10°, B.14°, B.15°
		B.8°–B.10°, B.29°–B.30°
Π, Σ	Poles of suppression of cycles	B.32°–B.39°
X, Y, U, V	Polarization predicates	B.25°–B.28°, B.32°–B.33°
A, B	Predicates with disturbed symmetry	B.25°–B.28°, B.36°
Predicates of the rigidity mechanism (R):		
Λ	Markers of the rigidity of quasisuccession	BR.40°– BR.44°
SP, SN	Indicators of sign coordinates	BR.42°– BR.44°

All the signature symbols are divided into 10 groups. The table characterizes their roles in the theory and indicates the corresponding axioms. The first eight groups of predicates correspond to the theory QS', while all ten groups form the signature of the rigid quasisuccessor theory QSR.

We proceed to the axiomatics of the theory. The first 39 axioms constitute the theory QS', and all the axioms constitute the rigid quasisuccessor theory QSR.

Axioms of the theory QS′

B.1°. $\sim$ is an equivalence relation and $\lhd$ is a successor relation on $\sim$-classes.

B.2°. H is an equivalence relation and M is a successor relation on H-classes.

B.3°. Each $\sim$-class intersects each H-class.

B.4°. $x \approx y \leftrightarrow x \sim y \,\&\, H(x, y)$.

B.5°. In every $\sim$-class, the predicate D distinguishes exactly one $\approx$-class.

B.6°. In every H-class, the predicate D distinguishes exactly one $\approx$-class.

B.7°. $x \sim y \,\&\, M(x, y) \,\&\, y \lhd z \,\&\, H(y, z) \;\rightarrow\; \big(D(x) \leftrightarrow D(z)\big)$.

B.8°. E is an equivalence relation.

B.9°. The intersection of each E-class and each $\approx$-class contains exactly two elements.

B.10°. $Q(x, y) \Leftrightarrow x \neq y \,\&\, x \approx y \,\&\, E(x, y)$.

B.11°. R is a successor relation.

B.12°. S is a successor relation.

B.13°. P is a successor relation.

B.14°. $R(x, y) \Rightarrow x \lhd y \,\&\, H(x, y) \,\&\, E(x, y)$.

B.15°. $S(x, y) \Rightarrow x \sim y \,\&\, M(x, y) \,\&\, E(x, y)$.

B.16°. $P(x, y) \Rightarrow x \approx y$.

B.17°. $Q(x, u) \,\&\, Q(y, v)$ implies $R(x, y) \Leftrightarrow R(u, v)$.

B.18°. $Q(x, u) \,\&\, Q(y, v)$ implies $S(x, y) \Leftrightarrow S(u, v)$.

B.19°. $Q(x, u) \,\&\, Q(y, v)$ implies $P(x, y) \Leftrightarrow P(u, v)$.

B.20°. $R(x, u) \,\&\, R(y, v)$ implies $S(x, y) \Leftrightarrow S(u, v)$.

B.21°. $R(x, u) \,\&\, R(y, v)$ implies $P(x, y) \Leftrightarrow P(u, v)$.

B.22°. $S(x, u) \,\&\, S(y, v)$ implies $P(x, y) \Leftrightarrow P(u, v)$.

B.23°. The predicates Π, Σ, A, B, U, V, X, and Y are mutually disjoint.

B.24°. $D(x) \Leftrightarrow \Pi(x) \vee \Sigma(x) \vee A(x) \vee B(x)$.

B.25°. $Q(x,y)$ implies $\Pi(x) \Leftrightarrow \Sigma(y)$.

B.26°. $Q(x,y)$ implies $A(x) \Leftrightarrow B(y)$.

B.27°. $Q(x,y)$ implies $U(x) \Leftrightarrow V(y)$.

B.28°. $Q(x,y)$ implies $X(x) \Leftrightarrow Y(y)$.

B.29°. In every $\approx$-class satisfying D, there is exactly one Π-element.

B.30°. In every E-class, there is exactly one Π-element.

B.31°. If $P(x,y)$, $E(x,u)$ & $\Pi(u)$, and $E(y,v)$ & $\Pi(v)$, then $u \lhd v$.

B.32°. If x and y are such that $E(x,y)$ and $x \sim y$, then $\Pi(x) \vee \Sigma(x) \vee U(x) \vee V(x) \Leftrightarrow \Pi(y) \vee \Sigma(y) \vee U(y) \vee V(y)$.

B.33°. If elements x and y are such that $E(x,y)$ and $H(x,y)$, then $\Pi(x) \vee \Sigma(x) \vee X(x) \vee Y(x) \Leftrightarrow \Pi(y) \vee \Sigma(y) \vee X(y) \vee Y(y)$.

B.34°. If $R(x,y)$, then $\Pi(x) \vee X(x) \Leftrightarrow \Pi(y) \vee X(y)$.

B.35°. If $S(x,y)$, then $\Pi(x) \vee U(x) \Leftrightarrow \Pi(y) \vee U(y)$.

B.36°. If $S(x,y)$ & $R(y,z)$, then $A(x) \Rightarrow A(z) \vee \Pi(z)$; moreover, $A(z) \Rightarrow A(x) \vee \Sigma(x)$.

B.37°. $P(x,y)$ implies $A(x) \vee \Pi(x) \Leftrightarrow A(y) \vee \Sigma(y)$.

B.38°. $P(x,y)$ & $R(y,z)$ implies $\Pi(x) \vee U(x) \Leftrightarrow \Pi(z) \vee U(z)$.

B.39°. $P(x,y)$ & $S(y,z)$ implies $\Pi(x) \vee X(x) \Leftrightarrow \Pi(z) \vee X(z)$.

The axioms of the theory QSR are formed by all the axioms just listed and the following five axioms:

BR.40°. The predicate Λ is well defined on $\sim$-classes.

BR.41°. $\Lambda(x_0)$ and $x_0 \lhd x_1 \lhd \ldots \lhd x_n$ imply $\neg\Lambda(x_n)$ for every natural number $n > 0$.

BR.42°. If elements x, y, and z satisfy the conditions $H(x,y)$ & $D(y)$ and $E(x,z)$ & $\Pi(z)$, then, under the condition $\Lambda(x)$ & $\Lambda(y)$ & $\Lambda(z)$, exactly one of the predicates SP and SN is true on x; otherwise, both predicates SP and SN are false on x.

BR.43°. $Q(x, y)$ implies $SP(x) \Leftrightarrow SN(y)$.

BR.44°. $\Pi(x) \lor A(x) \lor X(x) \lor U(x) \Rightarrow \neg SN(x)$.

In general, the theory QSR is not finitely axiomatizable because of BR.41°. This theory is not complete since the number of $\sim$-classes distinguished by the predicate Λ is not defined. However, this obstacle does not prevent us from using the theory QSR for construction of finitely axiomatizable and complete theories.

We give several simple corollaries of axioms of the theory QS'. For the sake of convenience, their enumeration continues the enumeration of axioms.

From B.1°–B.6° it follows that the predicate $\approx$ is an equivalence relation and the predicate D is well defined on $\approx$-classes. In other words, the following relation holds:

C.45°. $x \approx y \Rightarrow (D(x) \Leftrightarrow D(y))$.

Axioms B.1°–B.7° describe the connection between $\lhd$-successor and M-successor by means of $\approx$-classes satisfying D. Using them, one can prove the following two relations:

C.46°. $D(x) \;\&\; D(y) \Rightarrow (x \sim y \Leftrightarrow H(x, y))$.

C.47°. $D(x) \;\&\; D(y) \Rightarrow (x \lhd y \Leftrightarrow M(x, y))$.

Axiom B.10° implies the symmetry of the predicate Q, which, together with Axiom B.9°, allows us to prove the following relation:

C.48°. Q is a successor relation with 2-cycles.

Consequently, the Q-successor of any element coincides with its Q-predecessor. Furthermore, B.10° also leads to the relation

C.49°. $x \approx y \;\&\; E(x, y)$ implies $x = y \lor Q(x, y)$.

We emphasize one more useful assertion following from B.24° and B.29°:

C.50°. In every $\approx$-class, there is at most one Π-element.

We directly proceed to study the properties of the theories QS' and QSR. In Secs. 4.2–4.5, we consider the theory QS', study its properties, and describe models. Then we study the theory QSR.

4.2. The Impossibility of Cycles

Throughout Secs. 4.2–4.4, the symbol $\vdash$ means provability in the
theory QS'. In such proofs, digits indicate the numbers of Axioms B.1°–
B.39° and their corollaries C.45°–C.50°.
We begin with the principal lemma on quasisuccession.

Lemma 4.2.1. *In the theory QS', $\lhd$-cycles are impossible.*

PROOF. On the contrary, assume that, in a model $\mathfrak{N}$ of the theory
QS', there is a $\lhd$-cycle of length $s \geqslant 2$:

$$[a_0]_\sim \lhd [a_1]_\sim \lhd \ldots \lhd [a_{s-1}]_\sim \lhd [a_0]_\sim \qquad (4.2.1)$$
$$\neg(a_i \sim a_j) \text{ for } 1 \leqslant i < j < s$$

By B.5°, we can assume that the elements a_i, $i < s$, satisfy $D(x)$. By
C.46° and C.47°, we obtain

$$[a_0]_H \ M \ [a_1]_H \ M \ \ldots \ M \ [a_{s-1}]_H \ M \ [a_0]_H \qquad (4.2.2)$$
$$\neg H(a_i, a_j) \text{ for } 0 \leqslant i < j < s$$

In accordance with B.5° and B.29°, we find an element b_0^0 of class $[a_0]_\sim$
with the property $\Pi(x)$. Thus,

$$b_0^0 \approx a_0, \quad \Pi(b_0^0) \qquad (4.2.3)$$

Starting from the element b_0^0 and using B.11°, B.12°, and B.20°, we construct a "net" (perhaps, with repetitions) of elements b_i^j, $i, j \in \mathbb{Z}$, satisfying
the conditions

$$R(b_i^j, b_{i+1}^j), \quad i, j \in \mathbb{Z}$$
$$S(b_i^j, b_i^{j+1}), \quad i, j \in \mathbb{Z}$$

Using B.4°, B.14°, B.15°, and (4.2.1)–(4.2.3), we obtain $b_i^j \sim a_k$ for $i \equiv
k \pmod s$ and $H(b_i^j, a_l)$ for $j \equiv l \pmod s$. As a result, for all $i, j, k, l \in \mathbb{Z}$
we have

$$b_i^j \sim b_k^l \ \Leftrightarrow \ i \equiv k \pmod s$$
$$H(b_i^j, b_k^l) \ \Leftrightarrow \ j \equiv l \pmod s$$
$$b_i^j \lhd b_k^l \ \Leftrightarrow \ i + 1 \equiv k \pmod s$$
$$M(b_i^j, b_k^l) \ \Leftrightarrow \ j + 1 \equiv l \pmod s$$

By (4.2.3) and B.24°, we have $D(b_0^0)$. Applying B.7° several times, we obtain $D(b_i^i)$, $i = 0, 1, 2, \ldots$. By B.14° and B.15°,

$$E(b_i^j, b_k^l) \quad \text{for all} \quad i, j, k, l \in \mathbb{Z}$$

Using the above-mentioned properties of the "net" of elements b_i^j, we prove that $b_0^0 = b_s^0$ and $b_0^0 = b_0^s$. We use formal tree-like proofs:

$$34 \frac{\Pi(b_0^0), \; b_0^0 \, R \, b_1^0 \, R \ldots R \, b_s^0}{\Pi(b_s^0) \vee X(b_s^0),} \qquad 4 \frac{b_0^0 \sim b_s^0, \; H(b_0^0, b_s^0)}{\dfrac{\dfrac{\dfrac{b_0^0 \approx b_s^0, \; E(b_0^0, b_s^0)}{49}}{\Pi(b_0^0), \quad b_0^0 = b_s^0 \vee Q(b_0^0, b_s^0)}{25}}{\dfrac{\Pi(b_s^0) \vee \Sigma(b_s^0)}{\neg X(b_s^0)}23}} \\ \frac{\Pi(b_s^0), \quad \Pi(b_0^0), \; b_0^0 \approx b_s^0}{b_0^0 = b_s^0;}50$$

$$35 \frac{\Pi(b_0^0), \; b_0^0 \, S \, b_0^1 \, S \ldots S \, b_0^s}{\Pi(b_0^s) \vee U(b_0^s),} \qquad 4 \frac{b_0^0 \sim b_0^s, \; H(b_0^0, b_0^s)}{\dfrac{\dfrac{\dfrac{b_0^0 \approx b_0^s, \; E(b_0^0, b_0^s)}{49}}{\Pi(b_0^0), \quad b_0^0 = b_0^s \vee Q(b_0^0, b_0^s)}{25}}{\dfrac{\Pi(b_0^s) \vee \Sigma(b_0^s)}{\neg U(b_0^s)}23}} \\ \frac{\Pi(b_0^s), \quad \Pi(b_0^0), \; b_0^0 \approx b_0^s}{b_0^0 = b_0^s;}50$$

The coincidence of b_0^0 and b_0^s leads to pasting together other elements of the "net" along an R-successor. Thus, $b_i^0 = b_i^s$ for all $i \in \mathbb{Z}$. In particular, $b_s^0 = b_s^s$ for $i = s$. In view of the equality $b_0^0 = b_s^0$ proved above, we obtain $b_0^0 = b_s^s$.

By induction on i from 1 to $s - 1$, we prove that $B(b_i^i)$ holds. We consider an auxiliary element z such that $Q(b_i^i, z)$ holds. For Φ we take the assertion $\Pi(b_0^0)$ if $i = 1$ and the induction assumption $B(b_{i-1}^{i-1})$ if $1 < i < s$.

We now give a formal deduction:

$$7\ \frac{b_{i-1}^{i-1} \sim b_{i-1},\ M(b_{i-1}^{i-1}, b_{i-1}),\ b_{i-1}^i \lhd b_i^i,\ H(b_{i-1}, b_i^i)}{D(b_{i-1}^i) \Leftrightarrow D(b_i^i),} \qquad \frac{\varPhi}{D(b_{i-1}^i)}\ 24$$

$$\frac{}{D(b_i^i)}$$

$$10\ \frac{Q(b_i^i, z)}{b_i^i \approx z} \qquad\qquad\qquad\qquad\qquad \downarrow$$

$$4\ \frac{}{b_i^i \approx z} \qquad\qquad\qquad\qquad\qquad\qquad \downarrow$$

$$1\ \frac{b_i^i \sim z,\ \neg(b_0^0 \sim z)}{} \qquad\qquad 10\ \frac{Q(b_i^i, z)}{E(b_i^i, z)} \qquad \downarrow$$

$$30\ \frac{b_0^0 \neq z\ \&\ b_0^0 \neq b_i^i,\ \varPi(b_0^0),\ E(b_0^0, b_i^i),}{} \qquad\qquad \downarrow$$

$$25\ \frac{\neg \varPi(b_i^i)\ \&\ \neg \varPi(z),\ Q(b_i^i, z)}{} \qquad \downarrow$$

$$23\ \frac{\varPhi}{} \qquad \frac{\neg \varPi(b_i^i)\ \&\ \neg \varSigma(b_i^i)}{} \qquad \frac{D(b_i^i)}{A(b_i^i) \vee B(b_i^i)}\ 24$$

$$36\ \frac{\neg A(b_{i-1}^{i-1})\ \&\ \neg \varSigma(b_{i-1}^{i-1}),\ S(b_{i-1}^{i-1}, b_{i-1}),\ R(b_{i-1}, b_i^i)}{}$$

$$\frac{\neg A(b_i^i),\ \neg A(b_i^i) \Rightarrow B(b_i^i)}{B(b_i^i)}$$

We obtain $B(b_{s-1}^{s-1})$. By construction, $S(b_{s-1}^{s-1}, b_{s-1}^s)$, $R(b_{s-1}^s, b_s^s)$, and $b_0^0 = b_s^s$. We consider auxiliary elements x, y, and z such that $Q(b_{s-1}^{s-1}, x)$, $Q(b_{s-1}^s, y)$, and $Q(b_s^s, z)$. By B.17° and B.18°, we have $S(x, y)$ and $R(y, z)$. It remains to give a formal deduction:

$$26\ \frac{B(b_{s-1}^{s-1}),\ Q(b_{s-1}^{s-1}, x)}{}$$

$$36\ \frac{A(x),\ S(x, y),\ R(x, y)}{}$$

$$\frac{A(z) \vee \varPi(z),\ Q(b_s^s, z)}{B(b_s^s) \vee \varSigma(b_s^s)}\ 25,26$$

$$\frac{}{b_0^0 = b_s^s,\ \neg \varPi(b_s^s)}\ 23$$

$$\frac{}{\neg \varPi(b_0^0)}$$

We arrive at a contradiction with (4.2.3), which proves the impossibility of the cycle (4.2.1).   □

Lemma 4.2.2. *M-cycles are impossible in the theory QS'.*

PROOF. On the contrary, we suppose that, in a model $\mathfrak{N}$ of the theory QS', there is an M-cycle of length $s \geqslant 2$:

$$[a_0]_H \ M \ [a_1]_H \ M \ \ldots \ M \ [a_{s-1}]_H \ M \ [a_0]_H$$

By B.6°, we can assume that the representatives a_i, $i < s$, are D-elements. Using C.47°, we can construct a $\lhd$-cycle, which is impossible in view of Lemma 4.2.1. $\qquad\qquad\qquad\qquad\qquad\qquad\qquad\qquad\qquad\qquad\qquad\qquad\quad$ $\square$

Lemma 4.2.3. *The theory QS' has no finite models.*

4.3. Description of Models and Coordinates

In Fig. 4.3.1, we show the general form of a model, which follows from B.1°–B.7° in view of the absence of $\lhd$-cycles and M-cycles. This form is obtained by combining two mutually orthogonal successions: $\lhd$-successor on $\sim$-classes and M-successor on H-classes. Moreover, each $\sim$-class must intersect each H-class.

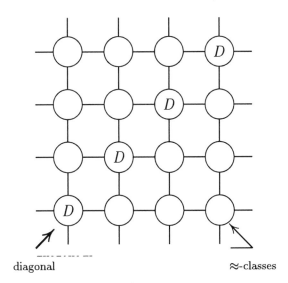

diagonal $\qquad\qquad\qquad\qquad\qquad\qquad$ $\approx$-classes

Fig. 4.3.1. General form of a model:

H-classes are marked by horizontal lines, $\sim$-classes are marked by vertical lines, $\uparrow$ is the $\langle M, H \rangle$-successor, $\rightarrow$ is the $\langle \triangle, \sim \rangle$-successor.

The predicate D distinguishes the *diagonal* of a model. With the help of the diagonal, a one-to-one correspondence between the set of $\sim$-classes and the set of H-classes is established. Moreover, in view of C.46° and C.47°, this correspondence is, in a sense, an "isomorphism" between $\lhd$-successor and M-successor. Therefore, in any model $\mathfrak{N}$ of the theory QS', the number of $\lhd$-chains coincides with the number of M-chains. This number is called the *length* of the model $\mathfrak{N}$ and is denoted by $\mathrm{Len}(\mathfrak{N})$. The intersection of a $\lhd$-chain and an M-chain will be called a *block*.

By a *net* we mean a minimal nonempty set of elements of a model that is closed with respect to R-connections and S-connections. By B.11°, B.12°, and B.20°, all nets have the form of a regular structure with simple squares. Let σ be a net. By B.17° and B.18°, the set σ' consisting of those elements that are Q-connected with σ is also a net. By the symmetry of Q-connection, we can consider σ and σ' together and, therefore, call them a *pair of nets*. All elements of any model are situated in the nets of this model. In turn, every set is included in the form presented in Fig. 4.3.1 in accordance with B.14° and B.15°. Consequently, in any model of the theory QS', all nets have strictly plane form without cycles.

In Fig. 4.3.2, we present five possible types of distributions of unary predicates on pairs of nets and give names of these types. Other types of distribution of unary predicates on pairs of nets are impossible. We note that elements satisfying Π, Σ, A, and B must be located on the diagonal D in view of B.24°.

We consider the equivalence E connecting pairs of nets of different blocks. By B.9° and B.10°, every E-class makes a cross section through the form (cf. Fig. 4.3.1) distinguishing a pair of Q-connected elements in each $\approx$-class so that the intersection of an E-class and the block consists of exactly one pair of Q-connected nets. By B.29° and B.30°, the number of E-classes agrees with the length of the model. Thus, the model as a whole is a 3-dimensional geometrical structure in which the mechanism of suppression of cycles is included. This mechanism is based on a special nonsymmetric distribution of unary predicates on pairs of nets Π-Σ.

By the above axioms, the predicate P is, in a sense, a structural isomorphism between pairs of nets that are located in neighboring E-classes of a block. The predicate P, together with the predicates R and S, forms a 3-dimensional geometrical structure which is called a *Q-connected pair of R–S–P-cubes*. As is indicated in Sec. 4.6 below, the predicate P plays a certain role in the rigidity mechanism.

We complete an informal description of models and pass to a strict definition of coordinates and the study of their properties.

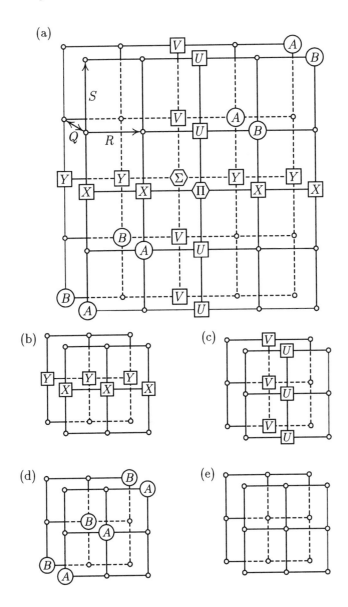

Fig. 4.3.2. Types of pairs of nets for models of the quasisuccessor theory:
(a) type Π–Σ; (b) type X–Y; (c) type U–V; (d) type A–B; (e) type $\varnothing$–$\varnothing$.

Let $\mathfrak{N}$ be an arbitrary model of the quasisuccessor theory. Denote by σa the R–S-net generated by an element a of $\mathfrak{N}$ and by $\sigma \mathfrak{N}$ the set of all such nets. We consider the quotient model $\mathfrak{N}/_\sim$ of the signature $\sigma' = \{\lhd, \sim\}$, which, by Lemma 4.2.1, is an ordinary successor relation without endpoints and cycles. We show how an arbitrary element $a \in |\mathfrak{N}|$ can be characterized by three coordinates with values in the indicated quotient model $\mathfrak{N}/_\sim$ and by the fourth sign coordinate with value in $\{-1, 1\}$.

We fix an arbitrary mapping

$$f \colon \sigma \mathfrak{N} \to \{-1, 1\} \tag{4.3.1}$$

such that

$$f(\sigma a) \neq f(\sigma b) \text{ if } Q(a, b) \tag{4.3.2}$$

$$f(\sigma a) = f(\sigma b) \text{ if } P(a, b) \tag{4.3.3}$$

For an element $a \in |\mathfrak{N}|$ we define the values of its coordinates as follows:

$r_1(a) = [a]_\sim$

$r_2(a) = [x]_\sim$ for x such that $H(a, x)$ & $D(x)$

$r_3(a) = [y]_\sim$ for y such that $E(a, y)$ & $\Pi(y)$

$s(a) = 1$ if $\sigma a \cap (U \cup V) \neq \varnothing$

           or $\sigma a \cap (U \cup V \cup X \cup Y) = \varnothing$ & $\sigma a \cap A \neq \varnothing$

           or $\sigma a \cap (U \cup V \cup X \cup Y \cup A \cup B) \doteq \varnothing$ & $f(\sigma a) = 1$

$s(a) = -1$ if $\sigma a \cap (X \cup Y) \neq \varnothing$

           or $\sigma a \cap (U \cup V \cup X \cup Y) = \varnothing$ & $\sigma a \cap B \neq \varnothing$

           or $\sigma a \cap (U \cup V \cup X \cup Y \cup A \cup B) = \varnothing$ & $f(\sigma a) = -1$

It is easy to see that the values of the coordinates $r_2(a)$ and $r_3(a)$ are independent of the choice of the representatives x and y; therefore, the definitions introduced are correct. We note that the choice function (4.3.1) defines the sign coordinate only on pairs of nets of type $\varnothing$–$\varnothing$. In other cases, the sign coordinate is defined from the distribution of unary predicates on pairs of nets.

We will use the following derivative relation, which can be expressed in terms of $\lhd$ in a weak second-order logic as follows:

$$x < y \Leftrightarrow (\exists k \in \mathbb{N} \setminus \{0\}) \; x \lhd^k y \tag{4.3.4}$$

where the notation $x \lhd^k y$ is used to designate the assertion that y is the kth $\lhd$-successor of the element x.

One can show that for an arbitrary sign coordinate, in the model of the quasisuccessor theory, the following relations hold:

$$Q(x, y) \Rightarrow s(x) \neq s(y) \tag{4.3.5}$$
$$R(x, y) \Rightarrow s(x) = s(y) \tag{4.3.6}$$
$$S(x, y) \Rightarrow s(x) = s(y) \tag{4.3.7}$$
$$P(x, y) \Rightarrow s(x) = s(y) \tag{4.3.8}$$
$$\Pi(x) \vee X(x) \vee U(x) \Rightarrow s(x) = 1 \tag{4.3.9}$$
$$\Sigma(x) \vee Y(x) \vee V(x) \Rightarrow s(x) = -1 \tag{4.3.10}$$
$$\text{If } A(x), \text{ then } s(x) = -1 \Leftrightarrow r_3(x) < r_1(x) \tag{4.3.11}$$
$$\text{If } B(x), \text{ then } s(x) = 1 \Leftrightarrow r_3(x) < r_1(x) \tag{4.3.12}$$

We show the representability of the universe of an arbitrary model of the quasisuccessor theory with respect to the coordinates introduced above.

Lemma 4.3.1. *If a sign coordinate is defined in a model $\mathfrak{N}$ of the quasisuccessor theory, then*
$$x = y \Leftrightarrow r_1(x) = r_1(y) \;\&\; r_2(x) = r_2(y) \;\&\; r_3(x) = r_3(y) \;\&\; s(x) = s(y)$$
for arbitrary elements x and y of this model

PROOF. The implication $\Rightarrow$ is obvious. In view of (4.3.5), the inverse implication is a direct consequence of the following auxiliary relation:
$$r_1(x) = r_1(y), \; r_2(x) = r_2(y), \; r_3(x) = r_3(y) \vdash x = y \vee Q(x, y)$$
We formalize and prove this relation:

$$x \sim y, \; H(x, u) \;\&\; D(u), \; H(y, v) \;\&\; D(v), \; u \sim v$$
$$E(x, w) \;\&\; \Pi(w), \; E(y, z) \;\&\; \Pi(z), \; w \sim z \vdash x = y \vee Q(x, y)$$

We give a formal proof:

$$46 \frac{u \sim v, D(u), D(v)}{H(u, v)}$$

$$2 \frac{H(x, u), H(y, v), H(u, v)}{\quad}$$

$$24 \frac{\Pi(w)}{\quad} \quad \frac{\Pi(z)}{\quad} 24$$
$$5 \frac{D(w), w \sim z, D(z)}{\quad}$$
$$50 \frac{w \approx z, \Pi(w), \Pi(z)}{\quad}$$

$$8 \frac{E(x, w), E(y, z), w = z}{E(x, y)}$$

$$4 \frac{H(x, y), x \sim y}{x \approx y,}$$

$$49 \frac{x \approx y, \quad E(x, y)}{x = y \vee Q(x, y)}$$

Lemma 4.3.2. *If a sign coordinate is defined in a model* $\mathfrak{N}$ *of the quasisuccessor theory, then for any three elements* $a_1, a_2, a_3 \in |\mathfrak{N}|$ *and any* $t \in \{-1, 1\}$ *there exists an element* $x \in |\mathfrak{N}|$ *such that its coordinates in* $\mathfrak{N}$ *are as follows:*

$$r_1(x) = [a_1]_\sim, \quad r_2(x) = [a_2]_\sim, \quad r_3(x) = [a_3]_\sim, \quad s(x) = t$$

PROOF. We successively find auxiliary elements u, v, w, z, x', and x'' such that

$$
\begin{aligned}
a_2 &\sim u \ \& \ D(u), & a_1 &\sim v \ \& \ H(u,v) \\
a_3 &\sim w \ \& \ D(w), & z &\approx w \ \& \ \Pi(z) \\
x' &\approx v \ \& \ E(x',z), & x'' &\approx v \ \& \ E(x'',z), \quad x' \approx x''
\end{aligned}
$$

The existence of such elements is guaranteed by B.3°, B.5°, B.9°, and B.29°. Let x stand for any of the elements x' and x''. The following formal deductions show that one of the elements x' and x'' is the required one:

$$
4\frac{x \approx w}{1\dfrac{x \sim v, \ a_1 \sim v}{\dfrac{x \sim a_1}{r_1(x') = r_1(x'') = [a_1]_\sim;}}}
\qquad
4\frac{x \approx y}{2\dfrac{H(x,v), \ H(u,v)}{\dfrac{H(x,u), \ D(u), \ u \sim a_2}{r_2(x') = r_2(x'') = [a_2]_\sim;}}}
$$

$$
4\frac{z \approx w}{1\dfrac{z \sim w, \ a_3 \sim w}{\dfrac{z \sim a_3, \ E(x,z), \ \Pi(z)}{r_3(x') = r_3(x'') = [a_3]_\sim;}}}
\qquad
8\dfrac{E(x',z), E(x'',z), \quad x' \approx v, \ x'' \approx v}{49\dfrac{E(x',x''), \quad x' \approx x''}{\dfrac{x' = x'' \vee Q(x',x''), \ x' \neq x''}{\dfrac{Q(x',x'')}{s(x') = t \vee s(x'') = t.}}}} 4
\tag{4.3.5}
$$

$\square$

Since any model of the quasisuccessor theory is infinite, from Lemmas 4.3.1 and 4.3.2 we obtain the following assertion.

Corollary 4.3.1. *Any model* $\mathfrak{N}$ *of the quasisuccessor theory and its quotient-model* $\mathfrak{N}/\!\sim$ *have the same cardinality.*

4.4. Connection Between Predicates and Coordinates

It turns out that the coordinates introduced agree with the language of the theory QS. The following lemma describes direct "analytic" representations of all predicates of this language in terms of the coordinates.

Lemma 4.4.1. *For any elements x and y of a model $\mathfrak{N}$ of the theory QS' with sign coordinate, the following relations hold:*

(a) $x \lhd y \Leftrightarrow r_1(x) \lhd r_1(y)$.

(b) $x \sim y \Leftrightarrow r_1(x) = r_1(y)$.

(c) $M(x, y) \Leftrightarrow r_2(x) \lhd r_2(y)$.

(d) $H(x, y) \Leftrightarrow r_2(x) = r_2(y)$.

(e) $x \approx y \Leftrightarrow r_1(x) = r_1(y) \,\&\, r_2(x) = r_2(y)$.

(f) $E(x, y) \Leftrightarrow r_3(x) = r_3(y)$.

(g) $Q(x, y) \Leftrightarrow r_1(x) = r_1(y) \,\&\, r_2(x) = r_2(y) \,\&\, r_3(x) = r_3(y) \,\&\, s(x) \neq s(y)$.

(h) $R(x, y) \Leftrightarrow r_1(x) \lhd r_1(y) \,\&\, r_2(x) = r_2(y) \,\&\, r_3(x) = r_3(y) \,\&\, s(x) = s(y)$.

(i) $S(x, y) \Leftrightarrow r_1(x) = r_1(y) \,\&\, r_2(x) \lhd r_2(y) \,\&\, r_3(x) = r_3(y) \,\&\, s(x) = s(y)$.

(j) $P(x, y) \Leftrightarrow r_1(x) = r_1(y) \,\&\, r_2(x) = r_2(y) \,\&\, r_3(x) \lhd r_3(y) \,\&\, s(x) = s(y)$.

(k) $D(x) \Leftrightarrow r_1(x) = r_2(x)$.

(l) $\Pi(x) \Leftrightarrow r_1(x) = r_2(x) = r_3(x) \,\&\, s(x) = 1$.

(m) $\Sigma(x) \Leftrightarrow r_1(x) = r_2(x) = r_3(x) \,\&\, s(x) = -1$.

(n) $A(x) \Leftrightarrow r_1(x) = r_2(x) \,\&\, r_1(x) \neq r_3(x) \,\&\, \big(s(x) = -1 \Leftrightarrow r_3(x) < r_1(x)\big)$.

(o) $B(x) \Leftrightarrow r_1(x) = r_2(x) \,\&\, r_1(x) \neq r_3(x) \,\&\, \big(s(x) = 1 \Leftrightarrow r_3(x) < r_1(x)\big)$.

(p) $U(x) \Leftrightarrow r_1(x) \neq r_2(x) \,\&\, r_1(x) = r_3(x) \,\&\, s(x) = 1$.

(q) $V(x) \Leftrightarrow r_1(x) \neq r_2(x)$ & $r_1(x) = r_3(x)$ & $s(x) = -1$.

(r) $X(x) \Leftrightarrow r_1(x) \neq r_2(x)$ & $r_2(x) = r_3(x)$ & $s(x) = 1$.

(s) $Y(x) \Leftrightarrow r_1(x) \neq r_2(x)$ & $r_2(x) = r_3(x)$ & $s(x) = -1$.

PROOF. Relations (a) and (b) are valid owing to the definition of the first coordinate, and relations (c) and (d) are derived from B.1°, B.2° and C.46°, C.47°. Relation (e) follows from (b), (d), and B.4°.

The implication $\Rightarrow$ in (f) follows from B.8° and B.30°. To prove the implication $\Leftarrow$ in (f) we formalize it as follows:

$$E(x, u) \; \& \; \varPi(u), \; E(y, v) \; \& \; \varPi(v), \; u \sim v \vdash E(x, y)$$

The proof has the form

$$
50 \frac{24 \dfrac{\varPi(u)}{D(u), \; u \sim v, \; D(v)} \dfrac{\varPi(v)}{5}}{\dfrac{\varPi(u), \; \varPi(v), \; u \approx v}{\dfrac{u = v, \; E(y, v), \; E(x, u)}{E(x, y)} 8}}
$$

Relation (g) follows from (e), (f), B.10°, and (4.3.5).

The implication $\Rightarrow$ in (h) follows from B.14° and relations (4.3.6), (a), (d), (f). The inverse implication $\Leftarrow$ in (h) is proved with the help of the implication $\Rightarrow$ in (h) in view of the uniqueness of an R-successor and Lemma 4.3.1.

The implication $\Rightarrow$ in (i) follows from B.15° and (4.3.7), (b), (c), and (f). The inverse implication $\Leftarrow$ in (i) is obtained, in view of the implication $\Rightarrow$ in (i), from the uniqueness of an S-successor and Lemma 4.3.1.

The implication $\Rightarrow$ in (j) follows from B.16°, B.30°, B.31°, (4.3.8), and (e). The inverse implication $\Leftarrow$ in (j) is obtained, in view of the implication $\Rightarrow$ in (j), from the uniqueness of a P-successor and Lemma 4.3.1.

We formalize the implication $\Rightarrow$ in (k) as follows: $D(x), H(x, y)$ & $D(y) \vdash x \sim y$. Now, we give the following proof:

$$
4 \frac{6 \dfrac{D(x), \; D(y), \; H(x, y)}{x \approx y}}{x \sim y}
$$

We formalize the inverse implication $\Leftarrow$ in (k) as follows: $H(x,y)$ & $D(y)$, $x \sim y \vdash D(x)$ and give the following proof:

$$4\frac{H(x,y),\ x \sim y}{45\dfrac{x \approx y,\ D(y)}{D(x)}}$$

By (4.3.8)–(4.3.12), to prove the remaining relations (l)-(s) it suffices to establish the following generalized relations:

(lm) $\Pi(x) \vee \Sigma(x) \Leftrightarrow r_1(x) = r_2(x) = r_3(x)$.

(no) $A(x) \vee B(x) \Leftrightarrow r_1(x) = r_2(x)$ & $r_1(x) \neq r_3(x)$.

(pq) $U(x) \vee V(x) \Leftrightarrow r_1(x) \neq r_2(x)$ & $r_1(x) = r_3(x)$.

(rs) $X(x) \vee Y(x) \Leftrightarrow r_1(x) \neq r_2(x)$ & $r_2(x) = r_3(x)$.

The proof of (no) can be omitted because it is a consequence of (k), (lm), and B.24°. We formalize the remaining relations (for the sake of convenience, we indicate the direction of implication in the notation):

(lm⇒) $\Pi(x) \vee \Sigma(x)$, $H(x,y)$ & $D(y)$, $E(x,z)$ & $\Pi(z) \vdash x \sim y$ & $x \sim z$.

(lm⇐) $H(x,y)$ & $D(y)$, $E(x,z)$ & $\Pi(z)$, $x \sim y$, $x \sim z \vdash \Pi(x) \vee \Sigma(x)$.

(pq⇒) $U(x) \vee V(x)$, $H(x,y)$ & $D(y)$, $E(x,z)$ & $\Pi(z) \vdash \neg(x \sim y)$ & $x \sim z$.

(pq⇐) $H(x,y)$ & $D(y)$, $E(x,z)$ & $\Pi(z)$, $\neg(x \sim y)$ & $x \sim z \vdash U(x) \vee V(x)$.

(rs⇒) $X(x) \vee Y(x)$, $H(x,y)$ & $D(y)$, $E(x,z)$ & $\Pi(z) \vdash \neg(x \sim y)$ & $y \sim z$.

(rs⇐) $H(x,y)$ & $D(y)$, $E(x,z)$ & $\Pi(z)$, $\neg(x \sim y)$, $y \sim z \vdash X(x) \vee Y(x)$.

We give proofs.
(lm⇒) Consider an auxiliary element w such that $Q(x,w)$ and give formal deductions:

$$10\frac{Q(x,w)}{10\dfrac{Q(x,w)}{4\dfrac{x \approx w,\quad x = z \vee w = z}{4\dfrac{x \approx z}{x \sim z}}}} \qquad \frac{10\dfrac{Q(x,w)}{E(x,w),\ E(x,z),\ \Pi(z),\ \Pi(x) \vee \Pi(w)}30 \qquad 25\dfrac{\Pi(x) \vee \Sigma(x),\ Q(x,w)}{}}{}$$

$$6\,\frac{H(x,y),\ D(y),\ D(x)}{x \approx y}\ \dfrac{\dfrac{\Pi(x) \vee \Sigma(x)}{}24}{}$$

$$\frac{}{x \sim y}$$

$(lm{\Leftarrow})$

$$5\,\frac{24\,\dfrac{\Pi(z)}{D(z),\ x \sim z,\ D(x)}\quad 45\,\dfrac{D(y),\ x \approx y}{}\ \dfrac{x \sim y,\ H(x,y)}{}4}{}$$

$$49\,\frac{x \approx z,\ E(x,z)}{}$$

$$25\,\frac{x = z \vee Q(x,z),\ \Pi(z)}{}$$

$$\frac{}{\Pi(x) \vee \Sigma(x)}$$

(pq$\Rightarrow$) Consider auxiliary elements t and w such that $x \sim t$, $E(x,t)$, $D(t)$, and $Q(t,w)$. The existence of such elements is provided by B.5°, B.9°, and B.10°. Give formal deductions:

$$32\,\frac{x \sim t, E(x,t), U(x) \vee V(x)}{\Pi(t) \vee \Sigma(t) \vee U(t) \vee V(t),}\quad 23,24\,\frac{D(t)}{\neg U(t)\ \&\ \neg V(t)}$$

$$10\,\frac{Q(t,w)}{t \approx w}\quad 30\,\frac{25\,\dfrac{\Pi(t) \vee \Sigma(t),\ Q(t,w)}{\Pi(t) \vee \Pi(w),\ \Pi(z),\ E(t,w),}\quad 8\,\dfrac{Q(t,w)\quad E(x,t), E(x,z)}{E(t,z)}}{t = z \vee w = z}$$

$$4\,\frac{z \approx t}{z \sim t,\ x \sim t}$$

$$1\,\frac{}{x \sim z}$$

$$24\,\frac{23\,\dfrac{U(x) \vee V(x)}{\neg \Pi(x)\ \&\ \neg \Sigma(x)\ \&\ \neg A(x)\ \&\ \neg B(x)}}{}$$

$$6\,\frac{\neg D(x), D(y), H(x,y)}{}$$

$$4\,\frac{H(x,y), \neg(x \approx y)}{}$$

$$\frac{}{\neg(x \sim y)}$$

$(pq\Leftarrow)$

$$32\frac{E(x,z),\ x\sim z,\ \Pi(z)}{\Pi(x)\vee\Sigma(x)\vee U(x)\vee V(x),}\qquad \frac{6\dfrac{4\dfrac{\neg(x\sim y)}{\neg(x\approx y),\ H(x,y),\ D(y)}}{\neg D(x)}}{\neg\Pi(x)\ \&\ \neg\Sigma(x)}24$$

$$U(x)\vee V(x)$$

$(rs\Rightarrow)$ Consider auxiliary elements t and w such that $H(x,t)$, $E(x,t)$, $D(t)$, and $Q(t,w)$. Such elements exist because of B.6°, B.9°, and B.10°. Give formal deductions:

$$33\frac{H(x,t),\ E(x,t),\ X(x)\vee Y(x)}{\Pi(t)\vee\Sigma(t)\vee X(t)\vee Y(t),}\qquad \frac{D(t)}{\neg X(t)\ \&\ \neg Y(t)}23,24$$

$$\Pi(t)\vee\Sigma(t)$$

$$10\frac{Q(t,w)}{t\approx w}\qquad 25\frac{\Pi(t)\vee\Sigma(t),\ Q(t,w)}{\Pi(t)\vee\Pi(w),\ \Pi(z),\ E(t,w),}\qquad \frac{Q(t,w)\quad \dfrac{E(x,t),\ E(x,z)}{E(t,z)}8}{E(t,w),\ E(t,z)}30$$

$$t=z\vee w=z$$

$$6\frac{24\dfrac{\Pi(z)}{D(z)},\ D(y),\ H(y,z)\quad \dfrac{4\dfrac{z\approx t}{H(z,t),\ H(x,t),\ H(x,y)}}{\ }2}{\ }$$

$$4\frac{y\approx z}{y\sim z}$$

$$23\frac{X(x)\vee Y(x)}{24\dfrac{\neg\Pi(x)\ \&\ \neg\Sigma(x)\ \&\ \neg A(x)\ \&\ \neg B(x)}{\dfrac{\neg D(x),\ D(y),\ H(x,y)}{\dfrac{H(x,y),\ \neg(x\approx y)}{\neg(x\sim y)}4}6}}$$

$(\mathrm{rs}\Leftarrow)$

$$24\,\dfrac{\Pi(z)}{D(z),\,D(y),\,y\sim z}$$

$$6\,\dfrac{\dfrac{\dfrac{\neg(x\sim y)}{H(x,y),\,D(y),\,\neg(x\approx y)}\,4}{24\,\dfrac{\neg D(x)}{\neg\Pi(x)\,\&\,\neg\Sigma(x),}} \qquad 5\,\dfrac{\dfrac{\dfrac{y\approx z}{H(y,z),\,H(x,y)}\,4}{\Pi(z),\,E(x,z),\,H(x,z)}\,2}{\Pi(x)\vee\Sigma(x)\vee X(x)\vee Y(x)}\,33}{X(x)\vee Y(x)}$$

$\square$

4.5. Main Properties of the Quasisuccessor Theory

Lemma 4.5.1. *Let $\mathfrak{M}$ be a model of the ordinary successor theory without endpoints and cycles of the signature $\sigma = \{\lhd^2\}$. There exists a model $\mathfrak{N}$ of the theory QS' such that the model $\mathfrak{M}$ is isomorphic to the quotient model $\mathfrak{N}/\!\sim$ with respect to the signature σ.*

PROOF. The relations of Lemma 4.4.1 make it possible to define the required model $\mathfrak{N}$ explicitly. The method is as follows. The signature successor predicate of the model $\mathfrak{M}$ will be written in the form $y = x + 1$. We define the corresponding derivative relation in a similar way as in (4.3.4) and write it as $x < y$. Starting from $\mathfrak{M}$, we construct a model $\mathfrak{N}$ of the theory QS. To this end, we put $|\mathfrak{N}| = |\mathfrak{M}|^3 \times \{-1, 1\}$. For arbitrary elements $x = \langle \alpha_1, \alpha_2, \alpha_3, \sigma \rangle$ and $y = \langle \beta_1, \beta_2, \beta_3, \tau \rangle$ of $|\mathfrak{N}|$ we set

$$x \lhd y \;\Leftrightarrow\; \beta_1 = \alpha_1 + 1$$
$$x \sim y \;\Leftrightarrow\; \beta_1 = \alpha_1$$
$$M(x,y) \;\Leftrightarrow\; \beta_2 = \alpha_2 + 1$$
$$H(x,y) \;\Leftrightarrow\; \beta_2 = \alpha_2$$
$$x \approx y \;\Leftrightarrow\; \beta_1 = \alpha_1 \;\&\; \beta_2 = \alpha_2$$
$$E(x,y) \;\Leftrightarrow\; \beta_3 = \alpha_3$$
$$Q(x,y) \;\Leftrightarrow\; \beta_1 = \alpha_1 \;\&\; \beta_2 = \alpha_2 \;\&\; \beta_3 = \alpha_3 \;\&\; \tau \neq \sigma$$
$$R(x,y) \;\Leftrightarrow\; \beta_1 = \alpha_1 + 1 \;\&\; \beta_2 = \alpha_2 \;\&\; \beta_3 = \alpha_3 \;\&\; \tau = \sigma$$

$$S(x, y) \;\Leftrightarrow\; \beta_1 = \alpha_1 \;\&\; \beta_2 = \alpha_2 + 1 \;\&\; \beta_3 = \alpha_3 \;\&\; \tau = \sigma$$
$$P(x, y) \;\Leftrightarrow\; \beta_1 = \alpha_1 \;\&\; \beta_2 = \alpha_2 \;\&\; \beta_3 = \alpha_3 + 1 \;\&\; \tau = \sigma$$
$$D(x) \;\Leftrightarrow\; \alpha_1 = \alpha_2$$
$$\Pi(x) \;\Leftrightarrow\; \alpha_1 = \alpha_2 = \alpha_3 \;\&\; \sigma = 1$$
$$\Sigma(x) \;\Leftrightarrow\; \alpha_1 = \alpha_2 = \alpha_3 \;\&\; \sigma = -1$$
$$A(x) \;\Leftrightarrow\; \alpha_1 = \alpha_2 \;\&\; \alpha_1 \neq \alpha_3 \;\&\; (\sigma = -1 \;\Leftrightarrow\; \alpha_3 < \alpha_1)$$
$$B(x) \;\Leftrightarrow\; \alpha_1 = \alpha_2 \;\&\; \alpha_1 \neq \alpha_3 \;\&\; (\sigma = 1 \;\Leftrightarrow\; \alpha_3 < \alpha_1)$$
$$U(x) \;\Leftrightarrow\; \alpha_1 \neq \alpha_2 \;\&\; \alpha_1 = \alpha_3 \;\&\; \sigma = 1$$
$$V(x) \;\Leftrightarrow\; \alpha_1 \neq \alpha_2 \;\&\; \alpha_1 = \alpha_3 \;\&\; \sigma = -1$$
$$X(x) \;\Leftrightarrow\; \alpha_1 \neq \alpha_2 \;\&\; \alpha_2 = \alpha_3 \;\&\; \sigma = 1$$
$$Y(x) \;\Leftrightarrow\; \alpha_1 \neq \alpha_2 \;\&\; \alpha_2 = \alpha_3 \;\&\; \sigma = -1$$

A routine verification shows that all the axioms of the theory QS' hold in the model $\mathfrak{N}$. It is easy to check that the mapping $\langle \alpha_1, \alpha_2, \alpha_3, \sigma \rangle \mapsto \alpha_1$ defines a $\lhd$-isomorphism between $\mathfrak{N}/\!\sim$ and $\mathfrak{M}$. □

Lemma 4.5.2. *Let $\mathfrak{N}'$ and $\mathfrak{N}''$ be models of the theory QS' with the same number of $\lhd$-chains. Then an arbitrary $\lhd$-isomorphism between the quotient models $\mu \colon \mathfrak{N}'/\!\sim \;\Rightarrow\; \mathfrak{N}''/\!\sim$ can be extended to an isomorphism $\mu^* \colon \mathfrak{N}' \;\Rightarrow\; \mathfrak{N}''$. Moreover, for a given μ the number of such extensions μ^* is equal to the number of ways of defining a sign coordinate in the model $\mathfrak{N}''$.*

PROOF. Let $\mu \colon \mathfrak{N}'/\!\sim \;\Rightarrow\; \mathfrak{N}''/\!\sim$ be an isomorphism between the quotient models. In the model $\mathfrak{N}'$, we introduce and fix a sign coordinate s'. Defining an arbitrary sign coordinate s'' in the model $\mathfrak{N}''$, we construct the natural mapping $\mu \colon |\mathfrak{N}'| \;\Rightarrow\; |\mathfrak{N}''|$ in accordance with the following rule: for $a \in |\mathfrak{N}'|$ and $b \in |\mathfrak{N}''|$ we set

$$\mu^*(a) = b \Leftrightarrow \underset{i=1}{\overset{3}{\&}} \big(r_i(b) = \mu r_i(a)\big) \;\&\; s''(b) = s'(a)$$

From Lemmas 4.3.1 and 4.3.2 it follows that μ^* is a bijection. By Lemma 4.4.1, all predicates of the theory QS' can be expressed in terms of the coordinates of their arguments; moreover, for every predicate the expressions are the same in $\mathfrak{N}'$ and in $\mathfrak{N}''$. Consequently, μ^* is an isomorphism.

It is easy to check that $s'' \mapsto \mu^*$ is a bijective correspondence between all the ways of defining a sign coordinate s'' in the model $\mathfrak{N}''$ and all isomorphisms μ^* induced by the isomorphism μ between the quotient models. □

Theorem 4.5.1. *The theory QS' is consistent, ω_1-categorical, complete, and model-complete.*

PROOF. The consistency of the theory QS' follows from Lemma 4.5.1. Let us prove that it is ω_1-categorical. Let $\mathfrak{N}'$ and $\mathfrak{N}''$ be two models of the theory QS' of cardinality $\alpha \geqslant \omega_1$. By Corollary 4.3.1, the quotient models $\mathfrak{N}'/\!\!\sim$ and $\mathfrak{N}''/\!\!\sim$ have cardinality α; therefore, they are isomorphic with respect to the signature $\sigma' = \{\vartriangleleft, \sim\}$. By Lemma 4.5.2, the models $\mathfrak{N}'$ and $\mathfrak{N}''$ are isomorphic.

By the Vaught theorem, the theory QS' is complete because it is ω_1-categorical and has no finite models. It is also model-complete, by the Lindström theorem, since the theory QS' is $\forall\exists$-axiomatizable and ω_1-categorical. $\square$

Theorem 4.5.2. *The theory QS' is almost strongly minimal and the Morley rank of QS' is 4. Moreover, the formula $\Pi(x)$ is strongly minimal and $|\mathfrak{N}| = \mathrm{acl}\big(\Pi(\mathfrak{N})\big)$ for any model $\mathfrak{N} \in \mathrm{Mod}(QS')$.*

PROOF. Axioms B.5°, B.24°, and B.29° imply that every $\sim$-class has exactly one Π-element so that the set Π is infinite. Let us prove that the formula $\Pi(x)$ is strongly minimal. Assume the contrary. Let there exist a model $\mathfrak{N} \in \mathrm{Mod}(QS')$ and a formula with constants in this model $\varphi(x, c_0, \ldots, c_{n-1})$ which divides $\Pi(\mathfrak{N})$ into two infinite parts. Introduce names of constants c_i, $i < n$, to the signature and consider the following collection of formulas in variables x and y:

$$\Sigma^*(x, y) = \{\Pi(x), \varphi(x, c_0, \ldots, c_{n-1})\} \cup \Sigma'(x)$$
$$\cup \{\Pi(y), \neg\varphi(y, c_0, \ldots, c_{n-1})\} \cup \Sigma''(y)$$

where $\Sigma'(x)$ means that the class $[x]_\sim$ is in $\vartriangleleft$-successor at a distance of more than k from the values of the coordinates $r_i(c_j)$ for all $i \in \{1, 2, 3\}$, $j < n$, and $k \in \mathbb{N}$. The set $\Sigma''(y)$ contains similar requirements for $[y]_\sim$. This assumption guarantees the local consistency of the set $\Sigma^*(x, y)$. We consider a model $\mathfrak{N}^*$ of the theory QS' with constants $c_0, \ldots, c_{n-1}$ such that, in this model, the set $\Sigma^*(x, y)$ is realized for, say, $x = a$, $y = b$, $a, b \in |\mathfrak{N}^*|$. Using Lemma 4.5.2 and taking into account that $r_i(x) = [x]_\sim$ and $s(x) = 1$ provided that x is a Π-element, it is possible to construct an automorphism μ of the model $\mathfrak{N}^*$ such that $\mu(c_i) = c_i$ for all $i < n$ and $\mu(a) = b$. But this contradicts the presence of $\varphi(x, \overline{c})$ and $\neg\varphi(y, \overline{c})$ in $\Sigma^*(x, y)$, which proves the strong minimality of the formula $\Pi(x)$.

In view of the properties of coordinates, the formula

$$\theta(x, y_1, y_2, y_3) = \mathop{\&}_{i=1}^{3} \big(\Pi(y_i) \ \& \ r_i(x) = [y_i]_\sim \big)$$

provides the validity of the relation $|\mathfrak{N}| = \mathrm{acl}\big(\Pi(\mathfrak{N})\big)$. Consequently, the theory QS' is almost strongly minimal and the rank of its universe is 3. Therefore, the Morley rank of the theory QS' is 4. $\qquad\square$

4.6. Rigidity Mechanism

We first study automorphisms of a model $\mathfrak{N} \in \mathrm{Mod}(QS')$ that are identical on the quotient model $\mathfrak{N}/\!\sim$. Such automorphisms are called *inner*. The group of all inner automorphisms of a model $\mathfrak{N}$ is denoted by $\mathrm{Aut}_0(\mathfrak{N})$.

By Lemma 4.5.2, the number of inner automorphisms of a model $\mathfrak{N}$ is equal to the number of ways of defining a sign coordinate in this model. In accordance with the description of the quasisuccessor theory, only pairs of nets of the form Π–Σ may be in models of length 1, whereas the pairs of the forms A–B, U–V, and X–Y can also occur in models of length 2. In both cases, a sign coordinate is uniquely defined in a model, without the attraction of the choice function (4.3.1). Therefore, such models have no nontrivial inner automorphisms.

We now consider a model $\mathfrak{N}$ of the theory QS' of an arbitrary length. In this model, we consider three $\lhd$-chains G_1, G_2, and G_3, which are not necessarily different. Then the set

$$K(G_1, G_2, G_3) = \{x \in |\mathfrak{N}| \mid r_1(x) \in G_1 \ \& \ r_2(x) \in G_2 \ \& \ r_3(x) \in G_3\} \tag{4.6.1}$$

is a pair of Q-connected R–S–P-cubes (we call it simply a cube for brevity). If $G_1 = G_2$ and $G_2 \neq G_3$, then, by Lemma 4.4.1, there is a pair of nets of type A–B in the cube (4.6.1). Therefore, in such a cube, a sign coordinate is uniquely defined without the attraction of the choice function (4.3.1). A similar situation arises in the other three cases if some of the $\lhd$-chains G_1, G_2, and G_3 coincide. If all the chains G_1, G_2, and G_3 are different, by Lemma 4.4.1, only pairs of nets of the type $\varnothing$–$\varnothing$ are in the cube (4.6.1). Hence, by the use of the function (4.3.1), such a cube leads to the arbitrariness of the choice of a sign coordinate (one of two possibilities). Thus,

$$\| \mathrm{Aut}_0(\mathfrak{N}) \| = 2^{k(k-1)(k-2)}, \quad k = \mathrm{Len}(\mathfrak{N})$$

We now turn to the theory QSR. Its models are given by models of the theory QS with additional predicates Λ, SP, and SN in view of BR.40°–BR.44°. By BR.40° and BR.41°, the predicate Λ distinguishes some family of $\sim$-classes, at most one class at each $\lhd$-chain. The number of classes that are distinguished by the predicate Λ on a model $\mathfrak{N}$ of the theory QSR is called the Λ-*weight* and is denoted by $\Lambda^+(\mathfrak{N})$. The number of $\lhd$-chains of this model that are free of the predicate Λ is called the *negative* Λ-*weight* and is denoted by $\Lambda^-(\mathfrak{N})$. The following relations are obvious:

$$\Lambda^+(\mathfrak{N}) \geqslant 0, \quad \Lambda^-(\mathfrak{N}) \geqslant 0, \quad \Lambda^+(\mathfrak{N}) + \Lambda^-(\mathfrak{N}) = \mathrm{Len}(\mathfrak{N})$$

By BR.42°–BR.44°, each of the predicates SP and SN distinguishes one pair of Q-connected elements in every cube (4.6.1) such that

$$G_1 \cap \Lambda \neq \varnothing, \quad G_2 \cap \Lambda \neq \varnothing, \quad G_3 \cap \Lambda \neq \varnothing \qquad (4.6.2)$$

It is clear that the Q-pair of elements of the cube (4.6.1) that is distinguished by the predicates SP and SN restricts inner automorphisms in such a manner as in the case in which the sign coordinate $s(x) = 1$ is assigned to an SP-element x and the sign coordinate $s(y) = -1$ is assigned to an SN-element y. Axioms BR.43° and BR.44° connect admissible values of the predicates SP and SN with the definition of a sign coordinate (cf. Sec. 4.3).

Summarizing, we can say that, in the theory QSR, the predicates SP and SN indicate a sign coordinate in the cubes (4.6.1) subject to the condition (4.6.2). In the rest of the cubes (4.6.1) with different chains G_1, G_2, and G_3, the arbitrariness is in defining a sign coordinate via the function (4.3.1). Thus, we have proved the following lemma.

Lemma 4.6.1. *The number of inner automorphisms of an arbitrary model $\mathfrak{N}$ of the theory QSR is defined by the formula*

$$\| \mathrm{Aut}_0(\mathfrak{N}) \| = 2^n$$

where

$$n = m(m + p - 1)(m + p - 2) + pm(m + p - 2) + p(p - 1)m$$
$$m = \Lambda^-(\mathfrak{N}), \quad p = \Lambda^+(\mathfrak{N})$$

In particular, it follows that $\mathfrak{N}$ has nontrivial inner automorphisms if $\mathrm{Len}(\mathfrak{N}) > 2$ *and* $\Lambda^-(\mathfrak{N}) > 0$.

4.7. Exterior Properties of the Basic Theory

In this section, we briefly describe properties of the theory QSR following from its description. We need these properties to use QSR as a basic theory in the universal construction (cf. Chapter 6).

Let $\mathfrak{N}$ and $\mathfrak{N}'$ be models of the theory QS' or the theory QSR. If $\mu\colon \mathfrak{N} \to \mathfrak{N}'$ is an isomorphic embedding and $\mu^*\colon \mathfrak{N}/\!\!\sim \to \mathfrak{N}'/\!\!\sim$ is an exterior isomorphic embedding induced by μ, then we say that μ is a *realization* of μ^*. A model $\mathfrak{N}$ of the theory QSR is called *filled* if on each ⊲-chain there are Λ-distinguished elements of $\mathfrak{N}$.

From the description of models of the theory QSR we obtain the following lemma.

Lemma 4.7.1. *The following assertions hold:*

(a) *let $\mathfrak{N}$ be an arbitrary model of the theory QS'. If a unary predicate Λ is defined in $\mathfrak{N}$, is well defined on $\sim$-classes, and distinguishes at most one $\sim$-class on any ⊲-chain, then it is possible to define the predicates SP and SN in $\mathfrak{N}$ in order to obtain a model of the theory QSR.*

(b) *Let $\mathfrak{N}'$ and $\mathfrak{N}''$ be models of the theory QSR. An arbitrary external isomorphic embedding $\mu^*\colon \mathfrak{N}'/\!\!\sim \to \mathfrak{N}''/\!\!\sim$ has a realization μ such that $\mu\colon \mathfrak{N}' \Rightarrow \mathfrak{N}''$ is an isomorphic embedding. If the mapping μ^* is surjective, then any such realization μ is surjective. Furthermore, if a model $\mathfrak{N}'$ is filled or satisfies the condition $\mathrm{Len}(\mathfrak{N}') \leqslant 2$, then μ is uniquely determined from a given μ^*. Otherwise, the embedding μ^* admits more than one realization.*

(c) *Let $\mathfrak{N}$, $\mathfrak{N}'$, and $\mathfrak{N}''$ be models of the theory QS' such that $\mathfrak{N} \subseteq \mathfrak{N}'$, $\mathfrak{N} \subseteq \mathfrak{N}''$ and let $\mu^*\colon \mathfrak{N}'/\!\!\sim \to \mathfrak{N}''/\!\!\sim$ be an external isomorphic embedding subject to the consistency condition $\mu^* \left([a]_{\sim}^{\mathfrak{N}'} \right) = [a]_{\sim}^{\mathfrak{N}''}$ for all $a \in |\mathfrak{N}|$. Then μ^* has a realization μ such that $\mu\colon \mathfrak{N}' \to \mathfrak{N}''$ is an isomorphic embedding and μ is identical on $\mathfrak{N}$.*

Chapter 5

Interpretations

Interpretations are of considerable importance for the universal construction. Therefore, in this chapter, we introduce terms and information concerning the notion of an interpretation. The main meaning of interpretations is to divide the proof of the main theorem into elementary stages. However, interpretations also play an important role in some other aspects of the study of finitely axiomatizable theories. In particular, by means of interpretations, finitely axiomatizable theories are reduced to the theory of the minimal signature. A special class of interpretations is used in the proof of the uniqueness theorem for a recursively saturated generalized Boolean algebra (cf. Chapter 10).

5.1. Interpretations in the Basic Construction

Let T_0 and T_1 be axiomatizable theories of signatures σ_0 and σ_1 respectively. We use the ordinary notion of an interpretation of the theory T_0 in the theory T_1 on the set distinguished by a unary predicate $U(x)$ [59]. A first-order definable relation can be taken for U. The domain of

135

an interpretation of T_0 in T_1 can be represented by elements of U with equality. The image of $\varphi \in \mathrm{FL}(\sigma_0)$ under an interpretation I is denoted by $I(\varphi)$.

- An interpretation I is called *effective* if the transformation $\varphi \mapsto I(\varphi)$ is effective, i.e., if it is given by a recursive function defined on Gödel numbers.

Let I be an interpretation of a theory T_0 in the domain $U(x)$ of a theory T_1 and let $\mathfrak{N}$ be a model of T_1. Using I, it is possible to define all predicates of the signature σ_0 on the first-order definable set $U(\mathfrak{N})$. As a result, we obtain the model $\mathfrak{M} = \langle U(\mathfrak{N}), \sigma_0 \rangle$ which is called the *kernel* of $\mathfrak{N}$ with respect to I and is denoted by $\mathbb{K}_I(\mathfrak{N})$ or $\mathbb{K}(\mathfrak{N})$. We assume that each interpretation I considered satisfies at most one of the following (equivalent) conditions:

$$T_0 \vdash \varphi \Leftrightarrow T_1 \vdash I(\varphi) \text{ for all } \varphi \in \mathrm{SL}(\sigma_0)$$
$$T_0 = \mathrm{Th}\{\mathbb{K}(\mathfrak{N}) \mid \mathfrak{N} \in \mathrm{Mod}(T_1)\}$$

We will use special relations between models of the theory T_1. They generalize the notion of a submodel to the case of interpretations:

$$\mathfrak{N} \sqsubseteq \mathfrak{N}' \Leftrightarrow \mathfrak{N} \subseteq \mathfrak{N}' \,\&\, U(\mathfrak{N}) = U(\mathfrak{N}') \cap |\mathfrak{N}| \,\&\, \mathbb{K}(\mathfrak{N}) \subseteq \mathbb{K}(\mathfrak{N}')$$
$$\mathfrak{N} \preccurlyeq^\circ \mathfrak{N}' \Leftrightarrow \mathfrak{N} \subseteq \mathfrak{N}' \,\&\, U(\mathfrak{N}) = U(\mathfrak{N}') \cap |\mathfrak{N}| \,\&\, \mathbb{K}(\mathfrak{N}) \preccurlyeq \mathbb{K}(\mathfrak{N}')$$

The model $\mathfrak{N}$ is called the *kernel submodel* of $\mathfrak{N}'$ if the first relation holds and the *kernel-elementary submodel* of $\mathfrak{N}'$ if the second relation holds. For models of the theory T_1 the notions of "kernel embeddings" and "kernel-elementary embeddings" also can be introduced.

- An interpretation I of a theory T_0 in the domain $U(x)$ of a theory T_1 is called $\exists$-*representable* if the following conditions hold:

  — the relation $U(x)$ and the negation of $U(x)$ are represented in the theory T_1 by $\exists$-formulas,

  — I-images of all signature predicates of the theory T_0 and their negations are represented in the theory T_1 by $\exists$-formulas.

The following lemma is obvious.

Lemma 5.1.1. *Let I be an $\exists$-representable interpretation of a theory T_0 in a theory T_1. Then $\mathfrak{N} \subseteq \mathfrak{N}'$ implies $\mathfrak{N} \sqsubseteq \mathfrak{N}'$ for any models $\mathfrak{N}$ and $\mathfrak{N}'$ of the theory T_1.*

- An interpretation I of a theory T_0 in a theory T_1 is called

  - *enveloping* if for any model $\mathfrak{N}$ of the theory T_1 and a submodel $\mathfrak{M} \subseteq \mathbb{K}(\mathfrak{N})$ such that $\mathfrak{M} \vDash T_0$ there is a submodel $\mathfrak{N}' \subseteq \mathfrak{N}$ such that $\mathfrak{N}' \vDash T_1$, $\mathfrak{N}' \sqsubseteq \mathfrak{N}$, and $\mathbb{K}(\mathfrak{N}') = \mathfrak{M}$,

  - *elementarily enveloping* if for any model $\mathfrak{N}$ of the theory T_1 and an elementary submodel $\mathfrak{M} \preccurlyeq \mathbb{K}(\mathfrak{N})$ there is a submodel $\mathfrak{N}' \subseteq \mathfrak{N}$ such that $\mathfrak{N}' \vDash T_1$, $\mathfrak{N}' \preccurlyeq^\circ \mathfrak{N}$, and $\mathbb{K}(\mathfrak{N}') = \mathfrak{M}$.

It is obvious that an enveloping interpretation is elementarily enveloping.

Using the compactness theorem, we obtain the following lemma.

Lemma 5.1.2. *Let an interpretation I of a theory T_0 in a theory T_1 be elementarily enveloping. Then for any model $\mathfrak{M}$ of the theory T_0 there exists a model $\mathfrak{N}$ of the theory T_1 such that $\mathfrak{M} \cong \mathbb{K}(\mathfrak{N})$.*

5.2. Isostone Interpretations

- An interpretation I of a theory T_0 in a theory T_1 is called *isostone* if the following conditions hold:

  (a) I is effective,

  (b) $\mathfrak{N}_0 \equiv \mathfrak{N}_1 \Leftrightarrow \mathbb{K}(\mathfrak{N}_0) \equiv \mathbb{K}(\mathfrak{N}_1)$ for any $\mathfrak{N}_0, \mathfrak{N}_1 \in \mathrm{Mod}\,(T_1)$.

The following lemma is proved by standard methods.

Lemma 5.2.1. *Let I be an isostone interpretation of a theory T_0 in a theory T_1. Then the mapping $\mu : \mathcal{L}(T_0) \to \mathcal{L}(T_1)$ defined by the formula $\mu([\varphi]_{T_0}) = [I(\varphi)]_{T_1}$, $\varphi \in \mathrm{SL}(\sigma_0)$, is a recursive isomorphism between the Lindenbaum algebras $\mathcal{L}(T_0)$ and $\mathcal{L}(T_1)$. Moreover, the isomorphism μ establishes a natural correspondence between complete extensions $T_0^* \supseteq T_0$ and $T_1^* \supseteq T_1$, i.e., $T_1^* = \mu(T_0^*)$ holds if and only if the equalities $T_0^* = \mathrm{Th}\,(\mathbb{K}(\mathfrak{N}))$ and $T_1^* = \mathrm{Th}\,(\mathfrak{N})$, which are equivalent to the relation $T_1^* = T_1 \cup I(T_0^*)$, are valid for some model $\mathfrak{N}$ of the theory T_1.*

In view of conditions (a) and (b) from the definition of an isostone interpretation, we can see that, in a sense, there is an isomorphism between the Stone spaces of Lindenbaum algebras. This explains the choice of the term "isostone interpretation."

Let I be an isostone interpretation of a theory T_0 in a theory T_1 and let L be a list of set-theoretic properties.

- An interpretation I is called an *L-interpretation* if the recursive isomorphism μ that is defined by I in accordance with Lemma 5.2.1 preserves all the properties in the list L.

The following lemma gives equivalent statements of condition (b) from the definition of an isostone interpretation.

Lemma 5.2.2. *For an interpretation I of a theory T_0 in a theory T_1, the following conditions are equivalent:*

(a) $\mathbb{K}(\mathfrak{N}) \equiv \mathbb{K}(\mathfrak{N}') \Leftrightarrow \mathfrak{N} \equiv \mathfrak{N}'$ *for any* $\mathfrak{N}, \mathfrak{N}' \in \mathrm{Mod}(T_1)$,

(b) $\mathbb{K}(\mathfrak{N}) \equiv \mathbb{K}(\mathfrak{N}') \Rightarrow \mathfrak{N} \equiv \mathfrak{N}'$ *for any* $\mathfrak{N}, \mathfrak{N}' \in \mathrm{Mod}(T_1)$,

(c) $\mathbb{K}(\mathfrak{N}) \cong \mathbb{K}(\mathfrak{N}') \Rightarrow \mathfrak{N} \equiv \mathfrak{N}'$ *for any* $\mathfrak{N}, \mathfrak{N}' \in \mathrm{Mod}(T_1)$,

(d) $\{I(\varphi) \mid \varphi \in \mathrm{SL}(\sigma_0)\}$ *generates* $\mathcal{L}(T_1)$.

PROOF. The implications (a) $\Rightarrow$ (b) $\Rightarrow$ (c) are obvious. The implication (c) $\Rightarrow$ (a) can be proved with the help of the Shelah theorem on the existence of isomorphic ultrapowers of elementarily equivalent models. The equivalence of (b) and (d) follows from Lemma 0.3.2. $\square$

5.3. Immersing Interpretations

We introduce a subclass of the class of isostone interpretations.

- An interpretation I of a theory T_0 in the domain $U(x)$ of a theory T_1 is called *immersing* if the following conditions hold:

   (a) I is effective,

   (b) for any models $\mathfrak{N}$, $\mathfrak{N}'$ of the theory T_1 and sets $X \subseteq U(\mathfrak{N})$, $Y \subseteq U(\mathfrak{N}')$, the mapping $f: \mathfrak{N} \upharpoonright X \to \mathfrak{N}' \upharpoonright Y$ is elementary if the mapping $f: \mathbb{K}(\mathfrak{N}) \upharpoonright X \to \mathbb{K}(\mathfrak{N}') \upharpoonright Y$ is elementary,

   (c) $\mathfrak{N} \preccurlyeq \mathfrak{N}' \Leftrightarrow \mathfrak{N} \preccurlyeq^{\circ} \mathfrak{N}'$ for any models $\mathfrak{N}$, $\mathfrak{N}'$ of T_1.

Lemma 5.3.1. *Every immersing interpretation is isostone.*

PROOF. Condition (b) from the definition of an isostone interpretation is a partial case of condition (b) from the definition of an immersing interpretation for $X = Y = \varnothing$. □

We state equivalent formulations of conditions (b) and (c) from the definition of an immersing interpretation. The proof of the following lemma is standard.

Lemma 5.3.2. *Let I be an isostone interpretation of a theory T_0 in the domain $U(x)$ of a theory T_1. Then the following assertions hold:*

(a) *for any models $\mathfrak{N}$, $\mathfrak{N}'$ of the theory T_1 and sets $X \subseteq U(\mathfrak{N})$, $Y \subseteq U(\mathfrak{N}')$, the mapping $f \colon \mathfrak{N} \restriction X \to \mathfrak{N}' \restriction Y$ is elementary if the mapping $f \colon \mathbb{K}(\mathfrak{N}) \restriction X \to \mathbb{K}(\mathfrak{N}') \restriction Y$ is elementary,*

(b) *for any $n < \omega$ and $\varphi \in \mathrm{FL}_n(\sigma_1)$ such that $T_1 \vdash \varphi(x_1, \ldots, x_n) \to U(x_1) \,\&\, \ldots \,\&\, U(x_n)$, there exists a formula $\psi \in \mathrm{FL}_n(\sigma_0)$ such that $T_1 \vdash \varphi \leftrightarrow I\psi$,*

(c) *if $\mathfrak{N}$ is a model of the theory T_1 and $p(x_1, \ldots, x_n)$ is a complete n-type of the theory $\mathrm{Th}(\mathbb{K}(\mathfrak{N}))$, then the set $Ip(x_1, \ldots, x_n)$ generates a complete n-type of the theory $\mathrm{Th}(\mathfrak{N})$.*

Let I be an interpretation of a theory T_0 in the domain $U(x)$ of a theory T_1. The formula $\varphi(x_1, \ldots, x_k)$ of the signature σ_1 is called an $\exists$-*primitive formula over kernel in the interpretation I* if

$$\varphi(x_1, \ldots, x_k) = (\exists x_{k+1}) \ldots (\exists x_n)\, \Phi(x_1, \ldots, x_n) \qquad (5.3.1)$$

where the formula $\Phi(x_1, \ldots, x_n)$ has the following special form: the set of subscripts of its variables can be divided into two disjoint parts:

$$\{1, \ldots, n\} = \{\alpha_1, \ldots, \alpha_s\} \cup \{\beta_1, \ldots, \beta_t\}, \quad \{\alpha_1, \ldots, \alpha_s\} \cap \{\beta_1, \ldots, \beta_t\} = \varnothing$$

so that $\Phi(x_1, \ldots, x_n)$ is the conjunction of three terms: $\Phi_1 \,\&\, \Phi_2 \,\&\, \Phi_3$, where

$\Phi_1(x_1, \ldots, x_n)$ is a primitive formula of the signature σ_1,
$\Phi_2(x_1, \ldots, x_n)$ is the conjunction of $U(x_i)$, $i \in \{\alpha_1, \ldots, \alpha_s\}$, and $\neg U(x_j)$, $j \in \{\beta_1, \ldots, \beta_t\}$,
$\Phi_3(x_1, \ldots, x_n)$ has the form $I\psi(x_{\alpha_1}, \ldots, x_{\alpha_s})$ for $\psi(x_{\alpha_1}, \ldots, x_{\alpha_s}) \in \mathrm{FL}(\sigma_0)$.

A formula $\varphi(x_1, \ldots, x_n)$ of the signature σ_1 is called *primitive over kernel in the interpretation I* if φ has the form (5.3.1) with $k = n$, i.e., if the existential quantifier is absent.

Lemma 5.3.3. *Let I be an isostone interpretation of a theory T_0 in a theory T_1. Then the following assertions are equivalent:*

(a) $\mathfrak{N} \preccurlyeq \mathfrak{N}' \Leftrightarrow \mathfrak{N} \preccurlyeq^\circ \mathfrak{N}'$ *for any models $\mathfrak{N}$ and $\mathfrak{N}'$ of the theory T_1,*

(b) *for any formula $\varphi(x_1, \dots, x_k)$ of the signature σ_1, a model $\mathfrak{N}$ of the theory T_1, and elements $a_1, \dots, a_k$ of $\mathfrak{N}$ such that $\mathfrak{N} \vDash \varphi(a_1, \dots, a_k)$, there exists a formula such that it is $\exists$-primitive over kernel in the interpretation I and*

$$\mathfrak{N} \vDash \psi(a_1, \dots, a_k), \quad T \vdash \psi(x_1, \dots, x_k) \to \varphi(x_1, \dots, x_k)$$

(c) *every formula $\varphi(x_1, \dots, x_k)$ of the signature σ_1 is equivalent in the theory T_1 to some finite disjunction of formulas that are $\exists$-primitive over kernel in the interpretation I,*

(d) *any complete type $p(x_1, \dots, x_k)$ of a completion T_1^* of the theory T_1 is generated by formulas that are primitive over kernel and occur in this type,*

(e) *for any models $\mathfrak{N}$ and $\mathfrak{N}'$ of the theory T_1 such that $\mathfrak{N} \preccurlyeq^\circ \mathfrak{N}'$, there exists a model $\mathfrak{N}''$ of the theory T_1 such that $\mathfrak{N}' \preccurlyeq^\circ \mathfrak{N}''$ and $\mathfrak{N} \preccurlyeq \mathfrak{N}''$.*

PROOF. For a model $\mathfrak{N}$ of the theory T_1 whose signature is enriched by constant symbols c_a, $a \in |\mathfrak{N}|$, we set

$$AD(\mathfrak{N}, U) = AD(\mathfrak{N}) \cup \{U(c_a) \mid a \in U(\mathfrak{N})\} \cup \{\neg U(c_a) \mid a \in \neg U(\mathfrak{N})\}$$

(a) $\Rightarrow$ (b). Assume that (a) holds and consider $\varphi \in FL_k(\sigma_1)$, a model $\mathfrak{N}$ of the theory T_1, and elements $a_1, \dots, a_k$ of $\mathfrak{N}$ such that $\mathfrak{N} \vDash \varphi(a_1, \dots, a_k)$. We enrich the signature σ_1 with constant symbols for elements of model $\mathfrak{N}$ and consider the set of formulas $\Sigma = T_1 \cup AD(\mathfrak{N}, U) \cup I(FD(\mathbb{K}(\mathfrak{N})))$. In view of (a), the canonical embedding of $\mathfrak{N}$ in any model $\mathfrak{N}^* \in \text{Mod}(\Sigma)$ is elementary. Therefore, the sentence $\varphi(a_1, \dots, a_k)$ is deducible from Σ. Consequently, it is deducible from a finite part $\Sigma' \subseteq \Sigma$. We can assume that Σ' contains $U(c)$ or $\neg U(c)$ for any constant symbol c from Σ'. By assumption, the interpretation I is isostone. By Lemma 5.2.2(d), the part of the set Σ' that is relative to T_1 can be transformed to the equivalent form $I(\psi)$, $\psi \in SL(\sigma_0)$. As a result, we obtain an equivalent set of formulas Σ''. Then the required formula ψ can be obtained by taking the conjunction of formulas from Σ'', replacing constant symbols by variables, and applying the $\exists$-quantifiers on those variables that are not associated with $a_1, \dots, a_k$.

(b) $\Rightarrow$ (c). Let $\varphi(x_1, \ldots, x_n)$ be a formula of the signature σ_1. We consider the set $\mathcal{P}$ of all those formulas $\psi(x_1, \ldots, x_n)$ that are $\exists$-primitive over kernel and such that the sentence $\psi(x_1, \ldots, x_n) \to \varphi(x_1, \ldots, x_n)$ is deducible in the theory T_1. In view of (b), the formula $\varphi(x_1, \ldots, x_n)$ is equivalent, in T_1, to the disjunction of all formulas in $\mathcal{P}$. By the compactness theorem, it is equivalent to some finite disjunction of formulas from $\mathcal{P}$, which proves (c).

(c) $\Rightarrow$ (d). The implication is obvious.

(d) $\Rightarrow$ (e). Assume that (d) holds and consider models $\mathfrak{N}$ and $\mathfrak{N}'$ satisfying (e). We enrich the signature σ_1 with constant symbols for elements of the model $\mathfrak{N}'$. Consider the set of formulas

$$\Sigma = FD(\mathfrak{N}) \cup AD(\mathfrak{N}', U) \cup I\big(FD(\mathbb{K}(\mathfrak{N}'))\big) \qquad (5.3.2)$$

and show that any finite subset $\Sigma' \subseteq \Sigma$ is consistent. We can suppose that Σ' is the conjunction of three formulas:

$$\varphi_1(\bar{c}_1, \bar{c}_2) \,\&\, \varphi_2(\bar{c}_1, \bar{c}_2, \bar{c}_3, \bar{c}_4) \,\&\, I\varphi_3(\bar{c}_1, \bar{c}_3) \qquad (5.3.3)$$

where $\varphi_1, \varphi_2 \in FL(\sigma_1)$, $\varphi_3 \in FL(\sigma_0)$, φ_2 is primitive over U, $\bar{c}_1 \in U(\mathfrak{N})$, $\bar{c}_2 \in |\mathfrak{N}| \setminus U(\mathfrak{N})$, $\bar{c}_3 \in U(\mathfrak{N}') \setminus U(\mathfrak{N})$, and $\bar{c}_4 \in |\mathfrak{N}'| \setminus (|\mathfrak{N}| \cup U(\mathfrak{N}'))$. A formula is called *primitive* over U if it is represented as the conjunction of a primitive formula and formulas of the form $U(a)$, $\neg U(b)$. Using (d), we replace $\varphi_1(\bar{c}_1, \bar{c}_2)$ in (5.3.3) by the stronger formula $I(\varphi_1'(\bar{e}_1)) \,\&\, \varphi_1''(\bar{e}_1, \bar{e}_2)$, which is primitive over kernel, where $\varphi_1' \in FL(\sigma_0)$, $\varphi_1'' \in FL(\sigma_1)$, and φ_1'' is primitive over U; moreover, $\bar{c}_1 \subseteq \bar{e}_1 \subseteq U(\mathfrak{N})$ and $\bar{c}_2 \subseteq \bar{e}_2 \subseteq |\mathfrak{N}| \setminus U(\mathfrak{N})$. The conjunction of four formulas obtained

$$I(\varphi_1'(\bar{e}_1)) \,\&\, \varphi_1''(\bar{e}_1, \bar{e}_2) \,\&\, \varphi_2(\bar{c}_1, \bar{c}_2, \bar{c}_3, \bar{c}_4) \,\&\, I(\varphi_3(\bar{c}_1, \bar{c}_3))$$

occurs in the consistent set $I\big(FD(\mathbb{K}(\mathfrak{N}'))\big) \cup AD(\mathfrak{N}') \subseteq FD(\mathfrak{N}')$. Consequently, the conjunction (5.3.3) is also consistent. By the compactness theorem, the set (5.3.2) is consistent. Consider a model $\mathfrak{N}^*$ for Σ. We restrict it to a model of the initial signature σ_1. As a result, we obtain a model $\mathfrak{N}''$ of the theory T_1 which satisfies (e) in accordance with the components of the set (5.3.2).

(e) $\Rightarrow$ (a). The implication from left to right in (a) is obvious. We establish the inverse implication in (a). Let $\mathfrak{N}_0$ and $\mathfrak{N}_1$ be models of the theory T_1 such that $\mathfrak{N}_0 \preccurlyeq^\circ \mathfrak{N}_1$. Using (e) many times, we construct the following chain of models of the theory T_1:

$$\mathfrak{N}_0 \preccurlyeq^\circ \mathfrak{N}_1 \preccurlyeq^\circ \mathfrak{N}_2 \preccurlyeq^\circ \ldots \preccurlyeq^\circ \mathfrak{N}_s \preccurlyeq^\circ \mathfrak{N}_{s+1} \preccurlyeq^\circ \ldots, \quad s < \omega \qquad (5.3.4)$$

where $\mathfrak{N}_s \preccurlyeq \mathfrak{N}_{s+1}$, $s < \omega$. Denote by $\mathfrak{N}_\omega$ the union of the models of this chain. On the one hand, $\mathfrak{N}_\omega$ is the union of the models in (5.3.4) with even subscripts. On the other hand, $\mathfrak{N}_\omega$ is the union of the models in (5.3.4) with odd subscripts. By construction, both chains are elementary. Therefore, all models constituting the chains are elementary submodels of the model $\mathfrak{N}_\omega$. As a result, we obtain $\mathfrak{N}_0 \preccurlyeq \mathfrak{N}_\omega$ and $\mathfrak{N}_0 \subseteq \mathfrak{N}_1 \preccurlyeq \mathfrak{N}_\omega$. Consequently, $\mathfrak{N}_0 \preccurlyeq \mathfrak{N}_1$. $\square$

Lemma 5.3.4. *If I is an $\exists$-representable immersing interpretation of a theory T_0 in a theory T_1, then the following assertions hold:*

(a) *if T_0 is model-complete, then T_1 is model-complete,*

(b) *if I is enveloping, then T_1 is model-complete if and only if T_0 is model-complete.*

PROOF. (a). Let $\mathfrak{N}$ and $\mathfrak{N}'$ be models of the theory T_1 such that $\mathfrak{N} \subseteq \mathfrak{N}'$. By Lemma 5.1.1, $\exists$-representability implies $\mathfrak{N} \sqsubseteq \mathfrak{N}'$. Therefore, $\mathbb{K}(\mathfrak{N}) \subseteq \mathbb{K}(\mathfrak{N}')$. Since the theory T_0 is model complete, we have $\mathbb{K}(\mathfrak{N}) \preccurlyeq \mathbb{K}(\mathfrak{N}')$. By condition (c) from the definition of an immersing interpretation, we obtain $\mathfrak{N} \preccurlyeq \mathfrak{N}'$, which proves the model completeness of T_1.

(b). As in (a), the model completeness of T_0 implies the model completeness of T_1. Conversely, let T_1 be a model-complete theory. We consider models $\mathfrak{M}_0$ and $\mathfrak{M}_1$ of the theory T_0 such that $\mathfrak{M}_0 \subseteq \mathfrak{M}_1$. Since the interpretation I is enveloping, it is elementarily enveloping. By Lemma 5.1.2, there exists a model $\mathfrak{N}_1$ of the theory T_1 such that $\mathfrak{M}_1 = \mathbb{K}(\mathfrak{N}_1)$. Since the interpretation I is enveloping, we can find a submodel $\mathfrak{N}_0 \subseteq \mathfrak{N}_1$ such that $\mathfrak{N}_0 \vDash T_1$, $\mathfrak{N}_0 \sqsubseteq \mathfrak{N}_1$, and $\mathfrak{M}_0 = \mathbb{K}(\mathfrak{N}_0)$. The model completeness of the theory T_1 implies $\mathfrak{N}_0 \preccurlyeq \mathfrak{N}_1$, whence $\mathfrak{M}_0 \preccurlyeq \mathfrak{M}_1$. Thus, the theory T_0 is model-complete. $\square$

5.4. Exact Interpretations

We begin with definitions.

- An interpretation I of a theory T_0 in a theory T_1 is called *model-bijective* if $\mathfrak{N} \mapsto \mathbb{K}(\mathfrak{N})$ is a one-to-one correspondence between isomorphism types of models of the theories T_0 and T_1, i.e., if $\mathrm{Mod}(T_0) = \{\mathbb{K}(\mathfrak{N}) \mid \mathfrak{N} \in \mathrm{Mod}(T_1)\}$ and $\mathfrak{N} \cong \mathfrak{N}' \Leftrightarrow \mathbb{K}(\mathfrak{N}) \cong \mathbb{K}(\mathfrak{N}')$, where $\mathfrak{N}, \mathfrak{N}' \in \mathrm{Mod}(T_1)$.

- An interpretation I of a theory T_0 in the domain $U(x)$ of a theory T_1 is called *exact* if the following conditions hold:

  (a) I is effective,

  (b) I is model-bijective,

  (c) every element a of any model $\mathfrak{N} \in \mathrm{Mod}\,(T_1)$ is first-order definable over $U(\mathfrak{N})$ by means of $\exists$-formulas,

  (d) for any model $\mathfrak{N}$ of the theory T_1, an automorphism $\mu\colon \mathbb{K}\,(\mathfrak{N}) \to \mathbb{K}(\mathfrak{N})$ of the kernel $\mathbb{K}\,(\mathfrak{N})$ can be extended to an automorphism $\mu^*\colon \mathfrak{N} \to \mathfrak{N}$ of the model $\mathfrak{N}$.

Lemma 5.4.1. *If I is an exact interpretation of a theory T_0 in the domain $U(x)$ of a theory T_1, then the following assertions hold:*

(a) $\|\mathfrak{N}\| < \omega \Rightarrow \|\mathbb{K}\,(\mathfrak{N})\| < \omega$ *for all* $\mathfrak{N} \in \mathrm{Mod}\,(T_1)$,

(b) $\|\mathfrak{N}\| \geqslant \omega \Rightarrow \|\mathbb{K}\,(\mathfrak{N})\| = \|\mathfrak{N}\|$ *for all* $\mathfrak{N} \in \mathrm{Mod}\,(T_1)$,

(c) I *is isostone,*

(d) I *is immersing,*

(e) I *is an* MQL-*interpretation.*

REMARK 5.4.1. If the corresponding complete extensions $T_0^* \supseteq T_0$ and $T_1^* \supseteq T_1$ have finite models, then, in (e), we regard the transmission of MQL-properties from T_0^* to T_1^* in the sense that the theories $T_0^* \oplus SI$ and $T_1^* \oplus SI$ have the same properties listed in MQL; the theory SI is defined in Remark 0.6.1.

PROOF OF LEMMA 5.4.1. (a), (b). The assertions follow from the fact that the universe has no two-cardinal property over $U(x)$.

(c). The assertion follows from Lemma 5.2.2(c).

(d). We first verify condition (b) from the definition of an immersing interpretation. Let $\mathfrak{N}_0$ be a model of the theory T_1 and let $p(x_1, \ldots, x_n)$ be a complete n-type of the theory $\mathrm{Th}\,(\mathbb{K}\,(\mathfrak{N}_0))$. It is obvious that the set $I(p(x_1, \ldots, x_n))$ is consistent in the theory $T_1^* = \mathrm{Th}\,(\mathfrak{N}_0)$. We prove that the type is complete. Assume the contrary. Then there is a countable model $\mathfrak{N}$ of the theory T_1^* and two sequences of elements $(b_1, \ldots, b_n)$ and $(c_1, \ldots, c_n)$ in $U(\mathfrak{N})$ which realize the type $p(x_1, \ldots, x_n)$ in $\mathbb{K}\,(\mathfrak{N})$, but whose types are different in $\mathfrak{N}$.

We construct a countable elementary extension $\mathfrak{N}'$ of the model $\mathfrak{N}$ such that the extension constructed is a homogeneous model. The kernel $\mathbb{K}(\mathfrak{N}')$ is also a homogeneous model. Therefore, there exists an automorphism μ: $\mathbb{K}(\mathfrak{N}') \to \mathbb{K}(\mathfrak{N}')$ such that $\mu(b_i) = c_i$, $i = 1, \ldots, n$. In view of condition (d) from the definition of an exact interpretation, the automorphism μ must be continued to an automorphism of the model $\mathfrak{N}'$, which is impossible because of the difference between the types of these sequences in $\mathfrak{N}$, and consequently, in $\mathfrak{N}'$. The contradiction obtained proves that $I(p(x_1, \ldots, x_n))$ is a complete type of the theory $\mathrm{Th}(\mathfrak{N})$. By Lemma 5.3.2, condition (b) from the definition of an immersing interpretation is valid. To verify condition (c) we use Lemma 5.3.3. Let $\mathfrak{N}$ be a model of the theory T_1. Consider the sequence

$$c_1, \ldots, c_n \in |\mathfrak{N}| \tag{5.4.1}$$

By condition (c) from the definition of an exact interpretation, the sequence (5.4.1) can be extended to the finite set

$$a_1, \ldots, a_s, \ b_1, \ldots, b_t \in |\mathfrak{N}| \tag{5.4.2}$$
$$a_1, \ldots, a_s \in U(\mathfrak{N}) \tag{5.4.3}$$
$$b_1, \ldots, b_t \in |\mathfrak{N}| \setminus U(\mathfrak{N}) \tag{5.4.4}$$

such that all the elements in (5.4.1) are first-order definable over the elements (5.4.3) by means of $\exists$-formulas in such a way that all the external $\exists$-quantifiers of these formulas are realized among the elements (5.4.3) and (5.4.4). Hence the type of the sequence (5.4.1) is completely characterized by the type of the sequence (5.4.3) in the theory $\mathrm{Th}(\mathbb{K}(\mathfrak{N}))$ and by some finite σ_1-diagram of the set (5.4.2). It suffices to include, in the σ_1-diagram, predicates of the $\exists$-formulas distinguishing the elements (5.4.1). As a result, we arrive at claim (b) of Lemma 5.3.3, which implies the validity of condition (c) from the definition of an immersing interpretation.

(e). We temporarily retract the above convention and admit interpretations such that the equality predicate goes to the first-order definable equivalences on U. Such interpretations are called *generalized*. Let $\mathfrak{M}$ be a model and let $s_1, \ldots, s_k$ be natural numbers. We introduce the notation

$$\mathfrak{M}\langle s_1, \ldots, s_k \rangle = \mathfrak{M} \cup |\mathfrak{M}|^{s_1} \cup \ldots \cup |\mathfrak{M}|^{s_k} \tag{5.4.5}$$

where the right-hand side of (5.4.5) means the model $\mathfrak{M}$ with added copies of the sets $|\mathfrak{M}|^s$ and necessary predicates for distinguishing and numbering these parts regarded as sets of sequences over $\mathfrak{M}$ of the corresponding length. By definition, the term in (5.4.5) with $s_i = 0$ represents a distinguished element in this model. For a theory T we introduce the notation

$$T\langle s_1, \dots, s_k \rangle = \mathrm{Th}\, \{ \mathfrak{M}\langle s_1, \dots, s_k \rangle \mid \mathfrak{M} \in \mathrm{Mod}\,(T) \}$$

It suffices to prove that for an exact interpretation I of the theory T_0 in the theory T_1 and some $s_1, \dots, s_k$ there exists a generalized model-bijective interpretation J of the theory T_1 in a theory of the form $T_0\langle s_1, \dots, s_k \rangle$ which, in a sense, is inverse to I. Namely, with respect to the interpretation J, the model $\mathfrak{M}$ is exactly the kernel of the model $I\mathfrak{M} = \mathfrak{M}\langle s_1, \dots, s_k \rangle$. The idea of the construction of such an interpretation J is obvious; therefore, we omit details. Assertion (e) is proved in Lemma 9.7.1 in another way.   □

Lemma 5.4.2. *Let T_0 and T_1 be theories of enumerable signatures such that T_0 is exactly interpreted in T_1. Then, to within omission of first-order definable predicates, the theory T_1 is finitely axiomatizable if and only if the theory T_0 is finitely axiomatizable.*

PROOF. The lemma is proved by standard methods. One can use the ideas indicated in the proof of Lemma 5.4.1(e).   □

- An interpretation I of a theory T_0 in a theory T_1 is called the *first-order definable invertible transformation of a signature* if I is effective and there exists an interpretation J of T_1 in T_0 such that I and J are mutually invertible.

For first-order definable equivalent theories T_0 and T_1, we use the denotation $T_0 \cong T_1$. A particular case of a first-order definable invertible transformation of a signature is a *complete first-order definable enrichment*, i.e., the case in which $\sigma_0 \subseteq \sigma_1$, $T_0 \subseteq T_1$, and $\sigma_1 \setminus \sigma_0$ contains predicates that present all the first-order relations in T_1 with respect to σ_0. We note that a complete first-order definable enrichment is a model-complete theory.

5.5. Prequasiexact and Immersing Interpretations

Isostone interpretations form the widest class of interpretations relative to the construction from Theorem 0.6.1. The class of immersing interpretations is sufficiently wide, and such interpretations possess good model-theoretic properties. The class of exact interpretations is narrower, but exact interpretations preserve all the properties indicated in the list

MQL, which is important in the sequel. However, to prove Theorem 0.6.1 by using exact interpretations is impossible in view of Lemma 5.4.2. Therefore, our nearest purpose is to define an intermediate class of interpretations (so-called quasiexact interpretations) in order to construct the required construction. We give the definition of an interpretation in two parts. The first part indicates conditions that guarantee immersion. The conditions from the second part provide the transmission of MQL-properties.

Let I be an interpretation of a theory T_0 in the domain $U(x)$ of a theory T_1. A model $\mathfrak{N}$ of the signature σ_1 is called *primitive over a set* $X \subseteq |\mathfrak{N}|$ if $\mathfrak{N} \in \mathrm{Mod}(T_1)$ and for any model $\mathfrak{N}'$ of the theory T_1 any injective partial mapping μ from $\mathfrak{N} \restriction (U(\mathfrak{N}) \cup X)$ in $\mathfrak{N}'$ such that

$$\left(\forall \psi \in \mathrm{EL}(\sigma_1)\right)\left(\forall \bar{c} \in U(\mathfrak{N}) \cup X\right) \mathfrak{N} \vDash \psi(\bar{c}) \Rightarrow \mathfrak{N}' \vDash \psi(\mu \bar{c}),$$

$$\left(\forall d \in U(\mathfrak{N}) \cup X\right) \mathfrak{N} \vDash U(d) \Leftrightarrow \mathfrak{N}' \vDash U(\mu d),$$

$$\left(\forall \varphi \in \mathrm{FL}(\sigma_0)\right)\left(\forall \bar{c} \in U(\mathfrak{N})\right) \mathbb{K}(\mathfrak{N}) \vDash \varphi(\bar{c}) \Leftrightarrow \mathbb{K}(\mathfrak{N}') \vDash \varphi(\mu \bar{c}),$$

can be extended to a kernel-elementary embedding $\mu^*\colon \mathfrak{N} \to \mathfrak{N}'$, where $\mathrm{EL}(\sigma_1)$ stands for the set of all $\exists$-formulas of the signature σ_1. A model $\mathfrak{N}$ of the theory T_1 is called *primitive* if it is primitive over the set $\varnothing$.

Let $\mathfrak{N}$ and $\mathfrak{N}'$ be models of the theory T_1 such that $\mathfrak{N} \preccurlyeq^{\circ} \mathfrak{N}'$. We say that the model $\mathfrak{N}'$ possesses the *universality property over the submodel* $\mathfrak{N}$ if for any models $\mathfrak{N}_1$ and $\mathfrak{N}_1'$ of the theory T_1 such that $\mathfrak{N}_1 \preccurlyeq^{\circ} \mathfrak{N}_1'$ and $\|\mathfrak{N}_1'\| \leqslant \|\mathfrak{N}'\|$ any injective partial mapping $\mu\colon \mathfrak{N}_1' \restriction (U(\mathfrak{N}_1') \cup |\mathfrak{N}_1|) \to \mathfrak{N}'$ possessing the properties

(1) $\mu \restriction |\mathfrak{N}_1|$ is an isomorphic embedding of $\mathfrak{N}_1$ into $\mathfrak{N}$,

(2) μ preserves $U(x)$,

(3) $\mu \restriction U(\mathfrak{N}_1')$ is an elementary embedding of $\mathbb{K}(\mathfrak{N}_1')$ into $\mathbb{K}(\mathfrak{N}')$

is extended to a kernel-elementary embedding $\mu^*\colon \mathfrak{N}_1' \to \mathfrak{N}'$.

A model $\mathfrak{N}$ of the theory T_1 is said to be *perfect over a set* $X \subseteq |\mathfrak{N}|$ if for any subset $Y \subseteq X$ and subset $Z \subseteq |\mathfrak{N}|$ of cardinality $|Z| < \|\mathfrak{N}\|$ the model $\mathfrak{N}$ possesses the universality property over a submodel $\mathfrak{N}' \preccurlyeq^{\circ} \mathfrak{N}$ that contains the set $Y \cup Z$ and is primitive over Y. A model $\mathfrak{N}$ of the theory T_1 is called *perfect* if it is perfect over $\varnothing$.

We formulate the first part of the definition of a quasiexact interpretation.

- An interpretation I of a theory T_0 in the domain $U(x)$ of a theory T_1 is called *prequasiexact* if the theory T_1 has no finite models and the following conditions hold:

(a) I is effective,

(b) if $\mathfrak{N}$ and $\mathfrak{N}'$ are perfect models of the theory T_1 of the same cardinality, then any isomorphism between the kernels $\mu \colon \mathbb{K}(\mathfrak{N}) \to \mathbb{K}(\mathfrak{N}')$ can be extended to an isomorphism between the models $\mu^* \colon \mathfrak{N} \to \mathfrak{N}'$,

(c) if $\mathfrak{N}$, $\mathfrak{N}'$, $\mathfrak{N}_1$, and $\mathfrak{N}_1'$ are models of the theory T_1 such that

   — $\mathfrak{N} \preccurlyeq^\circ \mathfrak{N}'$ and $\mathfrak{N}_1 \preccurlyeq^\circ \mathfrak{N}_1'$,

   — $\mathfrak{N}'$ is perfect over $|\mathfrak{N}|$ and $\mathfrak{N}_1'$ is perfect over $|\mathfrak{N}_1|$,

   — $\|\mathfrak{N}'\| = \|\mathfrak{N}_1'\|$,

   — a mapping $\mu \colon U(\mathfrak{N}') \cup |\mathfrak{N}| \to U(\mathfrak{N}_1') \cup |\mathfrak{N}_1|$ is bijective, and the restriction $\mu \restriction |\mathfrak{N}|$ is an isomorphism between the models $\mathfrak{N}$ and $\mathfrak{N}_1$ and the restriction $\mu \restriction U(\mathfrak{N}')$ is an isomorphism between the kernels $\mathbb{K}(\mathfrak{N}')$ and $\mathbb{K}(\mathfrak{N}_1')$,

then μ is extended to an isomorphism $\mu^* \colon \mathfrak{N}' \to \mathfrak{N}_1'$,

(d) for any countable model $\mathfrak{N}$ of the theory T_1 there exists a countable elementary extension $\mathfrak{N}'$ that is perfect over $|\mathfrak{N}|$,

(e) if, in the elementary chain $\mathfrak{N}_0 \preccurlyeq \mathfrak{N}_1 \preccurlyeq \ldots \preccurlyeq \mathfrak{N}_k \preccurlyeq \ldots$, $k < \omega$, of models of the theory T_1 every model $\mathfrak{N}_k$ is perfect, then the union of all models of this chain is also a perfect model.

We give some simple properties of prequasiexact interpretations.

Lemma 5.5.1. *Let I be a prequasiexact interpretation of a theory T_0 in the domain $U(x)$ of a theory T_1. Then the following assertions hold:*

(a) *if $\mathfrak{N}$ is a primitive model, then $\|\mathfrak{N}\| = \|\mathbb{K}(\mathfrak{N})\| + \omega$,*

(b) *for any model $\mathfrak{N}$ of the theory T_1 and a set $X \subseteq Y \subseteq |\mathfrak{N}|$, the following is true: if $\mathfrak{N}$ is perfect over Y, then $\mathfrak{N}$ is perfect over X,*

(c) *for any countable model $\mathfrak{N}$ of the theory T_1, there exists a countable elementary extension $\mathfrak{N}'$ such that $\mathfrak{N}'$ is perfect and the kernel $\mathbb{K}(\mathfrak{N}')$ is a homogeneous model,*

(d) *for any countable models $\mathfrak{N}_0$ and $\mathfrak{N}_1$ of the theory T_1 such that $\mathbb{K}(\mathfrak{N}_0) \equiv \mathbb{K}(\mathfrak{N}_1)$, there exist countable elementary extensions $\mathfrak{N}_0'$ and $\mathfrak{N}_1'$ that are perfect models, and the kernels $\mathbb{K}(\mathfrak{N}_0')$ and $\mathbb{K}(\mathfrak{N}_1')$ are homogeneous models with the same set of types realized in these models.*

PROOF. (a), (b). The assertions follow from definitions.

(c). We construct the required model $\mathfrak{N}'$ as the union of models of the following elementary chain:

$$\mathfrak{N} = \mathfrak{N}_0 \preccurlyeq \mathfrak{N}_1 \preccurlyeq \mathfrak{N}_2 \preccurlyeq \ldots \preccurlyeq \mathfrak{N}_k \preccurlyeq \ldots, \quad k < \omega$$

To this end, we use a standard procedure of constructing a homogeneous model. We must take into account all types appearing with successive realization of all extensions of types over their subtypes. In this procedure, we must include the procedure of taking perfect elementary extensions infinitely many times in accordance with condition (d) from the definition of a prequasiexact interpretation. This does not prevent the realization of the homogeneity procedure. Therefore, the model $\mathfrak{N}'$ is homogeneous by construction and the kernel of $\mathfrak{N}'$ is a homogeneous model. On the other hand, the model $\mathfrak{N}'$ is perfect in view of condition (e) from the definition of a prequasiexact interpretation.

(d). Let $\mathfrak{N}_0$ and $\mathfrak{N}_1$ be countable models of the theory T_1 such that $\mathbb{K}(\mathfrak{N}_0) \equiv \mathbb{K}(\mathfrak{N}_1)$. Following the scheme of the proof of (c), we present two separate procedures of constructing their countable homogeneous elementary extensions $\mathfrak{N}_0'$ and $\mathfrak{N}_1'$. Those types over kernel that appear in one of the constructions will be transferred to another construction as follows. We assume that in constructing the models $\mathfrak{N}_0'$ the type $p(x_1, \ldots, x_n)$ appears such that $U(x_1) \& \ldots \& U(x_n) \in p$. It is obvious that the set of formulas $II^{-1}p(x_1, \ldots, x_n) \subseteq p(x_1, \ldots, x_n)$ is consistent in the theory $\mathrm{Th}(\mathfrak{N}_0)$ as well as in the theory $\mathrm{Th}(\mathfrak{N}_1)$ because it presents a complete type for the kernels $\mathbb{K}(\mathfrak{N}_0)$ and $\mathbb{K}(\mathfrak{N}_1)$, which are elementarily equivalent by assumption. In order to construct some other model $\mathfrak{N}_1'$, we must transfer some complete n-type $p'(x_1, \ldots, x_n) \supseteq II^{-1}p(x_1, \ldots, x_n)$, extending the above set in the second construction of the theory $\mathrm{Th}(\mathfrak{N}_1)$. In the same way, after transformation, types from the second construction must be transferred for the first construction. Thus, we attain equality of families of types realized in the kernels $\mathbb{K}(\mathfrak{N}_0')$ and $\mathbb{K}(\mathfrak{N}_1')$ which are homogeneous. On the other hand, we insert the procedure of taking perfect elementary extensions into the procedure of constructing the models $\mathfrak{N}_0'$ and $\mathfrak{N}_1'$. The above-mentioned models are perfect in view of condition (e) from the definition of a prequasiexact interpretation. $\square$

Lemma 5.5.2. *Let I be a prequasiexact interpretation of a theory T_0 in a theory T_1. Then the following assertions hold:*

(a) *I is an isostone interpretation,*

(b) I *is an immersing interpretation,*

PROOF. (a). We consider models $\mathfrak{N}_0$ and $\mathfrak{N}_1$ of the theory T_1 such that $\mathbb{K}(\mathfrak{N}_0) \equiv \mathbb{K}(\mathfrak{N}_1)$. We also consider elementarily equivalent countable models $\mathfrak{N}_0'$ and $\mathfrak{N}_1'$ such that $\mathfrak{N}_0 \equiv \mathfrak{N}_0'$, $\mathfrak{N}_1 \equiv \mathfrak{N}_1'$, and $\mathbb{K}(\mathfrak{N}_0') \equiv \mathbb{K}(\mathfrak{N}_1')$. In view of Lemma 5.5.1(d), for the models $\mathfrak{N}_0'$ and $\mathfrak{N}_1'$ we can construct countable elementary perfect extensions $\mathfrak{N}_0''$ and $\mathfrak{N}_1''$ whose kernels are homogeneous models and have the same families of types realized there. Therefore, the kernels $\mathfrak{N}_0''$ and $\mathfrak{N}_1''$ are isomorphic. Consequently, the models are isomorphic by condition (b) from the definition of a prequasiexact interpretation. Therefore, $\mathfrak{N}_0' \equiv \mathfrak{N}_1'$, whence $\mathfrak{N}_0 \equiv \mathfrak{N}_1$. Thus, condition (b) of Lemma 5.2.2 is satisfied. Consequently, the interpretation I is an isostone one.

(b). Using Lemma 5.3.2, we verify condition (b) from the definition of an immersing interpretation. Consider a model $\mathfrak{N}$ of the theory T_1 and two sequences $\bar{c}$ and $\bar{d}$ of elements of $U(\mathfrak{N})$ that realize the same type $p(x_1, \ldots, x_n)$ in the theory $T_0^* = \mathrm{Th}\,(\mathbb{K}(\mathfrak{N}))$. We show that the types of these sequences coincide in the theory $T_1^* = \mathrm{Th}\,(\mathfrak{N})$. To this end, we consider a countable elementary submodel $\mathfrak{N}' \preccurlyeq \mathfrak{N}$ containing both sequences. By Lemma 5.5.1(c), we construct a countable elementary extension $\mathfrak{N}''$ of the model $\mathfrak{N}'$, which will be a perfect model with the homogeneous kernel. Since $\mathbb{K}(\mathfrak{N}'')$ is homogeneous and the types of sequences $\bar{c}$ and $\bar{d}$ coincide in this model, there exists an automorphism of $\mu \colon \mathbb{K}(\mathfrak{N}'') \to \mathbb{K}(\mathfrak{N}'')$ that transforms $\bar{c}$ into $\bar{d}$ and can be extended to an automorphism of the model $\mu^* \colon \mathfrak{N}'' \to \mathfrak{N}''$ in view of condition (b) from the definition of a prequasiexact interpretation. Therefore, the types of sequences $\bar{c}$ and $\bar{d}$ coincide in the model $\mathfrak{N}''$; thereby, they coincide in the model $\mathfrak{N}$. Thus, condition (c) of Lemma 5.3.2 is satisfied, which proves the validity of condition (b) from the definition of an immersing interpretation. We now prove the validity of (c). The implication from left to right is trivial. To prove the inverse implication we consider models $\mathfrak{N}$ and $\mathfrak{N}'$ of the theory T_1 such that $\mathfrak{N} \preccurlyeq^0 \mathfrak{N}'$, i.e.,

$$\mathfrak{N} \subseteq \mathfrak{N}' \ \& \ U(\mathfrak{N}) = U(\mathfrak{N}') \cap |\mathfrak{N}| \ \& \ \mathbb{K}(\mathfrak{N}) \preccurlyeq \mathbb{K}(\mathfrak{N}') \qquad (5.5.1)$$

It suffices to consider the case of countable models $\mathfrak{N}$ and $\mathfrak{N}'$. For the sake of simplicity, the set $U(\mathfrak{N})$ is denoted by X. In view of (5.5.1), the identity mapping $\mu \colon \mathbb{K}(\mathfrak{N}') \restriction X \to \mathbb{K}(\mathfrak{N}) \restriction X$ is elementary. In view of condition (b) from the definition of an immersing interpretation, the mapping $\mu \colon \mathfrak{N}' \restriction X \to \mathfrak{N} \restriction X$ is also elementary. Therefore, there exists a countable elementary extension $\mathfrak{N}_1$ of the model $\mathfrak{N}$ and an elementary embedding $\mu_1 \colon \mathfrak{N}' \to \mathfrak{N}_1$ that coincides with μ on X. However, the image

$\mu_1(\mathfrak{N})$, in general, cannot coincide with a submodel $\mathfrak{N}$ of $\mathfrak{N}_1$. However, $\mu_1(\mathfrak{N}) \preccurlyeq \mathfrak{N}_1$. Indeed, by condition (d) from the definition of a prequasiexact interpretation, we can construct a countable perfect elementary extension $\mathfrak{N}''$ of the model $\mathfrak{N}_1$, which, by Lemma 5.5.1(b), is a perfect model over $\mathfrak{N}$ and over $\mu_1(\mathfrak{N})$ as well. In view of condition (c) from the definition of a prequasiexact interpretation, there exists an automorphism of λ of the model $\mathfrak{N}''$ that is identical on the kernel of $\mathfrak{N}''$ and transforms $\mathfrak{N}$ into $\mu_1(\mathfrak{N})$. Therefore, $\mathfrak{N} \preccurlyeq \mathfrak{N}_1 \preccurlyeq \mathfrak{N}''$ implies $\mu_1(\mathfrak{N}) \preccurlyeq \mathfrak{N}''$, which, together with $\mu_1(\mathfrak{N}') \preccurlyeq \mathfrak{N}_1 \preccurlyeq \mathfrak{N}''$, leads to the relation $\mu_1(\mathfrak{N}) \preccurlyeq \mu_1(\mathfrak{N}')$, which implies $\mathfrak{N} \preccurlyeq \mathfrak{N}'$. Thus, condition (c) of the definition of an immersing interpretation is valid.   □

5.6. Envelopes and Quasiexact Interpretations

To introduce the notion of a quasiexact interpretation, we need some auxiliary notions of a primitive envelope and a saturated envelope. Let I be an interpretation of an axiomatizable theory T_0 in the domain $U(x)$ of a theory T_1.

- We say that the interpretation I has the *quasioperator of a primitive envelope* if the following conditions hold:

  (a) for any model $\mathfrak{M} \in \mathrm{Mod}\,(T_0)$, there exists a primitive model $\mathfrak{N} \in \mathrm{Mod}\,(T_1)$ such that $\mathbb{K}\,(\mathfrak{N}) \cong \mathfrak{M}$,

  (b) for any model $\mathfrak{N}$ of the theory T_1 and a set $X \subseteq |\mathfrak{N}|$, there exists a submodel $\mathfrak{N}' \preccurlyeq^\circ \mathfrak{N}$ such that $\mathfrak{N}' \vDash T_1$, $\mathbb{K}\,(\mathfrak{N}') = \mathbb{K}\,(\mathfrak{N})$, $X \subseteq |\mathfrak{N}'|$, and $\mathfrak{N}'$ is primitive over X,

  (c) if a model $\mathfrak{N}$ of the theory T_1 is primitive over $X \subseteq |\mathfrak{N}|$, then $\mathfrak{N}' = \mathfrak{N}$ for any submodel $\mathfrak{N}' \preccurlyeq^\circ \mathfrak{N}$ such that $U(\mathfrak{N}) \cup X \subseteq |\mathfrak{N}'|$ and $\mathfrak{N}' \vDash T_1$.

- We say that the interpretation I has the *operator of a perfect envelope* if conditions (b) and (c) from the definition of a prequasiexact interpretation as well as the following condition hold:

  (d) for any model $\mathfrak{N}$ of the theory T_1 and any cardinal $\alpha \geqslant \|\mathfrak{N}\|$, there exists a kernel-elementary extension $\mathfrak{N}'$ of the model $\mathfrak{N}$ of cardinality $\|\mathfrak{N}'\| = \alpha$ such that $\mathfrak{N}' \vDash T_1$, $\mathbb{K}\,(\mathfrak{N}) = \mathbb{K}\,(\mathfrak{N}')$, and the model $\mathfrak{N}'$ is perfect over $|\mathfrak{N}|$.

Lemma 5.6.1. *If a prequasiexact interpretation I of a theory T_0 in a theory T_1 has the quasioperator of a primitive envelope and the operator of a perfect envelope, then the following assertions hold:*

(a) *if $\mathfrak{N}$ is a primitive model and $\mathfrak{N}'$ is an arbitrary model of the theory T_1, then any elementary embedding $\mu\colon \mathbb{K}(\mathfrak{N}) \to \mathbb{K}(\mathfrak{N}')$ can be extended to a kernel-elementary embedding $\mu^*\colon \mathfrak{N} \to \mathfrak{N}'$,*

(b) *if $\mathfrak{N}$ and $\mathfrak{N}'$ are primitive models, then any isomorphism between the kernels $\mu\colon \mathbb{K}(\mathfrak{N}) \to \mathbb{K}(\mathfrak{N}')$ can be extended to an isomorphism between the models $\mu^*\colon \mathfrak{N} \to \mathfrak{N}'$,*

(c) *if $\mathfrak{N}$ is a model of the theory T_1, $X \subseteq |\mathfrak{N}|$, and submodels $\mathfrak{N}'$, $\mathfrak{N}''$ of $\mathfrak{N}$ are primitive over X and satisfy the conditions $\mathbb{K}(\mathfrak{N}') = \mathbb{K}(\mathfrak{N}'') = \mathbb{K}(\mathfrak{N})$, $X \subseteq |\mathfrak{N}'|$, $X \subseteq |\mathfrak{N}''|$, then there exists an isomorphism $\mu\colon \mathfrak{N}' \to \mathfrak{N}''$, which is identical on $U(\mathfrak{N}') \cup X$,*

(d) *for any model $\mathfrak{M} \in \mathrm{Mod}(T_0)$ and an infinite cardinal $\alpha \geqslant \|\mathfrak{M}\|$, there exists a perfect model $\mathfrak{N} \in \mathrm{Mod}(T_1)$ such that $\mathbb{K}(\mathfrak{N}) \cong \mathfrak{M}$,*

(e) *if $\mathfrak{N}$ and $\mathfrak{N}'$ are models of the theory T_1 such that the model $\mathfrak{N}'$ is perfect and $\|\mathfrak{N}\| \leqslant \|\mathfrak{N}'\|$, then any elementary embedding $\mu\colon \mathbb{K}(\mathfrak{N}) \to \mathbb{K}(\mathfrak{N}')$ can be extended to a kernel-elementary embedding $\mu'\colon \mathfrak{N} \to \mathfrak{N}'$,*

(f) *if $\mathfrak{N}$, $\mathfrak{N}'$, $\mathfrak{N}_1$, and $\mathfrak{N}_1'$ are models of the theory T_1 such that $\mathfrak{N} \preccurlyeq^\circ \mathfrak{N}'$, $\mathfrak{N}_1 \preccurlyeq^\circ \mathfrak{N}_1'$, $\mathbb{K}(\mathfrak{N}) = \mathbb{K}(\mathfrak{N}')$, $\mathbb{K}(\mathfrak{N}_1) = \mathbb{K}(\mathfrak{N}_1')$, the model $\mathfrak{N}'$ is perfect over $|\mathfrak{N}|$, and the model $\mathfrak{N}_1'$ is perfect over $|\mathfrak{N}_1|$, then any isomorphism $\mu\colon \mathfrak{N} \to \mathfrak{N}_1$ can be extended to an isomorphism $\mu^*\colon \mathfrak{N}' \to \mathfrak{N}_1'$ provided that $\|\mathfrak{N}'\| = \|\mathfrak{N}_1'\|$.*

PROOF. (a). Let $\mu\colon \mathbb{K}(\mathfrak{N}) \to \mathbb{K}(\mathfrak{N}')$ be an elementary embedding. By Lemma 5.5.2(b), the mapping $\mu\colon \mathfrak{N} \restriction U(\mathfrak{N}) \to \mathfrak{N}'$ is elementary. In particular, it preserves $U(x)$ and all $\exists$-formulas. Therefore, μ can be extended to a kernel-elementary embedding $\mu\colon \mathfrak{N} \to \mathfrak{N}'$ in accordance with the definition of a primitive model.

(b). In view of (a), the isomorphism $\mu\colon \mathbb{K}(\mathfrak{N}) \to \mathbb{K}(\mathfrak{N}')$ is extended to a kernel-elementary embedding $\mu\colon \mathfrak{N} \to \mathfrak{N}'$ which is a mapping "onto," i.e., μ is an isomorphism in view of condition (c) from the definition of an interpretation with the quasioperator of a primitive envelope.

(c). By Lemma 5.5.2, the interpretation I is an immersing one. Therefore, the models $\mathfrak{N}'$ and $\mathfrak{N}''$ are elementary submodels of the model $\mathfrak{N}$. The identity embedding $\mu_0\colon \mathfrak{N}' \upharpoonright \left(U(\mathfrak{N}') \cup X\right) \to \mathfrak{N}''$ is elementary. In particular, it preserves $U(x)$ and all $\exists$-formulas. Since $\mathfrak{N}'$ is primitive over X, the embedding μ_0 can be extended to a kernel-elementary embedding $\mu\colon \mathfrak{N}' \to \mathfrak{N}''$, which is surjective in view of condition (c) from the definition of an interpretation with the quasioperator of a primitive envelope and the primitivity of $\mathfrak{N}''$ over X.

(d). The assertion follows from condition (a) from the definition of an interpretation with the quasioperator of a primitive envelope, condition (d) from the definition of an interpretation with the operator of a perfect envelope, and Lemma 5.5.1(b).

(e). Let $\mathfrak{N}_0$ be a primitive submodel of the model $\mathfrak{N}$ such that $\mathbb{K}\left(\mathfrak{N}_0\right) = \mathbb{K}\left(\mathfrak{N}\right)$. Such a submodel exists in view of condition (b) from the definition of an interpretation with the quasioperator of a primitive envelope. Let $\mathfrak{N}_0'$ be a primitive submodel of the model $\mathfrak{N}'$ such that $\mathbb{K}\left(\mathfrak{N}_0'\right) = \mathbb{K}\left(\mathfrak{N}'\right)$. In view of (a), the elementary embedding $\mu\colon \mathbb{K}\left(\mathfrak{N}\right) \to \mathbb{K}\left(\mathfrak{N}'\right)$ can be extended to a kernel-elementary embedding $\mu^*\colon \mathfrak{N}_0 \to \mathfrak{N}_0'$ which, in turn, can be extended to a kernel-elementary embedding $\mu'\colon \mathfrak{N} \to \mathfrak{N}'$ since the perfect model $\mathfrak{N}'$ has the universality property over $\mathfrak{N}_0'$.

(f). This assertion is a partial case of condition (c) from the definition of a prequasiexact interpretation.   □

We consider the operators introduced on the whole. Let a quasiexact interpretation I have the quasioperator of a primitive envelope and let $\mathfrak{M}$ be a model of the theory T_0. From the above definitions and Lemma 5.6.1(b), it follows that a primitive model $\mathfrak{N}$ of the theory T_1 such that $\mathbb{K}\left(\mathfrak{N}\right) \cong \mathfrak{M}$ exists and is uniquely defined by the isomorphism type of the model $\mathfrak{M}$. For a primitive model $\mathfrak{N}$, we use the denotation $\mathbb{P}\left(\mathfrak{M}\right)$ or $\mathbb{P}\mathfrak{M}$ for simplicity. By definition, the equality $\mathfrak{M} = \mathbb{K}\left(\mathbb{P}\mathfrak{M}\right)$ holds and the model $\mathbb{P}\mathfrak{M}$ is primitive.

We consider a model $\mathfrak{N}$ of the theory T_1. We assume that for a set $X \subseteq |\mathfrak{N}|$ a submodel $\mathfrak{N}' \preccurlyeq^\circ \mathfrak{N}$ is given such that $\mathbb{K}\left(\mathfrak{N}'\right) = \mathbb{K}\left(\mathfrak{N}\right)$; moreover, it is a model of the theory T_1, contains X, and is primitive over X. In general, the definition of the quasioperator of a primitive envelope admits the existence of submodels of the model $\mathfrak{N}$ with the kernel $\mathbb{K}\left(\mathfrak{N}\right)$ that are different from $\mathfrak{N}'$ and are primitive over the same set X. However, all such models are isomorphic in the sense of claim (c) of Lemma 5.6.1. The quasioperator of a primitive envelope is called the *operator of a primitive envelope* if the primitive submodel $\mathfrak{N}' \preccurlyeq^\circ \mathfrak{N}$ is uniquely defined for any $\mathfrak{N} \in \mathrm{Mod}\left(T_1\right)$ and $X \subseteq |\mathfrak{N}|$.

We assume that a quasiexact interpretation I has the operator of a perfect envelope. Let $\mathfrak{N}$ be a model of the theory T_1 and let $\mathfrak{N}'$ be a model of cardinality $\alpha \geqslant \|\mathfrak{N}\|$. Moreover, let $\mathfrak{N}'$ be a perfect extension of the model $\mathfrak{N}$ such that $\mathbb{K}(\mathfrak{N}') = \mathbb{K}(\mathfrak{N}) \cong \mathfrak{M}$, $\mathfrak{M} \in \mathrm{Mod}(T_0)$. In view of condition (d) from the definition of an interpretation with the operator of a perfect envelope and Lemma 5.6.1(f), the model $\mathfrak{N}'$ is defined up to an isomorphism by α and $\mathfrak{N}$. The model $\mathfrak{N}'$ is denoted by $\mathbb{S}(\mathfrak{N}, \alpha)$, and the operator $(\mathfrak{N}, \alpha) \to \mathbb{S}(\mathfrak{N}, \alpha)$ is called the *operator of a perfect envelope* for the interpretation I. If $\alpha = \|\mathfrak{N}\|$, then we use the notation $\mathbb{S}(\mathfrak{N})$ or $\mathbb{S}\mathfrak{N}$ instead of $\mathbb{S}(\mathfrak{N}, \alpha)$ for brevity. We note that a perfect model $\mathfrak{N}$ of a given cardinality α is uniquely defined by the isomorphism type of the kernel in view of condition (b) from the definition of a prequasiexact interpretation.

For a model $\mathfrak{N}$ of the theory T_1 we introduce the notation

$$\mathrm{Stan}(\mathfrak{N}) = U(\mathfrak{N}) \cup \{x \in \mathfrak{N} \mid \mathrm{acl}(x) \setminus \mathrm{acl}(\varnothing) \neq \varnothing\}$$
$$\mathrm{Nost}(\mathfrak{N}) = |\mathfrak{N}| \setminus \mathrm{Stan}(\mathfrak{N})$$

Elements of the set $\mathrm{Stan}(\mathfrak{N})$ are called *standard* and elements of the set $\mathrm{Nost}(\mathfrak{N})$ are referred to as *nonstandard* elements of the model $\mathfrak{N}$. A type $p(x)$ is called a *standard* or *nonstandard* 1-*type* if it is realized by a standard element or a nonstandard element respectively.

- An interpretation I of a theory T_0 in the domain $U(x)$ of a theory T_1 is called *quasiexact* if the theory T_1 has no finite models and the following conditions hold:

(Q.1) I is prequasiexact,

(Q.2) I has the quasioperator of a primitive envelope,

(Q.3) I has the operator of a perfect envelope,

(Q.4) a model $\mathfrak{N}$ of the theory T_1 is primitive if and only if each of its countable kernel-elementary submodels is primitive,

(Q.5) for any model $\mathfrak{N}$ of the theory T_1, the number of types $p(x)$ over $\mathfrak{N}$ that are realized by nonstandard elements has cardinality at most $\|\mathfrak{N}\|$,

(Q.6) for any model $\mathfrak{N}$ of the theory T_1 and a 1-type $p(x)$ of the theory $T_1^* = \mathrm{Th}(\mathfrak{N})$ realized by a standard element, there exists a type $q(\overline{y})$ of the theory $T_0^* = \mathrm{Th}(\mathbb{K}(\mathfrak{N}))$ and a formula $\varphi(x, \overline{y}) \in \mathrm{FL}(\sigma_1)$ such that $T_1^* \vdash \varphi(x, \overline{y}) \to U(\overline{y})$ and the type $p(x)$ is generated by the set of formulas

$$p'(x) = \{(\exists \overline{y})\, \varphi(x, \overline{y})\, \&\, \psi(\overline{y}) \mid \psi \in Iq(\overline{y})\}$$

moreover, for any model $\mathfrak{N}' \equiv \mathfrak{N}$ and a sequence $\bar{c} \in \mathbb{K}(\mathfrak{N}')$ realizing the type $q(\bar{y})$, the number of different extensions of the formula $\varphi(x, \bar{c})$ to complete types over $|\mathfrak{N}'|$ is at most the cardinality of the model $\mathfrak{N}'$,

(Q.7) if a model $\mathfrak{N}$ of the theory T_1 of an uncountable cardinality α is α^+-homogeneous, then $\|\mathbb{K}(\mathfrak{N})\| = \|(\mathfrak{N})\|$,

(Q.8) the algorithmic dimension of any countable primitive model $\mathfrak{N}$ of the theory T_1 coincides with the algorithmic dimension of the kernel $\mathbb{K}(\mathfrak{N})$,

(Q.9) a countable perfect model $\mathfrak{N}$ of the theory T_1 is strongly constructivizable if and only if the kernel $\mathbb{K}(\mathfrak{N})$ is strongly constructivizable.

(Q.10) if a model $\mathfrak{N}$ of the theory T_1 is not primitive, then $\mathfrak{N}$ has a proper submodel $\mathfrak{N}' \preccurlyeq^0 \mathfrak{N}$ of cardinality $\|\mathfrak{N}'\| = \|\mathfrak{N}\|$,

(Q.11) if a model $\mathfrak{N}$ of the theory T_1 is not primitive, then there exists a nontrivial automorphism μ of $\mathfrak{N}$ that is identical on the kernel of $\mathfrak{N}$,

(Q.12) every element a of a primitive model $\mathfrak{N}$ is first-order definable over $U(\mathfrak{N})$,

(Q.13) there exists an ordinal $\alpha < \omega_1$ such that for any model $\mathfrak{M}$ of the theory T_0 of cardinality $\|\mathfrak{M}\| \leqslant \omega_\beta$ the number of isomorphism types of the model $\mathfrak{N}$ of the theory T_1 of cardinality $\|\mathfrak{N}\| = \omega_\beta$ with the kernel $\mathbb{K}(\mathfrak{N}) \cong \mathfrak{M}$ is at most $J_\alpha(|\omega + \beta|)$.

An ordinal δ is called the *spectral index* of an interpretation I if I satisfies (Q.13) with $\alpha = \delta$; in addition, such a value δ is the least among all possible values.

The following lemma indicates some properties of quasiexact interpretations that are a direct consequence of definitions. The proof is standard and is based on the definition of a quasiexact interpretation and the compactness theorem.

Lemma 5.6.2. *Let I be a quasiexact interpretation of a theory T_0 in the domain $U(x)$ of a theory T_1. Then the following assertions hold:*

(a) *if $\mathfrak{N}, \mathfrak{N}' \in \mathrm{Mod}(T_1)$ are models such that $\mathfrak{N} \preccurlyeq^0 \mathfrak{N}'$ and the model $\mathfrak{N}'$ is primitive, then the model $\mathfrak{N}$ is also primitive,*

(b) *if $\mathfrak{N}$ is not a primitive model of the theory T_1, then there exists a countable submodel $\mathfrak{N}' \preccurlyeq^0 \mathfrak{N}$ that is not a primitive model,*

(c) $\operatorname{Stan}(\mathfrak{N}) = \operatorname{Stan}(\mathfrak{N}') \cap |\mathfrak{N}|$ *for any models* $\mathfrak{N}$ *and* $\mathfrak{N}'$ *of the theory* T_1 *such that* $\mathfrak{N} \preccurlyeq^\circ \mathfrak{N}'$,

(d) *if* $p(x_1, \ldots, x_n)$ *is a complete type in some completion* T_1^* *of the theory* T_1 *and* $1 \leqslant i \leqslant n$, *then for all realizations of the type* $p(\bar{x})$ *the variable* x_i *corresponds only to either standard elements or nonstandard elements.*

5.7. Theorem on Transmission of Properties

This section is devoted to the proof of the key theorem concerning transmission of properties by quasiexact interpretations.

Theorem 5.7.1. *Let* I *be a quasiexact interpretation of a theory* T_0 *in a theory* T_1. *Then the following assertions hold:*

(a) $\operatorname{Mod}(T_0) = \{\mathbb{K}(\mathfrak{N}) \mid \mathfrak{N} \in \operatorname{Mod}(T_1)\}$,

(b) I *is elementarily enveloping,*

(c) I *is isostone,*

(d) I *is immersing,*

(e) *if* $\mu \colon \mathcal{L}(T_0) \to \mathcal{L}(T_1)$ *is a recursive isomorphism defined by* I *in accordance with Lemma 5.2.1 and* $T_0^* \supseteq T_0$, $T_1^* \supseteq T_1$ *are completions, which are corresponding in view of the isomorphism* μ, *then the following assertions hold:*

(e$_1$) *if* T_0^* *is a theory of a finite model, then the theories* $T_0^* \oplus SI$ *and* T_1^* *have the same descriptions within the framework of the list* MQL, *where* SI *is an* ω_1*-categorical successor theory with a single initial element and without cycles,*

(e$_2$) *if* T_0^* *has no finite models, then the theories* T_0^* *and* T_1^* *have the same description within the framework of the list* MQL.

PROOF (a). The assertion follows from condition (b) of the theorem, in view of Lemma 5.1.2.

(b). In view of (Q.2), the assertion follows from condition (a) from the definition of an interpretation with the quasioperator of a primitive envelope and Lemma 5.6.1(a).

(c), (d). The assertions follow from Lemma 5.5.2 in view of (Q.1).

(e). It is convenient to divide the assertion into several assertions and consider successively all the properties indicated in the list MQL. To the end of the proof, the completions $T_0^* \supseteq T_0$ and $T_1^* \supseteq T_1$ are assumed to be fixed. They are corresponding in the interpretation I, i.e.,

$$T_0^* = \text{Th}\,(\mathbb{K}\,(\mathfrak{N})), \quad T_1^* = \text{Th}\,(\mathfrak{N}) \tag{5.7.1}$$

for some model $\mathfrak{N}$ of the theory T_1. We restrict ourselves to the case in which the kernel $\mathbb{K}\,(\mathfrak{N})$ is infinite. If it is finite, the proof is simpler.

Thus, under the assumption that in (5.7.1) the kernel $\mathbb{K}\,(\mathfrak{N})$ of the model $\mathfrak{N}$ is infinite, for every p in MTL the theory T_1^* has the property p if and only if the theory T_0^* has the property p. This follows from Lemmas 5.7.1–5.7.9 and Propositions 5.7.1–5.7.8 below.   □

THE FIRST-ORDER DEFINABILITY AND STABILITY

Lemma 5.7.1. *Let α be an infinite cardinal. Then the theory T_1^* is stable with respect to a cardinality α if and only if the theory T_0^* is stable with respect to the same cardinality α.*

PROOF. If T_0^* is unstable with respect to a cardinality α, then T_1^* is also unstable with respect to the same cardinality. We assume that the theory T_0^* is stable with respect to the cardinality α. To prove the α-stability of the theory T_1^*, it suffices to establish the fact that for an arbitrary model $\mathfrak{N}$ of the theory T_1^* of cardinality α the number of 1-types over a model $\mathfrak{N}$ has cardinality at most α. This is guaranteed by (Q.5) and (Q.6) if we take into account that all 1-types over $\mathfrak{N}$ are divided into two disjoint classes in accordance with Lemma 5.6.2(d).

Thus, the α-stability of the theory T_0^* is equivalent to the α-stability of the corresponding theory T_1^*.   □

Lemma 5.7.1 leads to the following proposition.

Proposition 5.7.1. *The interpretation I preserves all the properties indicated in Theorem 0.6.1(a).*

PRIME, ATOMIC, AND MINIMAL MODELS

We first characterize a general relation between prime models and atomic models of the theories T_0 and T_1.

Lemma 5.7.2. *A model $\mathfrak{N}$ of the theory T_1 is prime if and only if the following conditions hold:*

(a) $\mathfrak{N}$ *is primitive,*

(b) $\mathbb{K}(\mathfrak{N})$ *is a prime model.*

PROOF. Let a model $\mathfrak{N}$ satisfy conditions (a), (b) and let a model $\mathfrak{N}_1$ of the theory T_1 be elementarily equivalent to $\mathfrak{N}$. In view of (b), there is an elementary embedding of the kernel $\mathbb{K}(\mathfrak{N})$ into the kernel $\mathbb{K}(\mathfrak{N}_1)$. By Lemma 5.6.1(a), we can extend it to a kernel-elementary embedding of the model $\mathfrak{N}$ into the model $\mathfrak{N}_1$. This embedding is elementary since the interpretation I is immersing.

To prove the converse assertion, we assume that $\mathfrak{N}$ does not satisfy condition (a) and consider a primitive model $\mathfrak{N}'$ of the theory T_1 such that $\mathfrak{N} \equiv \mathfrak{N}'$. In view of Lemma 5.6.2(a), an elementary embedding of $\mathfrak{N}$ into $\mathfrak{N}'$ is impossible. Otherwise, the image $\mu(\mathfrak{N})$ is a primitive model. But in this case, the model $\mathfrak{N}$ itself is primitive. Consequently, the model $\mathfrak{N}$ cannot be prime. We assume that condition (b) fails. Since the model $\mathbb{K}(\mathfrak{N})$ is not prime, there exists a model $\mathfrak{M}$ that is elementarily equivalent to $\mathbb{K}(\mathfrak{N})$. However, $\mathbb{K}(\mathfrak{N})$ is not embedded elementarily in $\mathfrak{M}$. Let $\mathfrak{N}'$ be a model of the theory T_1 such that $\mathbb{K}(\mathfrak{N}') \cong \mathfrak{M}$. Then $\mathfrak{N} \equiv \mathfrak{N}'$, but $\mathfrak{N}$ is not embedded elementarily into $\mathfrak{N}'$. Therefore, $\mathfrak{N}$ is not prime. □

Lemma 5.7.3. *A model* $\mathfrak{N}$ *of the theory* T_1^* *is atomic if and only if the following conditions hold:*

(a) $\mathfrak{N}$ *is primitive,*

(b) $\mathbb{K}(\mathfrak{N})$ *is an atomic model.*

PROOF. Let a model $\mathfrak{N}$ satisfy conditions (a), (b) and let $\bar{a}$ be a sequence of elements $\mathfrak{N}$ of finite length. We consider a countable elementary submodel $\mathfrak{N}' \preccurlyeq \mathfrak{N}$ containing $\bar{a}$. By Lemma 5.6.2(a), $\mathfrak{N}'$ is primitive. By construction, the kernel of $\mathfrak{N}'$ is an atomic model. Since the model $\mathfrak{N}'$ is countable, it is prime by Lemma 5.7.2. Therefore, $\bar{a}$ realizes the principal type in $\mathfrak{N}'$. Consequently, it realizes the principal type in the elementary extension $\mathfrak{N}$ of $\mathfrak{N}'$. We have proved that the model $\mathfrak{N}$ is atomic.

Conversely, let a model $\mathfrak{N}$ not satisfy condition (a). In view of Lemma 5.6.2(b), we can construct a countable elementary nonprimitive submodel $\mathfrak{N}' \preccurlyeq \mathfrak{N}$ of the model $\mathfrak{N}$. By Lemma 5.7.2, the model $\mathfrak{N}'$ is neither prime nor atomic. Therefore, the model $\mathfrak{N}$ is not atomic. If (a) is satisfied but (b) fails, we consider a sequence $\bar{a} \in |\mathbb{K}(\mathfrak{N})|$ realizing a nonprincipal type of the theory T_0^* and construct a countable elementary submodel $\mathfrak{N}' \preccurlyeq \mathfrak{N}$ containing $\bar{a}$. By Lemma 5.6.2(a), $\mathfrak{N}'$ is primitive. By Lemma 5.7.2, $\mathfrak{N}'$ cannot be prime. Consequently, the model $\mathfrak{N}$ is not atomic. □

Proposition 5.7.2. *The interpretation I preserves all the properties indicated in Theorem 0.6.1(b).*

PROOF. We assume that the theory T_0^* has a prime model $\mathfrak{M}$. By Lemma 5.7.2, the primitive model $\mathfrak{N} = \mathbb{P}(\mathfrak{M})$ is a prime model of the theory T_1^*. If the theory T_0^* has no prime models, then the theory T_1^* has no prime models by Lemma 5.7.2. The relation (Q.8) guarantees the preservation of the value of the algorithmic dimension for prime models of the theories T_0^* and T_1^*. The preservation of the number of atomic models with respect to each uncountable cardinality is guaranteed by the operator of a primitive envelope $\mathfrak{M} \mapsto \mathbb{P}(\mathfrak{M})$ which establishes a one-to-one correspondence between isomorphism types of atomic models of the theories T_0^* and T_1^* by Lemma 5.7.3 and preserves their cardinalities by Lemma 5.5.1(a).   □

Lemma 5.7.4. *A model $\mathfrak{N}$ of the theory T_1 is minimal if and only if the following conditions hold:*

(a) $\mathfrak{N}$ *is primitive,*

(b) $\mathbb{K}(\mathfrak{N})$ *is a minimal model.*

PROOF. Let a model $\mathfrak{N}$ of the theory T_1 of cardinality α satisfy conditions (a), (b) and let $\mathfrak{N}'$ be an elementary submodel of the model $\mathfrak{N}$ of cardinality $\|\mathfrak{N}'\| = \alpha$. By Lemma 5.6.2(a), the model $\mathfrak{N}'$ is primitive. By the choice of $\mathfrak{N}'$, we have $\mathbb{K}(\mathfrak{N}') \preccurlyeq \mathbb{K}(\mathfrak{N})$ and $\|\mathbb{K}(\mathfrak{N}')\| = \alpha$. Consequently, $\mathbb{K}(\mathfrak{N}') = \mathbb{K}(\mathfrak{N})$ in view of (b). As a result, we obtain $\mathfrak{N}' = \mathfrak{N}$ in view of (Q.2) and property (c) of the definition of the quasioperator of a primitive envelope. Thus, $\mathfrak{N}$ is minimal.

Conversely, we assume that a model $\mathfrak{N}$ of the theory T_1 of cardinality α does not satisfy condition (a). Then the model $\mathfrak{N}$ cannot be minimal in view of (Q.10). If (a) holds but (b) fails, then, taking into account Lemma 5.5.1(a), we have $\|\mathbb{K}(\mathfrak{N})\| = \|\mathfrak{N}\| = \alpha$; moreover, the kernel $\mathbb{K}(\mathfrak{N})$ has a proper elementary submodel $\mathfrak{M}$ of the same cardinality α. We consider a primitive model $\mathfrak{N}'$ such that $\mathbb{K}(\mathfrak{N}') \cong \mathfrak{M}$. Using Lemma 5.6.1(a), we find a kernel-elementary embedding $\mu\colon \mathfrak{N}' \to \mathfrak{N}$ such that $\mu\big(\mathbb{K}(\mathfrak{N})\big) = \mathfrak{M}$. Then the image $\mu(\mathfrak{N}')$ is a proper elementary submodel of the model $\mathfrak{N}$ of cardinality α. Consequently, the model $\mathfrak{N}$ is not minimal.   □

Proposition 5.7.3. *The interpretation I preserves all the properties indicated in Theorem 0.6.1(c).*

PROOF. By Lemma 5.7.5, the operator of a primitive envelope $\mathfrak{M} \mapsto \mathbb{P}(\mathfrak{M})$ establishes a one-to-one correspondence between isomorphism types of minimal models of the theories T_0^* and T_1^*. Furthermore, the operator $\mathbb{P}$ leaves cardinalities unchanged and, by (Q.8), preserves the value of the algorithmic dimension in the countable case. □

HOMOGENEOUS AND SATURATED MODELS

Lemma 5.7.5. *The following assertions hold:*

(a) *if $\alpha \geqslant \omega$ is a cardinal and a model $\mathfrak{N}$ of the theory T_0 is α-homogeneous, then the kernel $\mathbb{K}(\mathfrak{N})$ is an α-homogeneous model,*

(b) *if a countable model $\mathfrak{M} \in \mathrm{Mod}\,(T_0)$ is homogeneous, then its primitive envelope $\mathfrak{N} = \mathbb{P}(\mathfrak{M})$ is a homogeneous model,*

(c) *if a model $\mathfrak{M}$ of the theory T_1 of cardinality $\alpha \geqslant \omega$ is α^+-homogeneous, then its primitive envelope $\mathfrak{N} = \mathbb{P}(\mathfrak{M})$ is an α^+-homogeneous model.*

PROOF. (a). The assertion follows from condition (b) of the definition of an immersing interpretation and the obvious fact that the restrictions of automorphisms of the model $\mathfrak{N}$ to the domain $U(\mathfrak{N})$ are automorphisms of the kernel $\mathbb{K}(\mathfrak{N})$.

(b). Let $\mathfrak{N}$ be a primitive model with the kernel $\mathbb{K}(\mathfrak{N})$. In view of condition (b) from the definition of an immersing interpretation, all kernel-elementary embeddings are elementary. By Lemma 5.6.1(a), the model $\mathfrak{N}$ is prime over $U(\mathfrak{N})$. Consequently, any type $p(\overline{x})$ of the theory $\mathrm{Th}\,(\mathfrak{N})$ is uniquely defined by a single formula $\varphi(\overline{x}, \overline{y})$ and the I-image of some type $q(\overline{y})$ of the theory $\mathrm{Th}\,(\mathbb{K}(\mathfrak{N}))$. If the kernel $\mathbb{K}(\mathfrak{N})$ of a countable model $\mathfrak{N}$ is homogeneous, then the model $\mathfrak{N}$ is homogeneous.

(c). The assertion is proved by standard methods. The condition (Q.12) used in the proof provides a rigid connection of the envelope and the kernel. □

Proposition 5.7.4. *The interpretation I preserves all the properties indicated in Theorem 0.6.1(d).*

PROOF. If $\mathfrak{M}$ is a countable strongly constructivizable homogeneous or α^+-homogeneous model of the theory T_0^*, then, by assertions (a) and (b) of Lemma 5.7.5, the model $\mathbb{P}(\mathfrak{M})$ possesses the same properties in the theory T_1^*. The transmission of the above-mentioned properties from T_1^* to T_0^* follows from Lemma 5.7.5(a).

If a model $\mathfrak{M}$ of cardinality α is an α^+-homogeneous model of the theory T_0^*, then, by Lemma 5.7.5(c), the model $\mathbb{P}(\mathfrak{M})$ has the same cardinality α and is an α^+-homogeneous model of the theory T_1^*. To prove the converse assertion, we assume that $\mathfrak{N}$ is an α^+-homogeneous model of the theory T_1^* of cardinality α. In the case of an uncountable α, the kernel $\mathbb{K}(\mathfrak{N})$ has cardinality α in view of (Q.7). In the case of a countable α, the kernel $\mathbb{K}(\mathfrak{N})$ is countable by the assumption that T_0^* has no finite models. It remains to note that the kernel $\mathbb{K}(\mathfrak{N})$ is an α^+-homogeneous model of the theory T_0^* in view of Lemma 5.7.5(a). $\square$

Lemma 5.7.6. *A model* $\mathfrak{N}$ *of the theory* T_1 *of cardinality* α *is saturated if and only if the following conditions hold:*

(a) $\mathfrak{N}$ *is perfect,*

(b) *the cardinality of the kernel* $\mathbb{K}(\mathfrak{N})$ *is* α,

(c) $\mathbb{K}(\mathfrak{N})$ *is a saturated model.*

PROOF. Let a model $\mathfrak{N}$ of cardinality α satisfy conditions (a)–(c). To prove that $\mathfrak{N}$ is saturated, it suffices to show that for any subset X of $\mathfrak{N}$ of cardinality $|X| < \alpha$ any 1-type $p(x)$ over X is realized in $\mathfrak{N}$. We can only consider the case in which the restriction $\mathfrak{N}' = \mathfrak{N} \restriction X$ is an elementary submodel of the model $\mathfrak{N}$. We first realize the type $p(x)$ in some elementary extension $\mathfrak{N}_1$ of the model $\mathfrak{N}'$ of cardinality $\|\mathfrak{N}_1\| < \alpha$. Since the kernel $\mathbb{K}(\mathfrak{N})$ is saturated, it is possible to construct an elementary embedding $\mu\colon \mathbb{K}(\mathfrak{N}_1) \to \mathbb{K}(\mathfrak{N})$ that is identical on $\mathbb{K}(\mathfrak{N}')$. The model $\mathfrak{N}$ is perfect; therefore, it possesses, by definition, the universality property over a submodel $\mathfrak{N}'$ since the cardinality of the submodel is less than α. Since $\mathfrak{N}' \preccurlyeq^\circ \mathfrak{N}_1$, $\|\mathfrak{N}_1\| < \|\mathfrak{N}\|$, there exists a kernel-elementary embedding of $\mathfrak{N}_1$ into $\mathfrak{N}$ that is identical on $\mathfrak{N}'$ and extends μ. As a result, the type p over $\mathfrak{N}'$ is realized in $\mathfrak{N}$, whence $\mathfrak{N}$ is saturated. The proof of the converse assertion is trivial. $\square$

Proposition 5.7.5. *The interpretation* I *preserves all the properties indicated in Theorem 0.6.1(e).*

PROOF. We assume that theory T_0^* has a saturated model $\mathfrak{M}$ of cardinality α. Using Lemma 5.6.1(d), we construct a perfect model $\mathfrak{N}$ of cardinality α with the model kernel $\mathbb{K}(\mathfrak{N}) \cong \mathfrak{M}$. By Lemma 5.7.6, $\mathfrak{N}$ is a saturated model of the theory T_1^* of cardinality α. Conversely, if the theory T_0^* has no saturated model of cardinality α, then the theory T_1^* also has no such models in view of Lemma 5.7.6. The preservation of the strong constructivizability of a countable saturated model follows from (Q.9). $\square$

MODELS WITHOUT AUTOMORPHISMS

We consider models with first-order definable and almost first-order definable (i.e., algebraic) elements.

Lemma 5.7.7. *A model* $\mathfrak{N}$ *of the theory* T_1 *is a model with first-order definable* (*almost first-order definable*) *elements if and only if the following conditions hold*:

(a) $\mathfrak{N}$ *is primitive*,

(b) *all elements of the kernel* $\mathbb{K}(\mathfrak{N})$ *are first-order definable* (*almost first-order definable*) *in* $\mathfrak{N}$.

PROOF. We first consider the case of the first-order definability. If conditions (a) and (b) hold, then elements of $\mathfrak{N}$ are first-order definable in view of (Q.12).

To prove the converse assertion, we assume that condition (a) fails for $\mathfrak{N}$. Using (Q.2) and property (b) from the definition of the quasioperator of a primitive envelope, we find a primitive submodel $\mathfrak{N}' \preccurlyeq^{\circ} \mathfrak{N}$ such that $\mathbb{K}(\mathfrak{N}') = \mathbb{K}(\mathfrak{N})$. By condition, the model $\mathfrak{N}$ is not primitive. Therefore, $\mathfrak{N}'$ is a proper submodel of the model $\mathfrak{N}$. Consequently, none of the elements $a \in |\mathfrak{N}| \setminus |\mathfrak{N}'|$ is first-order definable in $\mathfrak{N}$. Let $\mathfrak{N}$ satisfy (a) but not (b). We assume the contrary, i.e., $\mathfrak{N}$ is a model with first-order definable elements. Then $\mathfrak{N}$ is prime. By Lemma 5.7.2, a model $\mathbb{K}(\mathfrak{N})$ is also prime. In our case, $\mathbb{K}(\mathfrak{N})$ is not a model with first-order definable elements. Therefore, there exists an element $a \in \mathbb{K}(\mathfrak{N})$ whose orbit O_a with respect to the actions of automorphisms of this model contains more than one element. By Lemma 5.6.1(a), any automorphism of the kernel $\mathbb{K}(\mathfrak{N})$ can be extended to an automorphism of the model $\mathfrak{N}$. Therefore, elements of O_a are also indiscernible in the model $\mathfrak{N}$ in the sense of the first-order definability, which contradicts the assumption that all elements are first-order definable in $\mathfrak{N}$.

In the case of the almost first-order definability of elements, it suffices to repeat verbally the above reasoning with the words "first-order definable" replaced by the words "almost first-order definable" and the condition $|O_a| \geqslant 2$ replaced by the condition $|O_a| \geqslant \omega$. $\qquad \square$

Proposition 5.7.6. *The interpretation* I *preserves all the properties indicated in Theorem* 0.6.1(f).

PROOF. By Lemma 5.7.7, the operator of a primitive envelope $\mathbb{P}$ establishes a one-to-one correspondence between isomorphism types of models of the theories T_0^* and T_1^* with first-order definable (almost first-order

definable) elements; moreover, the operator also preserves the cardinality
and the value of the algorithmic dimension.   □

Lemma 5.7.8. *A model* $\mathfrak{N}$ *of the theory* T_1 *is rigid if and only if
the following conditions hold:*

(a) $\mathfrak{N}$ *is primitive,*

(b) $\mathbb{K}(\mathfrak{N})$ *is a rigid model.*

PROOF. Let a model $\mathfrak{N}$ satisfy conditions (a), (b) and let μ be an automorphism of $\mathfrak{N}$. Then the restriction $\mu' = \mu \restriction U(\mathfrak{N})$ is an automorphism
of the kernel $\mathbb{K}(\mathfrak{N})$. By condition (b), μ' is identical on $\mathbb{K}(\mathfrak{N})$. By (Q.12),
μ is identical on the whole model $\mathfrak{N}$. Consequently, the model $\mathfrak{N}$ is rigid.

To prove the converse assertion, we assume that condition (a) fails for
$\mathfrak{N}$. Then the model $\mathfrak{N}$ cannot be rigid in view of (Q.11). If (a) holds but
(b) fails for $\mathfrak{N}$, it is possible to extend a nontrivial automorphism μ' of the
kernel $\mathbb{K}(\mathfrak{N})$ to an automorphism μ of the whole model $\mathfrak{N}$ in accordance
with Lemma 5.6.1(b). Consequently, the model $\mathfrak{N}$ is not rigid.   □

Proposition 5.7.7. *The interpretation* I *preserves all the properties indicated in Theorem* 0.6.1(g).

PROOF. By Lemma 5.7.8, the operator of a primitive envelope $\mathbb{P}$ establishes a one-to-one correspondence between isomorphism types of rigid
models of the theories T_0^* and T_1^*; moreover, it preserves their cardinalities
and values of the algorithmic dimension.   □

THE SPECTRUM FUNCTION

Lemma 5.7.9. *The interpretation* I *preserves the nonmaximality
property of the spectrum function.*

PROOF. By the Shelah theorem [60], for any countable complete theory T the spectrum function satisfies the following alternative:

$$IM(\omega_\beta, T) = 2^{\omega_\beta} \text{ for all } \beta \geqslant 1 \tag{5.7.2}$$

$$IM(\omega_\beta, T) < J_{\omega_1}(|\omega + \beta|) \text{ for all } \beta \geqslant 1 \tag{5.7.3}$$

where the cardinal-valued function $J_\alpha(\beta)$ is defined in the Introduction.
By Lemma 5.6.1(d) and the condition (Q.13), which is assumed to be
satisfied for some fixed ordinal $\alpha < \omega_1$, the following relations between the

spectrum functions of the theories T_0^* and T_1^* are valid:

$$IM(\omega_\beta, T_0^*) \leqslant IM(\omega_\beta, T_1^*) \leqslant J_\alpha(|\omega + \beta|) \circ \sum_{\delta \leqslant \beta} IM(\omega_\delta, T_0^*) \qquad (5.7.4)$$

From (5.7.4), we conclude that the spectrum function of the theory T_1^* is maximal provided that the spectrum function of the theory T_0^* is maximal. Conversely, we assume that (5.7.3) holds for the theory T_0^*. Since $J_\alpha(\beta)$ is monotone in α and β, from (5.7.3) we obtain the estimate from above $IM(\omega_\alpha, T_1^*) < J_{\omega_1}(|\omega + \alpha|)$, which implies that the spectrum function of the theory T_1^* is not maximal. $\qquad \Box$

REMARK 5.7.1. Lemma 5.7.9 can be easily proved under the assumption that the continuum hypothesis is true. Under this assumption, it is not necessary to use the alternative (5.7.2), (5.7.3).

Proposition 5.7.8. *The interpretation I preserves all the properties indicated in Theorem 0.6.1(i).*

PROOF. The necessary relations are obtained in Lemma 5.7.9. $\qquad \Box$

5.8. Elementary Transformations of Theories

Owing to the transitivity of the semantic similarity of theories, we can represent the transformation $T \mapsto \mathbb{F}(T)$ used in the proof of Theorem 0.6.1 as a chain of several simpler transformations called *elementary transformations* (cf. Table 5.8.1). Every elementary transformation allows us, starting from a given theory T_0, to construct a new theory T_1 and an interpretation of T_0 in T_1 with the properties described in Table 5.8.1.

We indicate concrete representations of the transformation $T \mapsto F$ via elementary ones.

If a theory T is given by its recursively enumerable index and the signature is finite, then the passage from T to F can be represented as follows:

$$T \xrightarrow{\Phi G} T_1 \xrightarrow{GF} T_2 \xrightarrow{\Phi G} T_3 \xrightarrow{GL} F \qquad (5.8.1)$$

If a theory T is given by its recursively enumerable index and the signature is infinite, then the passage from T to F can be represented as follows:

$$T \xrightarrow{AI} T_0 \xrightarrow{IG} T_1 \xrightarrow{GF} T_2 \xrightarrow{\Phi G} T_3 \xrightarrow{GL} F \qquad (5.8.2)$$

If a theory T is given by its weak recursively enumerable index, then (independently of the signature) the passage from T to F can be represented as follows:

$$T \xrightarrow{AI} T_0 \xrightarrow{IG} T_1 \xrightarrow{GF} T_2 \xrightarrow{\Phi G} T_3 \xrightarrow{GL} F \qquad (5.8.3)$$

Table 5.8.1. Elementary transformations

	Initial axiomatizable theory	Resulting axiomatizable theory	Interpretation type
AI	Arbitrary axiomatizable theory	Axiomatizable theory of an infinite signature	Complete first-order definable enrichment
IG	Arbitrary axiomatizable theory of an infinite signature	Axiomatizable theory extending GRE	Quasiexact $\exists$-representable interpretation
ΦG	Arbitrary axiomatizable theory of a finite signature	Axiomatizable theory extending GRE	Exact $\exists$-representable interpretation
GF	Arbitrary axiomatizable theory extending GRE	Finitely axiomatizable model-complete theory of a finite rich signature	Quasiexact $\exists$-representable interpretation
GL	Arbitrary axiomatizable theory extending GRE	Axiomatizable theory of a given finite rich signature	Exact $\exists$-representable interpretation

From the descriptions of elementary transformations (cf. Secs. 5.9 and 5.10), we see that the above-mentioned passages are realized effectively with respect to recursively enumerable indices and Gödel numbers of axioms of finitely axiomatizable theories. Weak recursively enumerable indices are used only in the first passage in (5.8.3). At first, from the weak recursively enumerable index of the theory T the transformation AI effectively constructs the recursively enumerable index of the theory T_0. After that, the same steps must be made as in the chain (5.8.2).

Using Table 5.8.1, we can verify the accordance between the properties of the intermediate theories obtained; each of them serves as the initial theory for the following transformation. We note that the transformation GF provides the finite axiomatizability and model-completeness of the theory. By Lemma 5.4.2, Lemma 5.4.1(d), and Lemma 5.3.4(a), these important properties are preserved along the chain up to the end.

Thus, Theorem 5.8.1 is reduced to the assertion that all the elementary transformations exist.

Theorem 5.8.1. *Let X be a transformation in the first column in Table 5.8.1. If a theory T_0 satisfies the conditions indicated in Table 5.8.1, then there exists a theory T_1 and an interpretation I of T_0 in T_1 which satisfy the conditions indicated in Table 5.8.1 for X. Furthermore, the recursively enumerable index of the theory T_1 and the interpretation I are effectively constructed from the recursively enumerable index of the theory T_0 and, in the case $X = AI$, from the weak recursively enumerable index of the theory T_0.*

We outline the proof. The elementary transformation AI is characterized in Table 5.8.1. The transformations IG, ΦG, and GL are described in Secs. 5.9 and 5.10. The description of the main elementary transformation GF is complex and occupies Chapter 6.

5.9. The Elementary Transformation *IG*

The transformation IG is a modification of a standard method of reducing signatures (cf., for example, [6]). We consider the infinite enumerable signature

$$\sigma_0 = \{P_0^{m_0}, P_1^{m_1}, \dots, P_s^{m_s}, \dots ; \ s < \omega\} \tag{5.9.1}$$

where $m_s > 0$ for all natural numbers s and $f(s) = m_s$ is a general recursive function.

Theorem 5.9.1. *Let T_0 be an axiomatizable theory of the signature (5.9.1). Starting from the recursively enumerable index of the theory T_0, it is possible to construct effectively an axiomatizable theory T_1 of the signature $\sigma = \{\Gamma^2\}$ that extends the graph theory GRE and a quasiexact $\exists$-representable interpretation I of the theory T_0 in the theory T_1.*

PROOF. We describe the transformation of a model $\mathfrak{M}$ of the theory T_0 into some model $\mathfrak{N} = \mathbb{P}\mathfrak{M}$ of the graph theory GRE. The theory of the class of models of the form $\mathbb{P}\mathfrak{M}$, $\mathfrak{M} \in \text{Mod}(T_0)$, turns out to be the required theory T_1. Let $\mathfrak{M}$ be a model of the theory T_0 of the signature (5.9.1). We define the model $\mathfrak{N} = \mathbb{P}\mathfrak{M}$. Let $|\mathfrak{N}|$ be equal to the union $A \cup B \cup C$ of the disjoint sets A, B, and C such that A has the same cardinality as $|\mathfrak{M}|$, B will be specified later, and $C = \{c_0, c_1, \ldots, c_6\}$ consists of seven elements. We fix a bijective mapping $f \colon |\mathfrak{M}| \to A$. We note that A, B, and C are definable in the model $\mathfrak{N}$ by means of the formulas $U(x)$, $V(x)$, and $W(x)$ indicated below.

We pass to the definition of Γ-connections on the set $|\mathfrak{N}|$. By means of f, we temporarily transfer the structure of the model $\mathfrak{M}$ to the set A. The truth and falsity of σ_0-predicates in A is coded by special configurations in the domain B (Fig. 5.9.1). We consider an m_k-placed predicate $P_k^{m_k} \in \sigma_0$ and elements $a_1, a_2, \ldots, a_{m_k} \in A$. If $P_k(a_1, a_2, \ldots, a_{m_k})$ is true, then, using elements of the set B and Γ-connections, over the sequence $(a_1, a_2, \ldots, a_{m_k})$ we construct a configuration of the form (t) (cf. Fig. 5.9.1).

If $P_k(a_1, a_2, \ldots, a_{m_k})$ is false, then over the above sequence we construct a configuration of the form (f).

Every configuration consists of the body and supports. The number of supports is equal to the number of places of the corresponding predicate. For the predicate P_k, the length of each transformation of the configurations is assumed to be equal to $k + 2$. Variants of the form are defined by the parameter $\alpha \in \{0, 1\}$, which indicates the falsity (f) or the truth (t) of the predicate on the given sequence. Antisymmetry of the form of the configuration provides the order on the sequence of arguments. For example, the configurations (t) and (f) (cf. Fig. 5.9.1) correspond to the predicate $P = P_k^{m_k}$ for $k = 1$ and $m_k = 3$ respectively. Different coding configurations (i.e., configurations for different predicates or for a single predicate but for different sequences) must have no common elements of B but they can have common elements of A. With each predicate $P_k^{m_k} \in \sigma_0$ and each sequence of elements of A of length m_k, only one coding configuration of the form (t) or (f) must be associated depending on the truth or falsity of the predicate on a given sequence. We now define B as the set of elements of all possible coding configurations, except A-elements.

Coding configurations do not exhaust all Γ-connections in the framework of the set $A \cup B$. In this domain, we introduce additional Γ-connections that are important for the rigidity mechanism. However, they do not prevent the coding configurations from giving the first-order definable representation of the structure of the model $\mathfrak{M}$ on the set A.

The domain
of the unique
structure
with seven
elements

The domain
of coding
configurations

The domain
of interpretations

$\mathfrak{N} = \mathbb{P}(\mathfrak{M})$

$\mathfrak{M} \in \mathrm{Mod}\ (T_0)$

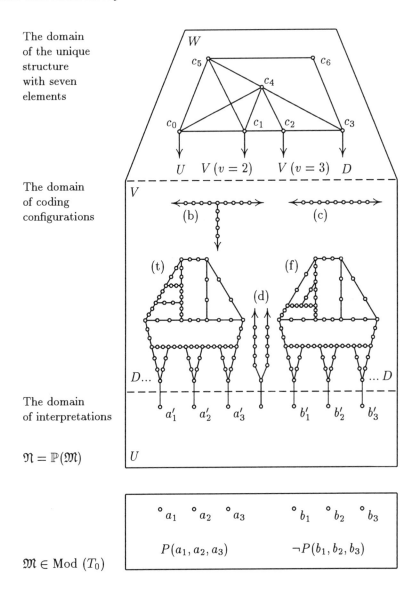

Fig. 5.9.1. Reduction of an arbitrary signature to graphs.

We introduce $D = \{x \in B \mid (\exists y \in A)\, \Gamma(x,y)\}$, $D[a] = \{x \in B \mid \Gamma(x,a)\}$, $a \in A$. By construction, $D \subseteq B$ and $D = \bigcup\{D[a] \mid a \in A\}$; moreover, different sets $D[a']$, $D[a'']$, $a' \neq a''$, have the same cardinality and are disjoint. By construction, coding configurations define no Γ-connections between elements of D.

ADDITIONAL Γ-CONNECTIONS

Additional Γ-connections are given on the set D so as to satisfy the following conditions:

(a) elements in the framework of a single set $D[a]$, $a \in A$, are not connected by the predicate Γ,

(b) the predicate Γ defines a one-to-one correspondence between any two sets $D[a']$ and $D[a'']$ for $a', a'' \in A$, $a' \neq a''$,

(c) $x, y, z \in D$ implies $\Gamma(x,y)\, \&\, \Gamma(y,z) \rightarrow \Gamma(x,z) \vee x = z$,

(d) if the following conditions are satisfied:

   (1) $P_k^{m_k}$ is a signature predicate of the theory T_0,

   (2) i is an index such that $1 \leqslant i \leqslant m_k$,

   (3) $(a_1, \ldots, a_{m_k})$ and $(b_1, \ldots, b_{m_k})$ are sequences of elements of A such that $a_i \neq b_i$, but a_j coincides with b_j for all $j \neq i$,

   (4) $\varkappa'$ and $\varkappa''$ are configurations over the above sequences for the predicate P_k,

   (5) $x \in D[a_i]$ is an element of the configuration $\varkappa'$ and is connected with the ith component a_i,

   (6) $y \in D[b_i]$ is an element of the configuration $\varkappa''$ and is connected with the ith component b_i,

   then the elements x and y are Γ-connected.

We have defined the predicate Γ on the set $A \cup B$. To define Γ-connections on the set C and between the sets C and $A \cup B$, on C, we introduce the graph structure shown in Fig. 5.9.1. We denote by $\Phi(z_0, \ldots, z_6)$ a quantifier-free formula describing the diagram of the Γ-structure with respect to an interpretation of z_i in terms of c_i. We introduce the notation

$$W_i(x) = (\exists z_0, \ldots, z_6)\big(\Phi(z_0, \ldots, z_6)\, \&\, x = z_i\big), \quad i < 7$$
$$W(x) = W_0(x) \vee \ldots \vee W_6(x)$$

We define Γ-connections between C and $A \cup B$ as follows. For $x \in A \cup B$ we set

$$\Gamma(c_0, x) \iff x \in A$$
$$\Gamma(c_1, x) \iff x \in (B \setminus D) \ \& \ (x \text{ has valence 2 in } B)$$
$$\Gamma(c_2, x) \iff x \in (B \setminus D) \ \& \ (x \text{ has valence 3 in } B)$$
$$\Gamma(c_3, x) \iff x \in D$$

In the rest of the cases, Γ is assumed to be false. We temporarily eliminate the introduced structure of the model $\mathfrak{M}$ in the set A. The description of the model $\mathfrak{N} = \mathbb{P}\mathfrak{M}$ of the signature $\sigma = \{\Gamma^2\}$ is complete. It is obvious that $\mathfrak{N}$ is a model of the graph theory GRE.

By construction, the realization of the formula Φ is unique; therefore, all seven elements c_i, $i < 7$, are first-order definable in the model $\mathfrak{N} = \mathbb{P}\mathfrak{M}$ by means of the formulas $W_i(x)$.

We now define T_1 as the theory of the class of models of the form $\mathbb{P}\mathfrak{M}$, $\mathfrak{M} \in \text{Mod}(T_0)$. The domain of interpretations of T_0 in T_1 is given by the formula $U(x) = \neg W(x) \ \& \ (\exists y)(W_0(y) \ \& \ \Gamma(y, x))$. With an atomic formula $P_k(x_1, x_2, \ldots, x_{m_k})$ of the theory T_0, we associate the formula of the signature σ_1 that asserts the occurrences of $U(x_i)$, $1 \leqslant i \leqslant m_k$, and the existence of a configuration of the form (t) over the sequence $(x_1, x_2, \ldots, x_{m_k})$ for the predicate P_k. We extend by induction this correspondence to the transformation $I : FL(\sigma_0) \to FL(\sigma_1)$, which is the required interpretation of T_0 in T_1. In addition, the formula

$$V(x) = \neg W(x) \ \& \ (\exists y)\big[\big(W_1(y) \vee W_2(y) \vee W_3(y)\big) \ \& \ \Gamma(y, x)\big]$$

defines the domain of coding configurations in the theory T_1 and the formula

$$VD(x) = V(x) \ \& \ (\exists y)\big[U(y) \ \& \ \Gamma(x, y)\big]$$

distinguishes the part of this domain for additional Γ-connections.

AXIOMS OF THE THEORY T_1

$1°$. The formulas $U(x)$, $V(x)$, and $W(x)$ divide the universe into three nonempty parts.

$2°$. The formula $W(x)$, true on exactly seven elements, presents a unique realization of the formula $\Phi(z_0, \ldots, z_6)$.

$3°$. Every element x of V which does not belong to VD has valence 2 or 3 in the domain V.

4°. An element x which does not belong to VD has valence 2 in the domain V if and only if $\neg W(x)$ & $(\exists y)\big(W_1(y)$ & $\Gamma(x,y)\big)$.

5°. An element x which does not belong to VD has valence 3 in the domain V if and only if $\neg W(x)$ & $(\exists y)\big(W_2(y)$ & $\Gamma(x,y)\big)$.

6°. If $V(x)$ & $V(y)$ & $\Gamma(x,y)$ and the element x is located in VD or has valence 3 in V, then y does not belong to VD and has valence 2 in the domain V.

7°. $U(x)$ and $U(y)$ imply $\neg\Gamma(x,y)$.

8°. $U(x)$ & $U(y)$ & $\Gamma(x,z)$ & $\Gamma(y,z)$ & $V(z)$ $\to$ $x=y$.

9°. For any predicate $P_k^{mk} \in \sigma_0$ and a sequence in U of the corresponding size, in V there exists exactly one coding configuration of the form (t) or (f) such that its size and form correspond to the predicate P_k^{mk} in accordance with the above description.

10°. Let V-elements $x_0,\dots,x_n$, $n<\omega$, form a Γ-chain, let $x_1,\dots,x_{n-1}$ have valence 2 in V, and let x_0, x_n belong to VD or have valence 3 in V. Then the elements are located in some coding configuration for one of the predicates P_k, $k \leqslant n-2$.

11°. In the domain VD, the predicate Γ is defined by conditions (a)–(d) from the description of additional Γ-connections.

12°. $I(\varphi)$ holds for every sentence $\varphi \in SL(\sigma_0)$ occurring in the system of axioms of the theory T_0.

It is obvious that the theory T_1 is axiomatizable and the recursively enumerable index of the theory T_1 can be effectively found from the recursively enumerable index of the theory T_0. In addition to models of the form $\mathbb{P}\mathfrak{M}$, $\mathfrak{M} \in \mathrm{Mod}\,(T_0)$, the theory T_1 has other models. In the domain of coding configurations, nonstandard configurations of types (b) and (c) can appear and the number of such configurations may be arbitrary (cf. Fig. 5.9.1). Furthermore, from every U-element of the model the number of double chains of type (d) starting at the U-element may be arbitrary. It is easy to see that the kernel operator in the interpretation I is the left inverse operator for the constructed operator $\mathfrak{M} \mapsto \mathbb{P}\mathfrak{M}$, i.e., these operators are connected by the relation $\mathbb{K}\,(\mathbb{P}\mathfrak{M}) \cong \mathfrak{M}$. On the other hand, $\mathbb{P}$ exactly is the operator of a primitive envelope for the interpretation I.

We study the role of Γ-connections defined by Axiom 11°. In accordance with conditions (a)–(d), the predicate Γ establishes a bijective correspondence between collections of edges directed from any two U-elements to the domain of coding configurations. Condition (d) guarantees that D-elements from standard configurations correspond to the same elements. Considering the remaining elements, we conclude that Γ is a bijective correspondence between chains of the form (d) starting from any two different U-elements, which restricts the number of models of the theory T_1 with given cardinality and kernel.

The description of the transformation IG is complete. It remains to prove that the theory T_1 and interpretation I obtained satisfy all the requirements of Theorem 5.9.1. The effectiveness of the construction of T_1 and I from the recursively enumerable index of the theory T_0 are guaranteed by construction. To verify that the interpretation I is quasiexact, we first describe primitive and saturated envelopes with respect to I.

Lemmas 5.9.1–5.9.3 follow from the construction.

Lemma 5.9.1. *A model $\mathfrak{N}$ of the theory T_1 is primitive over a set $X \subseteq |\mathfrak{N}|$ if and only if the following conditions hold:*

(a) *any configuration of the form (b) intersects X,*

(b) *any configuration of the form (c) intersects X,*

(c) *every nonempty family of nonstandard configurations of the form (d) intersects X provided that it is closed under Γ-connections in the domain VD.*

Lemma 5.9.2. *For a model $\mathfrak{N}$ of the theory T_1, the set $\mathrm{Nost}(\mathfrak{N})$ consists exactly of elements of all nonstandard configurations of the model $\mathfrak{N}$ and the set $\mathrm{Stan}(\mathfrak{N})$ contains all the remaining elements of the model $\mathfrak{N}$.*

Lemma 5.9.3. *A model $\mathfrak{N}$ of the theory T_1 of cardinality α is perfect over a set $X \subseteq |\mathfrak{N}|$ if and only if the following conditions hold:*

(a) *there exist α different configurations of the form (b) that do not intersect the set X,*

(b) *there exist α different configurations of the form (c) that do not intersect the set X,*

(c) *there exist α disjoint families of nonstandard configurations of the form (d) that are closed under Γ-connections in the domain VD; moreover, each of them does not intersect X.*

We show that the interpretation I is quasiexact. The verification of many conditions is simple and is left to the reader. We give arguments concerning the most delicate properties.

(d) (from the definition of a prequasiexact interpretation) For a given countable model $\mathfrak{N}$, the model $\mathfrak{N}'$, as the union of a countable chain of elementary extensions, realizes, at each next step, types describing diagrams of new nonstandard configurations in accordance with Lemma 5.9.3. Such descriptions are locally consistent since any family of finite fragments of nonstandard configurations is embedded in standard configurations. We have sufficiently many standard configurations with sufficiently large sizes because the signature σ_0 is infinite. Repeating the process ω times, we successively reach the validity of all the required conditions. The model obtained is the required one.

(Q.2) Since the stronger condition (Q.12) holds for T_1, the quasioperator of a primitive envelope is, in fact, the operator of a primitive envelope.

(Q.5), (Q.6) It is easy to see that the type $p(x)$ of the theory T_1 is the type of a nonstandard element if it contains the formula $V(x)$ & $\neg VD(x)$ and a collection of $\exists$-formulas describing neighborhoods of x in the form of some nonstandard configuration. On the other hand, the standard type $p(x)$ is the type of an element of the kernel, or one of the seven elements of the domain W, or an element located in a standard configuration of one of predicates. This fact allows us to establish (Q.5) and (Q.6) for standard and nonstandard types.

(Q.7) The number of nonstandard configurations is finite in α^+-homogeneous models of the theory T_1 of uncountable cardinality α.

(Q.8) Let $\mathfrak{N}$ be a primitive model with strongly constructivizable kernel. Then every strong constructivization ν of the kernel $\mathbb{K}(\mathfrak{N})$ is uniquely extended to a constructivization ν' of the whole model $\mathfrak{N}$. Indeed, the envelope of a primitive model can be explicitly constructed from the kernel. To realize it, we use Lemma 5.2.1, which implies that the elementary theory of the model $\mathfrak{N}$ is decidable if and only if the elementary theory of the kernel of $\mathfrak{N}$ is decidable. Conversely, let $\mathfrak{N}$ be a model of the theory T_1. Since the model $\mathfrak{N}$ is strongly constructivizable, it is obvious how to construct the strong constructivization of the kernel of $\mathfrak{N}$.

(Q.9) The proof is similar to that of (Q.8), but the difference is that an extension of the strong constructivization from the kernel to the corresponding perfect model is not unique. By standard methods, we can show that a strongly constructivizable countable perfect model is nonautostable.

(Q.10) To obtain $\mathfrak{N}'$ it suffices to eliminate one of the nonstandard configurations from $\mathfrak{N}$.

(Q.11) It is easy to see that all types of nonstandard configurations possess nontrivial automorphisms. At this point, the role of double form of supports of standard configurations becomes clear. Thus, the required automorphism of the model, which is not a primitive model, can easily be obtained by isomorphic permutation inside one nonstandard configuration of this model.

(Q.13) The required relation between isomorphism types of models of the theory T_1 and their kernels is described by Lemma 5.9.4.

Lemma 5.9.4. *Let* $\mathfrak{M}$ *be a model of the theory* T_0 *of cardinality* $\|\mathfrak{M}\| \leqslant \omega_\alpha$. *Then the number of isomorphism types of a model* $\mathfrak{N}$ *of the theory* T_1 *of cardinality* $\|\mathfrak{N}\| = \omega_\alpha$ *with the kernel* $\mathbb{K}(\mathfrak{N}) \cong \mathfrak{M}$ *is equal to* $\varepsilon = |\omega + \alpha|$. *In other words, the spectral index of the interpretation* I *is equal to* 0.

PROOF. Every model $\mathfrak{N}$ of the theory T_1 of cardinality ω_α such that $\mathbb{K}(\mathfrak{N}) \cong \mathfrak{M}$ is defined up to an isomorphism by the triple of ordinals δ_0, δ_1, δ_2, $0 \leqslant \delta_i \leqslant \omega + \alpha$, as follows. The component δ_0 characterizes the number of nonstandard fragments of the type (b) in the model $\mathfrak{N}$ so that if the cardinality of the set of all such fragments in $\mathfrak{N}$ is equal to β, then

$$\delta_0 = \begin{cases} \beta & \text{if } \beta < \omega \\ \omega + \gamma & \text{if } \omega \leqslant \beta \leqslant \omega_\alpha, \ \beta = \omega_\gamma \end{cases}$$

In the same way, δ_1 and δ_2 characterize the number of nonstandard fragments of types (c) and (d). As for (d), we mean the number of minimal families of configurations of the form (d) that are closed under Γ-connections in the domain VD. It is obvious that the assumption about the cardinality of the model $\mathfrak{N}$ leads to the following restriction on the triple of ordinals:

$$\|\mathfrak{M}\| < \omega_\alpha \Rightarrow |\delta_0 + \delta_1 + \delta_2| = |\omega + \alpha| \qquad (5.9.2)$$

From the description of the theory T_1 it follows that, in models $\mathfrak{N}$ of cardinality ω_α such that $\mathbb{K}(\mathfrak{N}) \cong \mathfrak{M}$ all the triples of ordinals satisfying (5.9.2) can be realized. Therefore, $\varepsilon \geqslant |\omega + \alpha + 1|^2 = |\omega + \alpha|$. On the other hand, by the condition $\mathbb{K}(\mathfrak{N}) \cong \mathfrak{M}$, the triple of ordinals characterizes the model $\mathfrak{N}$ of the theory T_1 uniquely up to an isomorphism. This fact gives the estimate from above $\varepsilon \leqslant |\omega + \alpha + 1|^3 = |\omega + \alpha|$. □

Thus, the interpretation I constructed is quasiexact.

The fact that the interpretation I is $\exists$-representable follows from the description of the construction IG. Indeed, the domain of interpretation is distinguished by the formula $U(x)$, which is equivalent to an $\exists$-formula and to a $\forall$-formula as well. With the atomic formula $P_k(x_1, x_2, \ldots, x_{m_k})$ of the theory T_0, we associate the $\exists$-formula of the theory T_1 asserting existence of a configuration (t) for a given predicate over a given sequence. With the negation of this atomic formula we associate the $\exists$-formula asserting the existence of a configuration (f) over a given sequence.

The proof of Theorem 5.9.1 is complete. Thereby, we have established the existence of the elementary transformation IG. □

5.10. The Elementary Transformations ΦG and GL

We first study the transformation ΦG, which is simpler than IG.

Theorem 5.10.1. *Let T_0 be an axiomatizable theory of a finite signature σ_0. Then from the recursively enumerable index of the theory T_0 it is possible to construct effectively an axiomatizable theory T_1 of the signature $\sigma_1 = \{\Gamma^2\}$ that extends the graph theory GRE and an exact $\exists$-representable interpretation I of T_0 in T_1.*

PROOF. As in the case of IG, we first describe the transformation of a model $\mathfrak{M}$ of the theory T_0 into some model $\mathfrak{N} = \mathbb{P}\,\mathfrak{M}$ of the graph theory GRE.

The theory T and the interpretation I are defined in the same way as in Sec. 5.9 for IG; moreover, we can manage by coding configurations with simple but not double contours of supports. A principal difference consists in the fact that the assertion that all elements of V are located only in standard coding configurations can be written as a single axiom. Since such an axiom does not admit the appearance of nonstandard fragments of configurations, additional Γ-connections are not necessary. It is obvious that the interpretation I obtained is exact and is $\exists$-representable. □

To consider GL we need the following lemma.

Lemma 5.10.1. *Let T_0 be an axiomatizable theory of the signature $\sigma_0 = \{\Gamma^2\}$ extending the graph theory GRE and let σ_1 be a finite* *rich signature. From the recursively enumerable index of the theory T_0 it is possible to construct effectively an axiomatizable theory T_1 of the signature σ_1 and an exact $\exists$-representable interpretation I of T_0 in T_1.*

PROOF. Depending on the type of the signature σ_1, we will construct the required theory T_1 and interpretation I in two different ways.

CASE 1: the signature σ_1 contains at least one predicate symbol or functional symbol of number of places at least 2. For T_1 we take the obvious representation of the theory T_0 of the signature σ_0 in the larger signature σ_1. The interpretation I is the first-order definable transformation of signature.

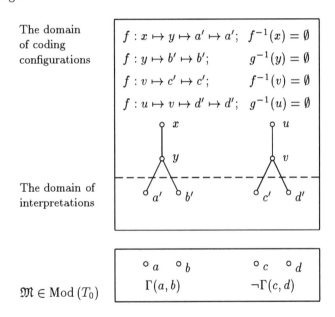

The domain of coding configurations

$$f : x \mapsto y \mapsto a' \mapsto a'; \quad f^{-1}(x) = \emptyset$$
$$f : y \mapsto b' \mapsto b'; \quad\quad g^{-1}(y) = \emptyset$$
$$f : v \mapsto c' \mapsto c'; \quad\quad f^{-1}(v) = \emptyset$$
$$f : u \mapsto v \mapsto d' \mapsto d'; \quad g^{-1}(u) = \emptyset$$

The domain of interpretations

$\mathfrak{M} \in \mathrm{Mod}\,(T_0)$

$\Gamma(a,b)$ $\neg\Gamma(c,d)$

Fig. 5.10.1. Reduction of graphs to two unary functions.

CASE 2: all predicate symbols and functional symbols occurring in σ_1 are unary. The signature contains at least two unary functional symbols since it is rich by condition. For the sake of definiteness, we denote these symbols by f and g. Only these functions are used in the construction. The rest of the signature symbols are trivially defined in T_1. The theory T_1 and the interpretation I are defined as in the cases of IG and ΦG. The main difference is that we have W here. The domain of the interpretation of T_0 in T_1 is defined by the formula $U(x) = \big(f(x) = x \,\&\, g(x) = x\big)$. The remaining elements in models of the theory T form the domain of coding configurations which are defined by the functions f and g (cf. Fig. 5.10.1). It is easy to see that the constructed theory T_1 regarded as a theory of the class of models $\mathbb{P}\mathfrak{M}$, $\mathfrak{M} \in \text{Mod}(T_0)$, is finitely axiomatizable and the interpretation I of T_0 in T is exact and is $\exists$-representable. $\square$

5.11. Reduction of the Signature of the Intermediate Construction

We are in a position to indicate a method of reduction to a finite rich signature, as is required in Theorem 3.1.1. It suffices to pass to the case of one binary predicate or two unary functions. The theory constructed in Chapter 3 possesses all the necessary properties except the condition on the signature. By construction, the signature of the theory $\mathbb{F}(m, s)$ contains 16 binary predicates and a finite number of unary predicates. Therefore, it is not hard to change the signature of $\mathbb{F}(m, s)$ so that the obtained signature contains 17 binary predicates. Then the theory $\mathbb{F}(m, s)$ can be reduced to a single binary predicate as follows. We apply the method of coding by configurations as in the case of ΦG but with the following modifications: the predicate Γ represents 16 predicates on $X \cup Y$ by means of configurations and represents V^2 on the domain W immediately. Using Lemma 3.16.1(c), it is easy to prove that the estimate (g) from Theorem 3.1.1 is valid and all other properties remain valid too.

In the case of two unary functions, the method is the same, but it is necessary to prepare 16 different configurations of a more complex form than is shown in Fig. 5.10.1 and code 16 binary predicates on $X \cup Y$ by means of these configurations (which are not symmetrical and antireflexive). The predicate $V(x, y)$ in the domain W is directly represented by one of these two functions.

Chapter 6

The Universal Construction

In this chapter, we describe the universal construction which, in fact, is the proof of Theorem 0.6.1. As already mentioned in Sec. 5.8, it suffices to construct the main elementary transformation GF. Although the main ideas somewhat repeat those of constructing the intermediate construction, we do not make any references to them, but discuss some connections and differences between these two constructions. We emphasize that the main difference is the use of different mechanisms for computing formulas. Such a mechanism is rather complex in the case of the universal construction.

6.1. A Sketch of the Proof of the Main Theorem

In this chapter, we consider the elementary transformation GF of the universal construction (cf. Table 5.8.1). The theory GRE was defined in the Introduction.

Theorem 6.1.1. *Let T be an axiomatizable theory of the signature $\sigma_0 = \{\Gamma^2\}$ which extends the graph theory GRE. Starting from the*

recursively enumerable index of the theory T, it is possible to construct
effectively a finitely axiomatizable model-complete theory F of a finite
signature σ_1 as well as a quasiexact $\exists$-representable interpretation I of
the theory T in the theory F.

We give a sketch of the proof of Theorem 6.1.1. For a Turing machine
$\mathcal{M}$ and $m \in \mathbb{N}$, we construct a finitely axiomatizable theory of the form

$$F(\mathcal{M}, m) = \mathrm{FRM}(m) + \mathrm{DEP} + \mathrm{CLC}(\mathcal{M}) + \mathrm{NOR} + \mathrm{RIG} \qquad (6.1.1)$$

whose axiomatics contains the description of $\mathcal{M}$ and m as the input parameters. Then we construct a special Turing machine $\mathcal{M}^*$ and a correcting
general recursive function $f(x)$ such that the desired finitely axiomatizable
theory F is represented in the form

$$F = GF(T) = F(\mathcal{M}^*, f(m_0)) \qquad (6.1.2)$$

m_0 is the recursively enumerable index of the theory T

Moreover, the function f is such that the numbers m and $f(m)$ are the
recursively enumerable indices of the same theory. In this chapter, we use
the same type of Turing machines as in Sec. 3.2, but with instructions of
the form $a_i q_j \longrightarrow a_s q_t L$ and $a_i q_j \longrightarrow a_s q_t R$. Instructions with division
of cells are not applied here. We will choose $\mathcal{M}$ and m in Sec. 6.12.

6.2. Signature

The signature σ_1 of the theory F contains 39 binary predicates and
$54 + d + e$ unary predicates, where d and e are parameters of the Turing
machine $\mathcal{M}$. As already mentioned, the Turing machine $\mathcal{M}$ and the parameters d and e are independent of the recursively enumerable index of
the theory T.

The signature of the theory F is presented in Table 6.2.1, where
predicates marked by the superscript 2 are binary predicates and the rest
are unary predicates. The predicates are divided into 10 groups. A brief
characteristic of each predicate and some additional information are given
in Table 6.2.1.

In accordance with our convention (cf. Sec. 4.7), all references to the
basic theory QSR are made only via the "external" predicates $\lhd$, $\sim$, and
Λ described in Table 4.1.1. The remaining predicates of the theory QSR
are assumed to occur in the signature σ_1. They are indicated by dots in
Table 6.2.1.

Table 6.2.1. The list of predicates of the theory F

Groups of predicates Predicates and their characterization	Depth	Dim
1. Predicate of the model kernel: $\quad$ U is the universe of the model kernel $\quad$ $\varGamma^2$ are edges (connections) of a graph		
2. Predicates of the skeleton mechanism: $\quad$ $\lhd^2$, $\sim^2$, ... the quasisuccession (crossing)		
3. Predicates of the frame and the geometric structure: X_1, X_2, X_3, X_4 are basic sets of dimensions E_1^2, E_2^2, E_3^2, E_4^2 are equivalences by dimensions S_1^2, S_2^2, S_3^2, S_4^2 are connections by the first dimension R_2^2, R_3^2, R_4^2 are connections by the second dimension P_3^2, P_4^2 are connections by the third dimensions T^2 is the connection by the fourth dimension	Lower index of predi- cate	
4. Predicates of the initial marking-out: $\quad$ O is the center of the computing block $\quad$ G is the general line $\quad$ M is the marking-out line $\quad$ K is the subdivision line $\quad$ I is the line generating tape cells	2 2 2 2 2	2 2 2 4 2
5. Predicates of the formation of the tree of sequences: $\quad$ B is the line of branching of sequences $\quad$ V is the line of extension of sequences $\quad$ V_0, V_1 are the indicators of the extension line $\quad$ Z is the screen of the results	2 2 4 2	3, 4 3, 4 4 2
6. Predicates of numbering and subdivision: $\quad$ N^2 is the numbering of branching of sequences $\quad$ H^2 is the subdivision equivalence $\quad$ EV^2 is the equivalence of branching $\quad$ EN^2 is the numbering equivalence	 4 3 3	

Table 6.2.1. The list of predicates of the theory F (continued).

Groups of predicates	Depth	Dim
Predicates and their characterization		
7. Predicates of the synthesis of formulas:		
D is the line with the initial messages	2	4
D_0, D_1, D_2 are the initial messages	4	4
Δ is the elementary line with the initial messages	2	3,4
Δ^* is the mark of the Δ-line	2	3
Δ_0, Δ_1, Δ_2 is elementary message	3	3,4
Σ is the elementary conjunction line	2	3
Σ_0, Σ_1 is the truth of the elementary conjunction	3	3
Ω is the matrix line	2	3
Ω^* is the mark of the Ω-line	2	3
Ω_0, Ω_1 is the truth of matrix	3	3
8. Predicates of the work of the Turing machine:		
A is a tape cell	2	2
A_0, ..., A_{d-1} are tape symbols	2	2
Q is the head line	2	2
Q_0, ..., Q_{e-1} are the internal state of the machine	2	2
L, R are the polarization of cells and the head	2	2
9. Predicates of model completeness:		
$D2$ is a 2-dimensional domain	2	2
$D3$ is a 3-dimensional domain	2	3
$D4$ is a 4-dimensional domain	2	4
$\Sigma0$, $\Sigma1$ are singular stations of the Σ-line	2	3
$\Omega0$, $\Omega1$ are singular stations of the Ω-line	2	3
10(R). Predicates of the rigidity mechanism:		
Λ, ... are marks of the rigidity of quasisuccession		
LR_3^2, LS_3^2, LG^2 are local isomorphisms of depth 3	3	
LR_4^2, LS_4^2, LP_4^2 are local isomorphisms of depth 4	4	
CN^2 is the coordination of branching numberings	3	
CV^2 is the coordination of branching indices	3	
CK^2 is the coordination of the subdivision lines	2	
CH^2 is the coordination of the subdivision classes	4	
CD^2 is the coordination of the fourth dimension	4	

6.3. Axiomatics. Frame

In this section, we describe axioms of the theory F, combining them in groups in accordance with their purposes. A prefix in the notation of groups indicates the corresponding term of (6.1.1).

(FRM.1) The model kernel axioms

$1°.$ $\Gamma(x, y) \rightarrow U(x) \,\&\, U(y)$.

$2°.$ $\neg\Gamma(x, x)$.

$3°.$ $\Gamma(x, y) \leftrightarrow \Gamma(y, x)$.

$4°.$ $(\exists x)(\exists y)\, \Gamma(x, y)$.

$5°.$ $(\exists x)(\exists y)\big[U(x) \,\&\, U(y) \,\&\, x \neq y \,\&\, \neg\Gamma(x, y)\big]$.

(FRM.2) Basis of the superstructure

$6°.$ The predicates U, X_1, X_2, X_3, and X_4 define a partition of the universe of a model.

$7°.$ All the signature predicates (except U, Γ, and N) are defined on $X = X_1 \cup X_2 \cup X_3 \cup X_4$ and are false outside X.

$8°.$ The predicates E_k, $k = 1, 2, 3, 4$, are equivalence relations on the set $X_1 \cup \ldots \cup X_4$ and are false outside this set.

$9°.$ $E_{k+1}(x, y) \rightarrow E_k(x, y)$, $k = 1, 2, 3$.

$10°.$ There are X_k-elements, $k = 1, 2, 3, 4$, in every E_k-class.

$11°.$ There are X_2-elements in every E_1-class.

(FRM.3) The skeleton mechanism

$12°.$ On the set $X = X_1 \cup X_2 \cup X_3 \cup X_4$, the predicates $\lhd$, $\sim$, and other predicates of the signature $\sigma_{QSPrime}$ define the family of pairwise disjoint models of the theory QS' which covers X. The set X_1 and sets of the form $[a]_{E_k} \cap X_{k+1}$, $a \in X_{k+1}$, $k = 1, 2, 3$, are exactly the universes of models of this family.

$13°.$ The set of X_k-elements, $k = 1, 2, 3, 4$, of the E_k-class is exactly the $\sim$-class of quasisuccession.

(FRM.4) Depth and dimension

Let x be an element of the set $X_2 \cup X_3 \cup X_4$. We will write

(a) $\dim(x) = 2$ if $[x]_{E_2} \subseteq X_2$,

(b) $\dim(x) = 3$ if $\dim(x) \neq 2$ and $[x]_{E_2} \subseteq X_2 \cup X_3$,

(c) $\dim(x) = 4$ if $\dim(x) \neq 2, 3$ and for every element $t \in X_3 \cap [x]_{E_2}$ there are X_4-elements in the class $[t]_{E_3}$.

14°. $D2(x) \leftrightarrow \dim(x) = 2$.

15°. $D3(x) \leftrightarrow \dim(x) = 3$.

16°. $D4(x) \leftrightarrow \dim(x) = 4$.

17°. $x \in X_2 \cup X_3 \cup X_4 \rightarrow D2(x) \vee D3(x) \vee D4(x)$.

We say that a signature predicate Π of the theory F has *depth* k if Π is defined on the set $X_k \cup \ldots \cup X_4$ and is well defined on E_k-classes.

We say that a unary signature predicate Π of the theory F has *dimension* l if $\Pi(x) \rightarrow \dim(x) = l$.

18°. Depths and dimensions of the signature predicates are indicated in Table 6.2.1.

(FRM.5) Geometric structure of depth 1

19°. $S_1(x, y) \leftrightarrow (\exists u)(\exists v)\big[E_1(x, u) \& E_1(y, v) \& X_1(u) \& X_1(v) \& u \lhd v\big]$.

(FRM.6) Geometric structure of depth 2

20°. $R_2(x, y) \leftrightarrow (\exists u)(\exists v)\big[E_2(x, u) \& E_2(y, v) \& X_2(u) \& X_2(v) \& u \lhd v\big]$.

21°. S_2 is the successor relation on E_2-classes; elements without S_2-predecessors are admissible.

22°. $S_2(x, y) \rightarrow S_1(x, y)$.

23°. If $R_2(x, y)$, $S_2(x, u)$, and $S_2(y, v)$, then one of the following relations holds:

    (a) $R_2(u, v)$,

    (b) $(\exists t_0)\, R_2(u, t_0) \& R_2(t_0, v)$,

(c) $(\exists t_0)(\exists t_1)\, R_2(u,t_0)\ \&\ R_2(t_0,t_1)\ \&\ R_2(t_1,v)$,

(d) $(\exists t_0)(\exists t_1)(\exists t_2)\, R_2(u,t_0)\ \&\ R_2(t_0,t_1)\ \&\ R_2(t_1,t_2)\ \&\ R_2(t_2,v)$.

In cases (b)–(d), the intermediate elements t_0, t_1, and t_2 have no S_2-predecessors.

24°. If $R_2(x,y)\ \&\ R_2(y,z)\ \&\ R_2(z,u)$, then at least one of the elements x, y, z, and u has an S_2-predecessor.

25°. In every E_1-class, the predicate O distinguishes a unique E_2-class.

6.4. General Structure of Models

6.4.1. Kernel and superstructure. An arbitrary model $\mathfrak{N}$ of the theory F consists of two parts (kernel and superstructure). On the set distinguished by the unary predicate U, we can define a graph with relation Γ in accordance with the axioms of (FRM.1). The graph is a model of the theory GRE. In fact, the predicate U is the domain of an interpretation I of the theory T in the theory F. For the interpretation I the *model kernel* is defined by the formula $\mathbb{K}(\mathfrak{N}) = \langle U(\mathfrak{N}); \Gamma \rangle$. Therefore, the predicate $\Gamma \in \sigma_1$ is the I-image of Γ from σ_0.

By the *superstructure* of a model we mean the family of quasisuccessions on the set $X = X_1 \cup X_2 \cup X_3 \cup X_4$ equipped with an additional structure. By Axiom 7°, all relationships between the model kernel and the superstructure can be realized by only the binary predicate N, called the *connection predicate*. The remaining predicates, except the predicates indicated in Axiom 7°, form the superstructure as a special computing mechanism. The latter codes sequences over the domain U, checks the truth of formulas in the model kernel, and controls the fact that all axioms of the theory T hold.

6.4.2. The skeleton mechanism and dimensions. By the axioms (FRM.2) and (FRM.3), on the superstructure (cf. Fig. 6.4.1(a)), there is the equivalence relation E_1 and a quasisuccessor chain which crosses classes with X_1-elements. We say that the quasisuccessor chain determines the *first dimension* of the superstructure. The succession appearing on E_1-classes is presented by the predicate S_1.

We turn to the internal structure of E_1-classes that are divided by the equivalence E_2 into small classes. In each E_1-class, a quasisuccessor chain passing through X_2-elements crosses E_2-classes and, thereby, determines the *second dimension* of the superstructure (Fig. 6.4.1(a)).

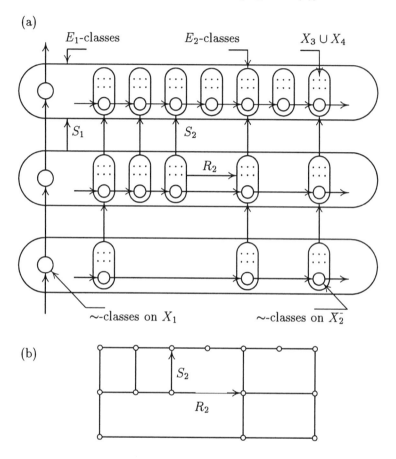

Fig. 6.4.1. The structure of the frame.

The superstructure also has *third dimension* and *fourth dimension*, which are determined by the crossing by the quasisuccessor chain passing through X_3-elements in E_3-classes and X_4-elements in E_4-classes respectively. Axiom 17° admits E_2-classes of dimensions 2–4 and inhibits, for example, E_2-classes of mixed dimension 3 and 4 (cf. Fig. 6.4.2).

6.4.3. Net-geometric structure. By Axiom $20°$, the predicate R_2 establishes connections between E_2-classes along the second dimension, i.e., in the horizontal direction, as is shown in Fig. 6.4.1(a). The predicate S_2 establishes connections between E_2-classes along the first dimension, i.e., in the vertical direction. Unlike R_2, the predicate S_2 is not the first-order definable one. By Axiom $23°$, R_2–S_2-connections between E_2-classes have the form of a "net" with various irregularities that are admissible in view of conditions (b)–(d) in Axiom $23°$. If only condition (a) in Axiom $23°$ holds within some locality of the superstructure, then we say that the locality has the *regular structure of R_2–S_2-connections*. Abstracting from other details of models, we can independently consider the structure of R_2–S_2-connections, representing E_2-classes as points of the R_2–S_2-net in the fragment of the superstructure (cf. Fig. 6.4.1(b)).

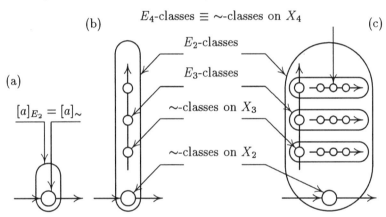

Fig. 6.4.2. The structure of E_2-classes of different dimensions:
(a) $\dim = 2$; (b) $\dim = 3$; (c) $\dim = 4$.

6.4.4. Depth and dimension. It is significant that we may study the superstructure taking into account only a part of the dimensions. This gives the possibility of studying several simple problems instead of the general one. Generally speaking, different dimensions are not equivalent. We may consider only the first k dimensions for $k = 1, 2, 3, 4$, and the corresponding object under consideration could be referred to as a *geometric structure of the superstructure up to depth k*. Increasing k, we include in the consideration more and more "delicate" details of the superstructure.

Every predicate, except the adjacency predicate, of the model kernel and the skeleton mechanism shows itself in the superstructure starting from some depth k, called the *depth of the predicate*. Unary predicates can also be characterized by the *dimension*, which shows the dimension of the superstructure in the domains of the predicates.

In fact, considering the superstructure up to depth k, we can regard E_k-classes as elementary (indivisible) objects. Therefore, we call an E_k-class a *k-point* or a *point* for brevity.

By the axioms in (FRM.5), the superstructure of depth 1 is a simple S_1-succession on 1-points without endpoints and cycles. The axioms in (FRM.6) contain the initial description of the superstructure of depth 2.

6.4.5. Blocks and zones. A minimal nonempty set of 2-points of a model $\mathfrak{N}$ satisfying the axioms is called a *block* if it is closed under R_2-connections and S_2-connections. Within the framework of a block, R_2–S_2-connections form a plane net, a possible fragment of which is depicted in Fig. 6.4.1. Generally speaking, the superstructure may consist of many blocks; moreover, some of the blocks can be connected by "remote" connections via common E_1-classes, as is shown in Fig. 6.4.3. A collection of blocks intersected by some E_1-class is called a *zone*. The superstructure of a model can consist of many zones, and every zone can consist of many blocks.

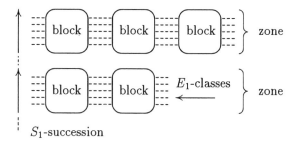

Fig. 6.4.3. Block-zone form of the superstructure.

6.5. Connections of Depth 2

In this section, we introduce some notions and terms concerning the combinatorial analysis of R_2–S_2-connections. Unary predicates of depth 2, except for the informational predicates A_i and Q_j, are very important in local structures of blocks. We divide these predicates into two lists:

$$\mathfrak{X} = \big\{O, G, M, K, I, B, V, Z, D, \Delta, \Delta^*, \Sigma, \Omega, \Omega^*, A, Q, L, R\big\}$$
$$\mathfrak{X}' = \big\{D2, D3, D4\big\}$$

We consider quantifier-free formulas of the form

$$O^{\alpha_1}(x) \ \& \ G^{\alpha_2}(x) \ \& \ \ldots \ \& \ R^{\alpha_{18}}(x), \quad \alpha_i \in \{0,1\}, \quad 1 \leqslant i \leqslant 18$$

constructed from predicates of the list $\mathfrak{X}$, where $\alpha_i = 1$ means positive occurrences and $\alpha_i = 0$ means negative occurrences. For such a formula we use the notation $\mathfrak{X}W(x)$, where $\mathfrak{X}$ is a special prefix and W is the word composed of those symbols of the list $\mathfrak{X}$ that have positive occurrences in the formula. Such a symbol has a unique occurrence in W, and the place of its occurence is of no importance. For example, $\mathfrak{X}OM(x)$ is the notation of the formula

$$O(x) \ \& \ {}^{\neg}G(x) \ \& \ M(x) \ \& \ {}^{\neg}K(x) \ \& \ \ldots \& {}^{\neg}R(x)$$

and $\mathfrak{X}(x)$ is the formula in which each occurrence of a predicate in the list $\mathfrak{X}$ is negative.

We omit the prefix $\mathfrak{X}$ in the notation $\mathfrak{X}W(x)$ if the word W has length at least 2. For example, instead of $\mathfrak{X}OM(x)$ we write $OM(x)$ for simplicity.

Let $\mathfrak{N}$ be a model satisfying all the axioms and let a_i, $i \leqslant k$, and b_j, $j \leqslant s$, be finite sequences of elements of $\mathfrak{N}$ such that

$$R_2(a_i, a_{i+1}), \quad i < k, \quad R_2(b_j, b_{j+1}), \quad j < s$$
$$S(a_i, b_{f(i)}), \quad i \leqslant k$$
$$f(0) = 0, \quad f(k) = s, \quad 1 \leqslant f(i+1) - f(i) \leqslant 4, \quad i < k$$

for some integer-valued function $f(x)$. This collection of elements, together with R_2–S_2-connections and an indication of the truth of predicates in $\mathfrak{X}' \cup \mathfrak{X}$, is called a *chain*, and the number s is called the *length* (of the chain). In the above situation, the elements b_j, where j is different from $f(0), \ldots, f(k)$, cannot have S_2-predecessors, in view of Axiom 23°.

We arrange chains by setting R_2-connections in the horizontal direction and S_2-connections in the vertical direction. Elements of chains are represented as boxes. True predicates of $\mathfrak{X}$ are written near the box and the dimension is written within the box. To indicate the truth of formulas of the form $\mathfrak{X}W(x)$, we write down the word W near the element.

In Catal 6.1, a number of elementary chains are presented. The chain Catal 6.1(i) is denoted by θ_i, $1 \leqslant i \leqslant 72$. By a *junction* of the chains θ' and θ'' of lengths m and n respectively, we mean the chain $\theta = \theta'\theta''$ of length

$m+n-1$ that is obtained by identification of the right endpoint of the chain θ' with the left endpoint of the chain θ'', provided that the descriptions of these endpoints are identical. Otherwise, the result of the operation $\theta'\theta''$ is undefined. For example, the junction of the elementary chains θ_7 and θ_{72} is defined. However, the junction of these chains in the inverse order is undefined. We say that the chains ζ_0 and ζ_1 have a *nontrivial overlapping* if there exist chains θ, θ', and θ'' such that the length of θ is greater than 1 and for some $i \in \{0, 1\}$ we have $\zeta_i = \theta'\theta\theta''$ & $\zeta_{1-i} = \theta$ or $\zeta_i = \theta'\theta$ & $\zeta_{1-i} = \theta\theta''$. For example, the elementary chains θ_1, θ_2, and θ_3 have nontrivial overlappings with the chains θ_{11}, θ_{14}, θ_{25}, and others. However, all overlappings between elementary chains are exhausted by similar obvious examples.

The following lemma can be proved by exhaustion of all possible cases.

Lemma 6.5.1. *There is no nontrivial overlapping among elementary chains θ_i, $4 \leqslant i \leqslant 72$.*

The junction operation is associative. Therefore, in the notation of a junction we omit parentheses.

The following lemma is a consequence of Lemma 6.5.1. It characterizes a uniqueness property for covers by elementary chains.

Lemma 6.5.2. *Let θ be a chain and let a, b, c, and d be four different points of its upper R_2-chain (from left to right); moreover, each of them has an S_2-predecessor. Let*

$$\theta' = \theta_\alpha \theta_\beta \ldots \theta_\gamma, \qquad \theta'' = \theta_\delta \theta_\varepsilon \ldots \theta_\nu$$

be constructed from elementary chains so that the chain θ' is isomorphic to the locality from a to c and the chain θ'' is isomorphic to the locality from b to d of the chain θ. Then, on the locality from b to c, the decompositions of θ' and θ'' into elementary chains are the same.

6.6. Axiomatics. Local Forms of Blocks

In this section, we study the group of combinatorial axioms. The results obtained will be used in the definition of the geometric form of blocks of depth 2 for models of the theory F.

(FRM.7) Local forms of blocks

26°. Let x and y satisfy $R_2(x, y)$. Then, in a neighborhood of x and y, the form of R_2–S_2-connections and the distribution of unary predicates of $\mathfrak{X} \cup \mathfrak{X}'$ are such that x and y, together with the neighborhood, are isomorphically covered by one of the elementary chains θ_i, $1 \leqslant i \leqslant 72$, and are located on the upper R_2-chain.

27°. From $S_2^3(y, x)$ & $R_2^4(y, z)$, it follows that $VD\Delta\Delta^*(x) \leftrightarrow OJ(z)$, where the superscript indicates the number of the successor in the chain of this relation.

28°. From $R_2^4(u, v)$ & $S_2^{19}(z, v)$, it follows that $\Omega\Omega^*(u)$ & $\mathfrak{X}Z(v) \leftrightarrow OZ(z)$.

To state the last axiom of this group, we introduce an integer-valued parameter $m > 0$ as in (6.1.1). A concrete value of m will be chosen in Sec. 6.17.

29°. Let x and y be such that y is the $(m+3)$th R_2-successor of x. Then $OZ(x) \leftrightarrow IQR(y)$. Furthermore, for z satisfying $IAL(z)$ there is a natural number $k < m + 3$ such that the kth R_2-predecessor t of z satisfies $OZ(t)$.

6.7. Description of Blocks

In this section, we describe all possible forms of blocks defined by the axioms in Sec. 6.6. Catal 6.2 presents 25 types of blocks. Our goal is to show that Catal 6.2 is complete, i.e., it contains exactly all types of blocks that are admitted by the axioms.

Table 6.7.1 characterizes various parameters used in the description of blocks in Catal 6.2 and the abstraction levels of considering blocks. In accordance with Table 6.7.1, there are several ways of addressing blocks in Catal 6.2. They differ by the abstraction level.

The abstraction level E characterizes blocks up to an elementary equivalence. This means that two blocks can join and form a new block under the same elementary embedding.

The abstraction level F characterizes blocks up to an isomorphism. The intermediate abstraction level $F3$ characterizes blocks up to an isomorphism with respect to the first three dimensions. The corresponding parameters τ and π are used in Catal 6.2 without sharpening.

Table 6.7.1. Parameters and the abstraction levels of blocks

Parameters	Characterization	Abstraction level
Notation	It indicates the characteristic fragment that distinguishes a given type of blocks among other types by its functional purpose in the theory F.	N
α, ν, ν'	It characterizes all possible variants of the form of the R_2–S_2-grid and the truth values of predicates in the list $\mathfrak{X} \cup \mathfrak{X}'$ for blocks with a given notation.	X
ε, i, j, δ, δ', δ''	They indicate states of the informational lines and characterize up to an elementary equivalence (without the rigidity predicates) all possible variants of blocks of a given R_2–S_2-form with a given distribution of predicates in the list $\mathfrak{X} \cup \mathfrak{X}'$.	E
τ, π	They indicate the length of fragments of a block by the third and fourth dimensions and characterize up to an isomorphism all possible variants of blocks of a given elementary type.	$F3$, F

We establish the completeness of Catal 6.2 at the abstraction level X. By B_i we denote a block of type Catal 6.2(i), $1 \leqslant i \leqslant 25$. We first investigate the structure of the computing block B_{25} by making more precise its brief description $O(m, \lambda)$, $m \geqslant 0$, $\lambda = \langle \lambda_i, i < \omega \rangle$, $\lambda_i \in \{L, R\}$ presented in Catal 6.2(25).

Lemma 6.7.1. *The following assertions hold:*

(a) *for any $m \in \mathbb{N}$ and λ, the block $O(m, \lambda)$ exists and is defined in a unique way up to an isomorphism at the abstraction level X,*

(b) *for any m and λ, in the block $O(m, \lambda)$, there is a unique 2-point satisfying $O(x)$,*

(c) *any two blocks of the forms $O(m, \lambda)$ and $O(m', \lambda')$ can be different only in the first quarter, i.e., in the domain of the computing controller. In the rest of the domains, they have the same structure of R_2–S_2-connections and that of unary predicates at the abstraction level X.*

PROOF. We describe how to line the block $O(m, \lambda)$ with elementary chains, using Sec. 6.5. Ignoring the parameters m and λ, we gather all the parts of the block, except the first quarter, and then we construct the first quarter of the block for given m and λ. We first place the elementary chain θ_{71} and, to the right of it, add a series of $(m + 1)$ chains θ_{21}, one chain θ_{69}, and a countable number of chains θ_{22}. Thereby, we have defined the initial state of the head on the tape and prepared the upper row of a free domain. The lower row is lined with the chain θ_5 and the series made up of a countable number of chains θ_1. The subsequent lower rows are lined in the same way. As a result, we obtain the free domain with the line G. We line the marking-out domain with vertical columns from bottom to top, moving to the left and using the chains θ_1, θ_4, θ_{27}, and θ_6. Finally, we fill the marking-out domain and the line K containing the points KB and KVD. Then we line the third quarter. As the first chain, we take θ_{72} in the first row and one of the chains θ_{57}, θ_{58}, θ_{59}, and θ_{68} in the following rows. We line a row from right to left, using the chains from Catal 6.1. At every step, only one chain is suitable. The configuration of lines lying in the third quarter is shown in Figs. 6.7.1 and 6.7.2. The uniqueness fails only if the chains θ_{30} and θ_{66} are suitable for the next step. In this case, the choice problem is solved by Axiom 27°. Namely, in the first case, we use the singular chain θ_{66}; thereby, we determine the beginning of the line $\Delta(3^*)$. In the second case, the chain θ_{30} must be taken. As a result, we line the third quarter of the block O, and the boundary line Z of the first quarter is ready. The domain of the computing controller is lined row-by-row from bottom to top. The parameter m appearing here determines the beginning of the Q-line. Furthermore, after points marking the execution of commands in the chains θ_{60} and θ_{63}, an arbitrariness arises in the choice of the chains θ_{61} and θ_{62} or θ_{64} and θ_{65}. The choice problem is solved by the next component λ_i of the sequence λ.

Thus, we have constructed the block $O(m, \lambda)$ satisfying (a) and (c). Since we begin lining the block with the chain θ_{71}, which is not used in any of the domains, the block also satisfies (b). $\square$

We prove the key assertion, which can be called the *combinatorial theorem of level* X.

Theorem 6.7.1. *Let $\mathfrak{N}$ be a model satisfying Axioms* 1°–25° *and let B be a block of $\mathfrak{N}$. Axioms* 26°–29° *hold on B if and only if the description of B at the abstraction level X is contained in* Catal 6.2.

PROOF. If a block B is contained in Catal 6.2, then the validity of Axioms 26°–29° on B is directly verified. To prove the converse assertion,

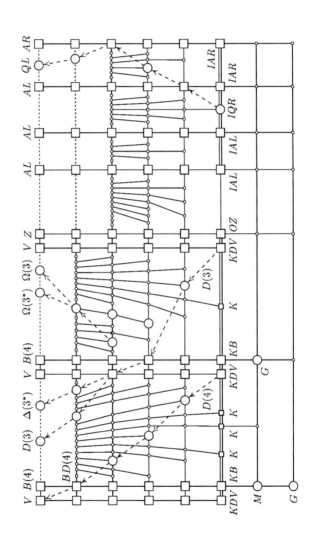

Fig. 6.7.1. The computing block $O(m.\lambda)$

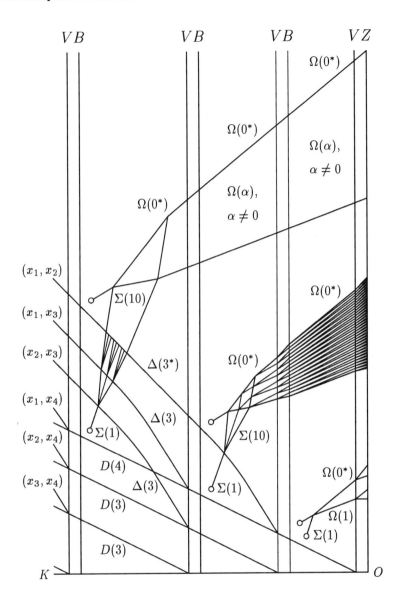

Fig. 6.7.2. Configuration of lines in the domain of synthesis of formulas.

we need some definitions and auxiliary claims given below. Using them, we consider all possible cases for the block $\mathcal{B}$ satisfying Axioms $26°$–$29°$ and, in each case, we show that $\mathcal{B}$ belongs to Catal 6.2.   □

Two realizations θ' and θ'' of elementary chains θ_i and θ_j in $\mathcal{B}$ are called *connected* if θ' and θ'' do not coincide, but have a common 2-point marked by predicates from the list $\mathfrak{X}$. Let θ_i be a chain in Catal 6.2(i), where i is one of the numbers 4–6, 8–22, 25, 26. A finite or infinite series of consequently connected realizations of a chain θ_i is called a *singular line of type* θ_i. For $i \neq 21$, a singular line of type θ_i is called *free* and that of type θ_{21} is called *dependent*. It is easy to see that each of the blocks B_4–B_{15} is a singular unbounded line passing through all the blocks. It is called the *characteristic line* of the block. Singular lines can be identified with the sequence of their 2-points marked by predicates from the list $\mathfrak{X}$. Therefore, a line of type θ_4 is called an *M-line*, a line of type θ_5 is called a *G-line*, and so on. Singular lines introduce irregularities in the structure of R_2–S_2-connections of the block in a regular way. For such lines, the integer-valued parameters α and β are naturally defined. They characterize the "evasion speed" of the line to the left and to the right of the regular net. In this case, the singular line is called an (α, β)-line.

Let θ_i be one of the elementary chains, where i is one of the numbers 28, 30, 33, 36, 38, 40, 43, 46, 48, 50, 52, 55, 58, 61, 62, 64, 65, 66, 69, 71. The realization of θ_i in the block $\mathcal{B}$ is called a *singular point of type* θ_i. Realizations of the chains θ_{66} and θ_{69} are called *dependent points* and the realization of the chain θ_{71} is called the *initial point*. The remaining 17 types of singular points are called *free points*. We note that each of the blocks B_{16}–B_{24} has exactly one singular point, called the *characteristic point* of the block. Generally speaking, the computing block also has the characteristic point of type θ_{71}. However, a block of type O will be considered independently.

The connections between singular lines and points in blocks are characterized by the following three lemmas. The first two lemmas are proved by direct verification.

Lemma 6.7.2. *If, in a block $\mathcal{B}$ satisfying Axioms $26°$–$29°$, there is a realization of an elementary chain θ_i, $i \geqslant 27$, that is not of the singular-point type, then, above or below, there exists a realization of a chain of the singular-point type.*

Lemma 6.7.3. *Let $\mathcal{B}$ be a block of a model satisfying Axioms $26°$–$29°$ and let L be a singular line of $\mathcal{B}$. Then, in both directions,* *the line L either continues infinitely or terminates at a point that is a singular point or a chain neighboring to a singular point.*

Lemma 6.7.4. *Let a block B contain two different singular lines. Then none of the lines can be extended infinitely.*

PROOF. It is obvious that two horizontal lines of type $K+K$, $I+I$, or $K+I$ cannot lie in a common block. One horizontal line and one vertical (or sloping) line must have contact. It remains to consider the case of two nonhorizontal singular lines. We assume that at the level of an E_1-class Y, the distance between such lines along the R_2-succession is k. In this case, in the successor E_1-class or in the preceding E_1-class, the distance is strictly less because of the construction of (α, β)-lines. Advancing further, we arrive at a point of contact of these lines; therefore, at least one of them must terminate.   □

We now complete the proof of Theorem 6.7.1. Let B be a block satisfying Axioms $26°–29°$. We show that the isomorphism type of the block B at the abstraction level X is contained in Catal 6.2.

A block B is called *pure* if all the predicates from the list $\mathfrak{X}$ are false at the points of B.

Lemma 6.7.5. *Let a block B satisfying Axioms $26°–29°$ be pure. Then B is a block of one of the following types: D2, D3, and D4.*

PROOF. A pure block is completely lined with one of the chains $\theta_1–\theta_3$, which gives the necessary form of the block.   □

Denote by η the equivalence on the set of elementary chains θ_i, $4 \leqslant i \leqslant 26$, such that the classes by η are as follows:

$$\{\theta_4, \theta_5\}, \ \{\theta_6, \theta_7\}, \ \{\theta_{19}, \theta_{21}, \theta_{23}\}, \ \{\theta_{20}, \theta_{22}, \theta_{24}\}$$

and, in addition, 13 one-element classes formed by the remaining elements of the above-mentioned set.

Lemma 6.7.6. *If, in a block B, two elementary chains of family θ_i, $4 \leqslant i \leqslant 26$, located in different classes by η are realized, then the block B contains a singular point.*

PROOF. By condition, in the block B there are two disjoint singular lines. By Lemma 6.7.4, a singular point must appear in this block.   □

Lemma 6.7.7. *If a block B is not pure and has no singular point, then it is one of the blocks B_i, $4 \leqslant i \leqslant 15$.*

PROOF. Since the block B contains no singular point, it is covered by elementary chains within the framework of the family θ_i, $1 \leqslant i \leqslant 26$, by Lemma 6.7.2. Lemma 6.7.6 restricts the number of variants for elementary chains realized in B, and all variants can be verified. As a result, we obtain exactly all the above-mentioned blocks.   □

Lemma 6.7.8. *If, in a block B, there is a dependent line and a dependent or initial singular point or connected realization of the chains θ_{58} and θ_{68}, then B is the computing block.*

PROOF. By Axiom 29°, a dependent singular line may be situated only in a neighborhood of a point satisfying $O(x)$ at a distance of at most $m + 2$. By Axioms 27° and 28°, the case of a dependent singular point θ_{66} and the pair of chains θ_{58} and θ_{68} leads to a point satisfying $O(x)$. The case of the chain θ_{69} is covered by Axiom 29°, and the case of the initial point is trivial.   □

Lemma 6.7.9. *Suppose that there is a unique singular point in a block B. Then B is a block of Catal 6.2 of one of the types B_i, $16 \leqslant i \leqslant 24$.*

PROOF. A singular point realized in B is free. Otherwise, by Lemma 6.7.8, there are other singular points in the block B. For the sake of definiteness, let a singular point have type θ_k. We describe how to line the block B with elementary chains starting from the point θ_k. First, we line a neighborhood of this singular point with suitable chains θ_i, $i \geqslant 27$. If $k = 58$, the chain θ_{68} cannot be used, in view of Lemma 6.7.8. Further, we must use only the elementary chains θ_1, θ_2, θ_3 as well as the chains θ_i, $4 \leqslant i \leqslant 26$. It is easy to check that this process is unambiguous. Thus, the form of the block B is defined by the type of the point θ_k, and the block turns out to be in the list B_i, $16 \leqslant i \leqslant 24$.   □

We continue to study an arbitrary block B. It remains to consider the case in which there is more than one singular point in B. We show that B contains a point satisfying $O(x)$.

Lemma 6.7.10. *Suppose that for a block B at least one of the following situations occurs:*

(a) *a Q-line and an I-line,*

(b) *a Z-line and an I-line,*

(c) *a Z-line and a Q-line,*

(d) a *Q-line and two different A-lines,*

(e) a *G-line and a K-line,*

(f) *two singular points on a common K-line,*

(g) a *B-line and a Z-line,*

(h) *two B-lines not extending each other.*

Then, in the block $\mathcal{B}$, there is a point satisfying $O(x)$.

PROOF. In cases (a)–(c), we must move down along the lines up to their break, which leads to the point O. In case (d), two A-lines terminate at an I-line, and we are in the position of case (a). In case (e), the point of contact of the G-line and the K-line gives the point O. Case (f) leads to case (e). In case (g), the break of a Z-line gives the point O. Case (h) leads to case (f).   □

We investigate the structure of a block $\mathcal{B}$ containing more than one singular point. To this end, we introduce the notion of a *loop*, i.e., a closed polygon (contour) whose boundary is formed by segments of singular lines. No singular points or singular lines are contained within a loop. Hence the interior of the domain of a loop has a regular structure of the R_2–S_2-net of dimension α, $2 \leqslant \alpha \leqslant 4$, which is called the *dimension* of the loop. By the *type* of a loop we mean the description at the abstraction level X that contains information about the type and number of sides of the loop as well as the type of corner points. Roughly speaking, the type of a loop is an abstract topological description of the loop regarded as a contour whose sides and angles are of given types but for which there is no information about the size of the loop.

We exclude from consideration dependent singular points and dependent singular lines which lead to the point O by Lemma 6.7.8. There are 17 types of angle points for constructing loops. As angle points, we also may regard points at which two adjacent M-lines terminate on G and two adjacent $A(R)$-lines terminate on I. There are 17 types of independent singular lines that can be taken as sides of loops. G-lines and I-lines are excluded from consideration since they cannot be sides of loops.

The number of types of loops is infinite because we can use points of types $\Sigma\Omega(3^*)$, $AQ(R, L)$, and $AQ(L, R)$ representing a "straight angle." It is natural to group the above-mentioned cases in series, which allows us to conclude that there are only a finite number of series and types of loops. Namely, 26 series and 108 different types of loops. To prove the following assertion, it is necessary to construct explicitly all the above-mentioned types of singular lines and singular points as well as all types and series of loops (combining them in groups in accordance with dimensions).

Lemma 6.7.11. *If a block contains more than one singular point, then the block contains a loop.*

PROOF. By the downward-directed crowd principle, in the block $\mathcal{B}$ there are two singular points joined by a segment of a singular line. Looking over all possible cases, we arrive at the assertion of the lemma. □

The following lemma can be proved by exhaustion with the help of the downward-directed crowd principle.

Lemma 6.7.12. *Let a and b be singular points of a block joined by a segment of a singular line of type $(-2, 3)$ or $(3, -2)$; moreover, there are no other singular points within this segment. Then the segment ab is a side of some loop located below the segment.*

Lemma 6.7.13. *If a block contains a loop, then the block contains a point satisfying $O(x)$.*

PROOF. We consider loops of different types, taking into account their size. It is easy to see that every loop contains a segment of the form $(-2, 3)$ or $(3, -2)$ such that the loop itself is located below the segment; moreover, such a segment is unique for a given loop. Such a segment is called the *upper segment* and its length (along the S_1-chain) is called the *size* of the loop. The notion of the *lower segment* of a loop is analogous. However, a loop can have no lower segments.

We apply the method of descent on the size of loops. A loop Z is called a *deadlock loop* if one of the items of Lemma 6.7.10 can be applied, which guarantees the existence of the point O in the block. Beginning with the loop $Z = Z_0$ defined by the assumptions of the lemma, we construct a sequence of loops Z_1, Z_2, ... that are not deadlock ones in $\mathcal{B}$. We assume that the loop Z_k has been constructed and it is not a deadlock loop. If Z_k has a lower segment, then the loop defined from this segment by Lemma 6.7.12 can be taken as Z_{k+1}. Otherwise, Z_k has the form of a triangle, the base upward, such that its lateral sides are lines of type $(1, 1)$. In this case, as Z_{k+1} we take the loop that adjoins the least lateral side of the loop Z_k in $\mathcal{B}$. Looking over all cases, we see that if the loop Z_{k+1} is not a deadlock one, then its size is strictly less than that of the loop Z_k. Since the sizes of loops decrease, this procedure has to finish, i.e., in the sequences Z_i, $i = 0, 1, 2, \ldots$, a deadlock loop appears, which proves the lemma. □

6.8. Peculiarities of Block-Zone Structure of Models

We continue to study the general structure of models of the theory F (cf. Sec. 6.4). Based on the description of blocks, we consider the block-zone form of the superstructure shown in Fig. 6.4.3. In Catal 6.2, of special interest is a block of type B_{25} with a computing mechanism that controls the model kernel. The remaining blocks presented in Catal 6.2 can be regarded as only "nonstandard fragments" of the computing block. Therefore, it is natural that a block of type B_{25} is referred to as *standard* and the rest of the blocks of the catalogue are called *nonstandard*.

In accordance with the description of blocks, every 2-point satisfying $O(x)$ generates a standard block in which the remaining 2-points do not satisfy $O(x)$. This makes clear the role of Axiom 25° which connects the number of standard blocks with the collection of 1-points of the model and arranges all such blocks displacing the central points along the S_1-succession. The number and location of the rest of the blocks are arbitrary. Conceptually, standard blocks deal with a common model kernel. That is why they have similar structures.

The many-times repeated computation in different blocks is dictated by the following reasons. In accordance with the description in Sec. 6.7, every standard block contains finite fragments, as large as is wished, of pure domains, singular points, and singular lines of all types, from the blocks in Catal 6.2. Owing to the position of standard blocks displaced along the S_1-succession, we obtain the saturation of every E_1-class by the above-mentioned fragments. As a result, we can neutralize the influence of nonstandard blocks on the elementary theory of the model obtained. The global combinatorial part of the construction has been described, and we now pass to details.

6.9. Axiomatics. Geometric Structure

(DEP.8) Geometric structure of depth 3

30°. $P_3(x, y) \leftrightarrow (\exists u)(\exists v)[E_3(x, u) \ \& \ E_3(y, v) \ \& \ X_3(u) \ \& \ X_3(v) \ \& \ u \triangleleft v]$.

31°. The relations R_3 and S_3 are defined on the set $X_3 \cup X_4$ and are well defined on E_3-classes.

32°. $R_3(x, y) \rightarrow R_2(x, y)$.

$33°$. Every x has at most one R_3-successor, i.e., $R_3(x,y)$ & $R_3(x,z)$ $\rightarrow$ $E_3(y,z)$.

$34°$. $(\exists y)\, R_3(x,y)$ $\leftrightarrow$ $\dim(x) \geqslant 3$ & $(\exists z)\big(R_2(x,z)$ & $\dim(z) \geqslant 3\big)$.

$35°$. Every x satisfying $X_3(x)$ & $\neg G(x)$ has at least one R_3-predecessor.

$36°$. If $X_3(x)$ & $B(x)$, then $P_3(y,z)$ implies $R_3(y,x)$ $\leftrightarrow$ $R_3(z,x)$.

$37°$. If $X_3(x)$ & $\neg B(x)$ & $\neg G(x)$, then x has a unique R_3-predecessor relative to E_3.

$38°$. If $R_3(x,u)$, $R_3(y,v)$, and $\neg B(u)$, then $P_3(x,y)$ $\leftrightarrow$ $P_3(u,v)$.

$39°$. The predicate S_3 is the successor relation on E_3-classes (elements without S_3-predecessors are admissible).

$40°$. $S_3(x,y)$ $\rightarrow$ $S_2(x,y)$.

$41°$. If x satisfies X_3, then $(\exists y)\, S_3(y,x)$ $\leftrightarrow$ $(\exists z)\big(S_2(z,x)$ & $\dim(z) \geqslant 3$ & $\neg G(z)\big)$.

$42°$. If $S_3(x,u)$ and $S_3(y,v)$, then $P_3(x,y)$ $\leftrightarrow$ $P_3(u,v)$.

$43°$. For any k, $1 \leqslant k \leqslant 3$, and u, y, x, z_0, z_1, $\ldots$, z_k, from $S_3(y,u)$, $R_3(x,y)$, $S_3(x,z_0)$, $R_3(z_0,z_1)$, $\ldots$, $R_3(z_{k-1},z_k)$, it follows that $E_2(u, z_k)$ $\rightarrow$ $E_3(u,z_k)$.

(DEP.9) Geometric structure of depth 4

$44°$. $T_4(x,y)$ $\leftrightarrow$ $X_4(x)$ & $X_4(y)$ & $x \lhd y$.

$45°$. The predicates R_4, S_4, and P_4 are defined on the set X_4 and are well defined on E_4-classes.

$46°$. If Π denotes any of the symbols R, S, and P, then $\Pi_4(x,y)$ $\rightarrow$ $\Pi_3(x,y)$.

$47°$. If Π denotes any of the symbols R, S, P and $\Pi_3(x,y)$ holds, then Π_4 is a one-to-one correspondence between 4-points of the classes $[x]_{E_3}$ and $[y]_{E_3}$.

$48°$. If Π denotes any of the symbols R, S, and P, then

$$T_4(x,y) \ \& \ T_4(u,v) \ \rightarrow \ \big(\Pi_4(x,u) \ \leftrightarrow \ \Pi_4(y,v)\big)$$

A quantifier-free formula φ in n free variables $x_1, \ldots, x_n$ is called an *RSP$_4$-formula* if it has the form φ_1 & $\ldots$ & φ_{n-1}, where every conjunctive term φ_i is an atomic formula of the form $\Pi_4(x_i, x_{i+1})$ or $\Pi_4(x_{i+1}, x_i)$ for some Π, where Π denotes one of the symbols R, S, and P.

49°. For every natural number n, $5 \leqslant n \leqslant 8$, and an RSP_4-formula $\varphi(x_1, \ldots, x_n)$ in n variables, the following relation holds:

$$\varphi(x_1, \ldots, x_n) \ \& \ E_3(x_1, x_n) \ \to \ E_4(x_1, x_n)$$

6.10. Commutative Stratifications and Branchings

The notions presented in this section explain the meaning of the axioms in Sec. 6.9. A set W of 2-points of a block $\mathcal{B}$ is called *connected* if any two points a and b can be joined by a finite chain $a = c_1, c_2, \ldots,$ $c_n = b$, $c_i \in W$, such that the neighboring points c_i and c_{i+1} are connected by one of the relations R_2 and S_2, i.e.,

$$\varphi_2(x, y) = R_2(x, y) \vee R_2(y, x) \vee S_2(x, y) \vee S_2(y, x) \tag{6.10.1}$$

A set W of 2-points is called *nonbranching* if it contains no points a and b such that $V(a)$, $B(b)$, $R_2(a, b)$.

A set W of 2-points of a block $\mathcal{B}$ is called a *region* of $\mathcal{B}$ provided that

(a) $a \in W \ \Rightarrow \ \dim(a) \geqslant 3 \ \& \ \neg G(a)$,

(b) W is connected,

(c) W is nonbranching,

(d) W is maximal under conditions (a)–(c).

It is obvious that lines of type V, B, and K serve as the boundary of regions. In a standard block, regions are infinite domains bounded by the line B to the left, by the line V to the right, and by the line K from below. Double V–B-lines separate neighboring regions; the components V and B belong to different regions. It is not hard to describe regions of the remaining blocks.

Let W be one of the regions of a block $\mathcal{B}$ and let a be a 3-point of W. We denote by W_a the set of all 3-points $b \in W$ joined with a by a finite chain $a = c_1, c_2, \ldots, c_n = b$, $c_i \in W$, two neighboring points of which c_i,

c_{i+1} are connected by one of the relations R_3 and S_3, i.e.,

$$\varphi_3(x,y) = R_3(x,y) \vee R_3(y,x) \vee S_3(x,y) \vee S_3(y,x) \qquad (6.10.2)$$

Lemma 6.10.1. *The intersection of W_a with each E_2-class from the region W is exactly a single 3-point.*

PROOF. From the connection between W and the axioms in (DEP.8) it follows that the set W_a has a nontrivial intersection with each of the E_2-classes from W. It remains to show that each of the above-mentioned intersections consists of at most one 3-point. Assume the contrary. In the region W, we can take a sequence of 3-points

$$c_1, c_2, \ldots, c_n \qquad (6.10.3)$$

in which any two neighboring points are connected by (6.10.2) and the endpoints c_1 and c_n lie in the same E_2-class but in different E_3-classes. We can assume that (6.10.3) defines a closed contour consisting of 2-points and surrounding a minimal number of elementary cells of the R_2–S_2-net. If (6.10.3) surrounds only a single elementary cell of the R_2–S_2-net, we arrive at a contradiction with Axiom 43°. Otherwise, the contour can be divided into two parts. Constructing an intermediate joining chain of 3-points, we obtain a new contour of the form (6.10.3) that has the same properties but surrounds a lesser number of cells of the R_2–S_2-net, which yields a contradiction. □

We investigate the structure of sets of the form W_a, which will be called 3-*layers* of the region W. By the axioms in (DEP.8) and Lemma 6.10.1, every layer W_a is a precise copy of the structure of the initial region. Roughly speaking, the general structure of the region W is defined by the relations R_2 and S_2, and a copy of this structure is defined by the relations R_3 and S_3 in every layer W_a. On the other hand, by Axioms 38° and 42°, the set of all P_3-successors of points in W_a is a layer of the region W. The same assertion is valid for the second P_3-successor, P_3-predecessor, and so on. By a 3-*block* of a given region, we mean the minimal collection of 3-layers that is closed under P_3-connections. The number of 3-blocks in a given region W is defined by the length of the P_3-successor (in other words, by the length of the third dimension within the framework of the region).

We say that in regions there is a *commutative stratification* of the geometric R_2–S_2-structure *with respect to the third dimension* given by the P_3-succession. As a result of the stratification, we obtain a 3-dimensional structure whose R_3–S_3-layers can be regarded as structure copies of the initial region. The use of the term "commutative" can be explained by the permutation principle contained in Axiom 43°. We investigate the form of the geometric structure at the join of regions. We consider a block $\mathcal{B}$ containing two adjacent regions W' and W separated by a $V-B$-line; moreover, the region W' is to the left of W. It is easy to see that, at the join, the axioms in (DEP.8) describe R_3-connections on a $V-B$-line in the form of *branching* from right to left. Every layer W_a of the region W is connected by the relation R_3 with a family of layers of the region W'. The family will be denoted by $R_3^{-1}(W_a)$ or RW_a. In view of Axiom 35°, the family RW_a is nonempty. Axiom 36° guarantees that the family RW_a is closed under P_3-succeeding and P_3-preceding layers in W'. Consequently, RW_a consists of an integer number of 3-blocks. This number is called the *branching index* of the layer W_a. The general form of a branching is presented in Fig. 6.10.1, which shows the cut of the 3-dimensional geometric $R_3-S_3-P_3$-structure by the R_3-P_3-plane in a neighborhood of the interface of regions. As indicated in Fig. 6.10.1, different layers of the region W can have different branching indices.

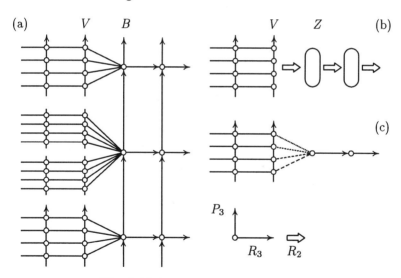

Fig. 6.10.1. The scheme of branchings.

We consider the $V-Z$-line, which can be referred to as the *branching line*. We can assume that the line Z, where the third dimension is absent, is a unique layer branching in the passage from Z to V (cf. Fig. 6.10.1 (b)). The branching is imaginary because it is not represented by the predicate

R_3 as in the case of ordinary branchings. Therefore, a line of type Z is referred to as a *singular region*.

We now characterize the geometric structure of depth 4 defined by the axioms in (DEP.9) in a domain of dimension dim $= 4$. In the standard block, such a domain is located between the K-line and a line of type $D(3)$. In other blocks, its location is defined in accordance with Catal 6.2.

Let Y be a connected set of all 2-points of a block $\mathcal{B}$ of dimension 4. By the above axioms, there is a commutative stratification of the 3-dimensional R_3–S_3–P_3-structure of the domain Y with respect to the fourth dimension, which is determined by the T_4-succession. For a 4-point $a \in Y$ and $k \in \{3, 4\}$, let $Y_a^{(k)}$ stand for the set of all k-points b in Y joined with the point a by a finite chain of k-points $a = c_1, c_2, \ldots, c_n = b$, $c_i \in Y$, such that any two neighboring points c_i and c_{i+1} are connected by one of the relations R_k, S_k, and P_k. We use the notation Y_a^* for $Y_a^{(3)}$ and Y_a for $Y_a^{(4)}$. A set of the form Y_a^* is called a 3-*component* of the domain Y of the block $\mathcal{B}$, and a set of the form Y_a is called a 4-*layer* of the component Y_a^*.

The following assertion is proved in a similar way as Lemma 6.10.1.

Lemma 6.10.2. *A set of 3-points of Y is the union of disjoint terms of the form Y_a^*, $a \in Y$. If a is a 4-point of Y, then the intersection of the layer Y_a and an E_3-class in Y_a^* is exactly one 4-point.*

Lemma 6.10.2 shows that every layer Y_b, $b \in Y_a^*$, presenting this structure by R_4–S_4–P_4-connections, is a copy of the 3-dimensional geometric R_3–S_3–P_3-structure of any component Y_a^*; moreover, the T_4-succession is taken as the fourth dimension with respect to which the stratification is realized. There is no branching here, and this stratification is commutative in the whole domain. At the same time, different 3-components Y_a^* of a block can have different lengths by the fourth dimension. By a 4-*block* we mean a collection of 4-layers of the domain Y that is closed under T_4-connections.

We have completed the characterization of the geometric structure of blocks formed by stratifications and branchings.

6.11. Axiomatics. Informational Lines

In the construction of the computing mechanism, lines of types $B(4)$, D, Δ, Σ, Ω, A, Q are used for the transportation of messages, while information is processed at points of contact of these lines. We introduce necessary notions and notation. Let Π be one of the signature predicates indicated in Table 6.2.1 as the informational line with messages Π_0, Π_1, ..., Π_{n-1}. Then Π is a predicate of depth 2 and Π_i, $i < n$, are predicates of depth $k \leqslant 4$. The predicate Π is well defined on E_2-classes and the predicates Π_i, $i < n$, are well defined on E_k-classes.

We say that Π is the *informational line with messages* Π_0, Π_1, ..., Π_{n-1}, writing this fact as follows:

$$\Pi = \Pi_0 \sqcup \Pi_1 \sqcup \ldots \sqcup \Pi_{n-1}$$

if $\Pi_0(x) \vee \Pi_1(x) \vee \ldots \vee \Pi_{n-1}(x) \rightarrow \Pi(x)$ and, along the Π-line, the following conditions hold:

(1) $\Pi_i(x) \rightarrow \neg \Pi_j(x)$, $i, j < n$, $i \neq j$.

(2) $\Pi(x) \,\&\, X_k(x) \rightarrow \Pi_0(x) \vee \Pi_1(x) \vee \ldots \vee \Pi_{n-1}(x)$.

(3) Let k-points a and b belong to neighboring E_2-classes of the same Π-line, i.e., $S_1(a, b)$. In the case $k \geqslant 3$, we also assume that the points a and b lie on a common k-layer of the stratification with respect to the kth dimension. If elements a and b are not found at a singular point, then $\Pi_i(a) \leftrightarrow \Pi_i(b)$ for all $i < n$.

(4) If $k \geqslant 3$ and a, b are placed at a common 2-point on neighboring k-layers (in the sense of P_3 or T_4), then $\Pi_i(a) \leftrightarrow \Pi_i(b)$ for all $i < n$.

We give axioms concerning the structure of informational lines used in the computing mechanism. The depth and dimension of predicates are assumed to be the same as is indicated in Table 6.2.1.

(CLC.10) Informational lines

$50°$. $V = V_0 \sqcup V_1$ for a line V in the 4-dimensional domain.

$51°$. $D = D_0 \sqcup D_1 \sqcup D_2$.

$52°$. $\Delta = \Delta_0 \sqcup \Delta_1 \sqcup \Delta_2$.

$53°$. $\Sigma = \Sigma_0 \sqcup \Sigma_1$.

$54°$. $\Omega = \Omega_0 \sqcup \Omega_1$.

$55°$. $A = A_0 \sqcup A_1 \sqcup \ldots \sqcup A_{d-1}$.

$56°$. $Q = Q_0 \sqcup Q_1 \sqcup \ldots \sqcup Q_{e-1}$.

6.12. Axiomatics. Numbering of Sequences and Subdivision

We consider the second quarter of a standard block $\mathcal{B}$. Owing to the marking-out mechanism, a countable number of branching lines will be generated to the left of the Z-line. By Sec. 6.10, the structure of depth 3 of the second quarter of the block $\mathcal{B}$ has the form of a branching tree that is used in the construction to represent finite sequences of elements over the model kernel. To this end, branchings of every 3-layer in the block $\mathcal{B}$ on the line K are connected with elements of the model kernel by the special *numbering predicate* N.

We now proceed to axiomatics.

(CLC.11) Numbering of branchings of sequences

$57°.$ $N(x, y) \rightarrow U(x)$ & $KVD(y)$ & $[X_3(y) \vee X_4(y)]$.

$58°.$ $E_3(y, z)$ implies $N(x, y) \leftrightarrow N(x, z)$.

$59°.$ $P_3(y, z)$ implies $N(x, y) \leftrightarrow N(x, z)$.

$60°.$ For any y such that $KVD(y)$ & $[X_3(y) \vee X_4(y)]$, there exists a unique element $x \in U$ such that $N(x, y)$.

$61°.$ For any $x \in U$ and z satisfying the formula $OJ(z)$, there exists an element y such that $N(x, y)$ & $R_2(y, z)$.

$62°.$ For any $x \in U$ and z satisfying the formula $KB(z)$ & $[X_3(z) \vee X_4(z)]$, there exists y such that $N(x, y)$ & $R_3(y, z)$.

$63°.$ $EV(y, z)$ if and only if at least one of the following relations holds:

(a) $(\exists u)[Z(u)$ & $R_2(y, u)$ & $R_2(z, u)$ & $\neg X_2(y)$ & $X_2(z)]$,

(b) $(\exists v)[B(v)$ & $R_3(y, v)$ & $R_3(z, v)]$.

$64°.$ $EN(y, z) \leftrightarrow EV(y, z)$ & $(\exists x)[N(x, y)$ & $N(x, z)]$.

The following two groups of axioms describe the so-called subdivision of the model kernel, i.e., input the initial messages about the model kernel in the computing mechanism of a standard block.

(CLC.12) Subdivision equivalence

$65°.$ The predicate H is an equivalence relation on the set $K \cap X_4$ and is false outside this set.

66°. The predicate H is well defined on E_4-classes.

67°. $H(x, y) \rightarrow E_1(x, y)$.

68°. Every H-class contained in the E_1-class $[z]_{E_1}$ intersects each of the E_3-classes of $[z]_{E_1} \cap K$.

69°. $K(x)$ & $R_4(x, y) \rightarrow H(x, y)$

70°. $K(x)$ & $P_4(x, y) \rightarrow H(x, y)$

71°. $K(x)$ & $T_4(x, y) \rightarrow H(x, y)$

72°. For every EN-class $[b]_{EN}$, there exists an H-class $[a]_H$ such that $EN(b, x) \rightarrow \big(V_1(x) \leftrightarrow H(a, x)\big)$ for all x,

73°. For every H-class $[a]_H$, there exists an EN-class $[b]_{EN}$ such that $H(a, x) \rightarrow \big(V_1(x) \leftrightarrow EN(b, x)\big)$ for all $x \in X_4$.

(CLC.13) Subdivision of the model kernel

74°. Let x_1, x_2, y, and z be such that $U(x_1)$, $U(x_2)$, $N(x_1, y)$, $N(x_2, z)$, $V_1(y)$, $H(y, z)$. Then $D_0(z) \leftrightarrow (x_1 = x_2)$, $D_1(z) \leftrightarrow \big(x_1 \neq x_2 \ \& \ \neg \Gamma(x_1, x_2)\big)$, $D_2(z) \leftrightarrow \Gamma(x_1, x_2)$.

The subdivision of the model kernel is defined in Sec. 6.15.

6.13. Axiomatics. Interaction of Messages

Axioms presented in this section describe the process of interaction of messages at singular points of informational lines.

(CLC.14) Interaction of Information

75°. At a point of type BD, a message is transformed in accordance with Catal 6.2(17). An elementary message appears as follows. Let x satisfy the formula $VD\Delta(x)$. Then for any $y \in [x]_{E_3} \cap X_3$ and $i < 3$

$$\Delta_i(y) \leftrightarrow (\exists t)\big[E_3(y, t) \ \& \ V_1(t) \ \& \ D_i(t)\big]$$

76°. At a point of realization of the chain θ_{66}, a message is transformed in the same way as at a point of type $BD(3)$ in accordance with Catal 6.2(17), but with the following changes.

(a) Instead of the line $\Delta(3)$, the line $\Delta(3^*)$ is generated.

(b) In addition, the line $\Omega(3^*)$ is generated with identically false message, as in the case of the point $B\Delta(3^*)$.

77°. A message on lines passes through a point of type $D\Delta$ in accordance with Catal 6.2(18).

78°. At a point of type $B\Delta$, a message on lines is transformed in accordance with Catal 6.2(19).

79°. At a point of type $\Sigma\Delta$, messages on lines interact in accordance with Catal 6.2(20).

80°. At a point of type $\Sigma\Omega$, messages on lines interact in accordance with Catal 6.2(21).

81°. At a point of type $B\Omega$, a message is transformed by the rule of quantifier realization as follows. Let x and y be elements such that $S_3(x, y)$, $B\Omega(x)$, and $B\Omega(y)$. Then $\Omega_1(x) \leftrightarrow (\exists z)[R_3(z, x)\,\&\,\Omega_1(z)]$, $\Omega_1(y) \leftrightarrow (\forall z)[R_3(z, x) \rightarrow \Omega_1(z)]$.

82°. At a point of type $Z\Omega$, as well as at a point of the connected realization of the chains θ_{58} and θ_{68}, a message is transformed as follows. Let x and y be such that $S_2(x, y)$, $Z\Omega(x)$, and $Z\Omega(y)$. Then $\Omega_1(x) \leftrightarrow (\exists z)[R_2(z, x)\,\&\,\Omega_1(z)]$, $\Omega_1(y) \leftrightarrow (\forall z)[R_2(z, x)\,\&\,\neg X_2(z) \rightarrow \Omega_1(z)]$. Moreover, a message on the display Z determines the initial values of two generated A-lines in accordance with Catal 6.2(23). More exactly, the value Ω_0 on the display generates a line of type $A(L, 0)$ and the value Ω_1 generates a line of type $A(L, 1)$.

83°. At a point of type AQ, a message on lines is transformed in accordance with Catal 6.2(24) and the program of the Turing machine $\mathcal{M}$. Combinations of the form $A_i(x)\,\&\,Q_j(x)$ are inhibited for those pairs i, j for which, in the situation $a_i q_j$, none of the instructions of the Turing machine $\mathcal{M}$ is applicable.

(CLC.15) Initial information for the Turing machine

84°. $OZ(x)\,\&\,R_2(x, y) \rightarrow A_2(y)$.

85°. $IAL(x)\,\&\,IAL(y)\,\&\,R_2(x, y) \rightarrow A_1(y)$.

86°. $IAR(y) \rightarrow A_2(y)$.

87°. $IQR(y) \rightarrow Q_0(y)$.

We now study the theory defined by Axioms $1°-87°$, and then we will continue to state axioms. We denote the basic part of the theory (6.1.1) by $F_0(\mathcal{M}, m)$, where $\mathcal{M}$ is the Turing machine described in Axiom $83°$ and m is the integer-valued parameter appearing in Axiom $29°$.

6.14. General Arrangement of the Computing Block

In Secs. 6.14–6.17, we give an informal description of all main mechanisms in the computing block $\mathcal{B}$ of a model $\mathfrak{N}$ of the theory $F_0(\mathcal{M}, m)$, where the parameters $\mathcal{M}$ and m will be specified below. Catal 6.2(25) contains a general scheme of domains of the computing block that present the main mechanisms of the block. The place of input of the initial message about the model kernel is also indicated there. The work of the computing mechanism in a standard block is regarded as a time process. The first dimension serves as time, i.e., the direction of the S_1-succession. The initial time is assigned with the central point of the computing block. Fixing the first two dimensions, we can regard a block as a plane object. Therefore, we often use the geometric terminology. In the computing block, there are the natural axes K, I, Z and the center O; therefore, we can speak about the coordinate quarters. We give general characteristics of the most important mechanisms working in the computing block.

The marking-out mechanism. It is located in the upper part of the third quarter. Its main purpose is to make the initial marking-out of a subdivision line, as shown in Fig. 6.7.1. As a result, we distinguish the 2-points located to the left of the center O at a distance of $n^2 - 1$, $n = 1, 2, 3, \ldots$. Over these points, the branching lines of the tree of sequences are generated.

Subdivision lines. By the subdivision of the kernel, we mean the division into elementary indivisible particles that play the role of the initial messages for those mechanisms that determine if a formula is true in the model kernel. The subdivision is realized on the axis separating the second and third quarters.

Tree of sequences. It is located in the second quarter of the computing block. The sequences of elements of the model kernel are presented by 3-points of the tree of sequences. With a sequence we associate the set of 3-points representing the set of 3-layers of some region closed with respect to P_3-successors and predecessors. The tree of sequences serves as a support of all those mechanisms that compute the values of formulas.

The mechanism for processing the initial messages. It is located in the lower part of the second quarter between the lines K and $D(3)$. Information about the truth of atomic formulas on elements of sequences is initial. A message on a subdivision line must be transported from the model kernel to the corresponding points of the tree of sequences. To this end, we need a special mechanism. Information that participates in its operation is initial.

The mechanism for computing formulas. It is located in the upper part of the second quarter between the lines $D(3)$ and Z. Information about the values of atomic formulas is initial. This information is obtained as a result of operation of the mechanism for processing the initial messages and is presented by special unary predicates on the tree of sequences. As intermediate information, we compute elementary conjunctions, matrices, and realizations of quantifiers. As a result, we obtain the truth values for all sentences of the model kernel. These values appear on a special display.

Display. It is the axis separating the first quarter from the second quarter in a standard block. The values of formulas presented by special unary predicates on the display are the initial messages for operating the computing controller.

Computing controller. It occupies the first quarter of the computing block and is a special Turing machine that has access to information about a theory of the model kernel kept on the display. The Turing machine program must control the fact that, in the model kernel, all axioms of the initial axiomatizable theory T hold. Otherwise, the program calls for a halt. A special axiom inhibits situations leading to halts and, thereby, guarantees the validity of all axioms of the theory T in the model kernel.

6.15. Tree of Sequences and Subdivision of the Model Kernel

In Secs. 6.15–6.17, we use the notions and notation introduced in Sec. 6.10. We consider a standard block $\mathcal{B}$ of a model $\mathfrak{N}$ of the theory $F_0(\mathcal{M}, m)$ with model kernel $\mathfrak{M} = \mathbb{K}(\mathfrak{N})$. We also discuss a system of coding sequences over $\mathfrak{M}$ in the clock $\mathcal{B}$. As already mentioned in Sec. 6.10, the second quarter of the block $\mathcal{B}$ is divided by the $V-B$-lines into a countable number of regions, which, from right to left, are denoted by W_0, W_1, W_2, $W_3, \ldots$, including a conditional region W_0 that is the line Z.

We consider the set $\mathcal{D} = \mathcal{D}(\mathcal{B})$ consisting of all 2-points of the region W_0 and all 3-points of the regions W_i, $i \geqslant 1$. The set $\mathcal{D}$, together with the geometric structure of depth 3 in the block $\mathcal{B}$, is a tree branching at the join of regions, including the conditional branching between W_0 and W_1. The form of the tree with the branching lines and D-lines is presented at the top of Fig. 6.15.1. At the bottom is the top view of the projection of this 3-dimensional domain onto the cut by the K-line.

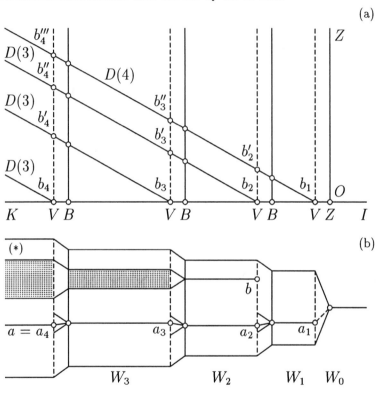

Fig. 6.15.1. The scheme of working the mechanism for the initial messages.

For the predicate N, to every point $a \in \mathcal{D}$ we assign a finite sequence $\varkappa = \varkappa(a)$ of elements of the model kernel as follows.

(1) For $a \in Z$, $\varkappa(a)$ is equal to the empty sequence $\langle \, \rangle$.

(2) For u, x, and z satisfying $U(u)$, $N(u, x)$, $R_2(x, z)$, and $OZ(z)$, we set $\varkappa(x) = \langle u \rangle$.

(3) for u, x, and z satisfying $U(u)$, $N(u, x)$, $R_3(x, z)$, and $KB(z)$, while $\varkappa(z) = \langle u_1, \ldots, u_{n-1} \rangle$, we set $\varkappa(z) = \langle u_1, \ldots, u_{n-1}, u \rangle$.

(4) If to a 3-point a of a region W different from Z, we ascribe the sequence $\varkappa(a) = \varkappa$, then for b belonging to the layer W_a, we set $\varkappa(b) = \varkappa$.

By the axioms in (CLC.11), we see that the imposed conditions yield a well defined inductive definition of the function $\varkappa$. By Axiom 59°, $P_3(a, b) \Rightarrow \varkappa(a) = \varkappa(b)$ for all $a, b \in \mathcal{D}$. We note that the operation $\varkappa$ acts on all possible finite sequences over the model kernel $\mathfrak{M}$; moreover, with sequences of length n we associate points of the nth region W_n.

We consider the mechanism for subdivision and the initial messages, the support of which is a 4-dimensional domain of the tree of sequences $Y = \{a \in \mathcal{D} \mid \dim(a) = 4\}$ located between the lines K and $D(3)$. We denote by V^* the set of points of Z in the E_2-class that adjoins the display. It is easy to see that every 3-component Y_a^* of the domain Y is a minimal set of 3-points that includes one P_3-chain from V^* and is closed under S_3-successors and R_3-predecessors in Y, in particular, through branchings. Moreover, every 4-layer Y_a in Y contains one P_4-chain in V^*, and its R_4-S_4-P_4-structure is an exact copy of the structure of R_3-S_3-P_3-connections of the component Y_a^*.

By Axioms 72° and 73°, there is a one-to-one correspondence realized by V_1-elements between all the H-classes on the K-line and all the numbering branchings $[x]_{EN}$ on this line. We consider one of the H-classes intersecting the K-line of the block $\mathcal{B}$, which is denoted by H_0. For the sake of definiteness, we assume that V_1-elements of the class H_0 are connected with the class $[a]_{EN}$, $a = a_4$, represented in Fig. 6.15.1. By the axioms in (CLC.12), the class H_0 intersects every E_3-class by the K-line of a given block and is closed under R_4-connections, P_4-connections, and T_4-connections. Thus, in the block $\mathcal{B}$, the H-class under consideration generates a family of 4-layers $H_0 Y$ that passes through all 3-points of the tree $\mathcal{D}$ and copies the structure of the tree by their R_4-S_4-P_4-connections. Moving from the element a in the direction of the R_4-succession up to the center O, we pass through the branching points $a = a_4$, a_3, a_2, and a_1 that define the sequences of the model kernel $\varkappa = \langle u_1, u_2, u_3, u_4 \rangle$, $N(u_i, a_i)$, $i = 1, 2, 3, 4$, by means of the numbering predicate N. The sequence $\varkappa$ is exactly a result of the operation $\varkappa(a)$.

The subdivision of the model kernel is realized as follows. In the 4-dimensional domain Y, Axiom 50° defines V as the informational line with messsages V_0 and V_1. Therefore, the indicator V_1 will be transported along the V-line by one of 4-layers of the family $H_0 Y$. Let a 4-point b

satisfy $KVD(b)$ and lie on the K-line in the class H_0. Since a and b are in the same H-class and $V_1(a)$ holds, in view of Axiom 74°, by the 4-layer Z_b along the D-line the initial message presented by one of the predicates D_i, $i \in \{0, 1, 2\}$, will be transported. Namely, we must take the predicate D_k such that $\varepsilon_k(u_4, u)$ holds, where

$$\varepsilon_0(x, y) = (x = y) \tag{6.15.1}$$
$$\varepsilon_1(x, y) = (x \neq y \ \& \ \neg\Gamma(x, y)) \tag{6.15.2}$$
$$\varepsilon_2(x, y) = \Gamma(x, y) \tag{6.15.3}$$

and u is an element of the model kernel with $N(u, b)$. Thus, from the point a in the direction of the S_4-succession along the line $b_4 b'_4$, we direct the indicator V_1, and from the point b along the D-line we direct D_k, for some $k \in \{0, 1, 2\}$.

Future trends depend on the possible meeting of the initial messsage and its indicator at branchings. If they do not meet (as in the case of the element b), then the intial message does not lead to any result. Otherwise (as in the case of the element $b = a_2$), by Axiom 75°, the elementary message Δ_i, $i \in \{0, 1, 2\}$, appears at a contact point of the V-line and the D-line. It can take place at the 3-points of the tree $\mathcal{D}$ that are related to the sequence $\varkappa(a)$. We note that the processes described are independently realized in different H-classes. In the general case, we arrive at the following assertion.

Lemma 6.15.1. *Let a point a be located on the nth (from the display) branching line of the tree $\mathcal{D}$, let $VD\Delta(a)$, and let a be the kth (from the K-line upward) point of type $VD\Delta$ on this V-line. We also assume that $\varkappa(a) = \langle u_1, \ldots, u_n \rangle$. Then $k < n$ and $\Delta_i(a) \leftrightarrow \varepsilon_i(u_k, u_n)$, for every $i < 3$, where the formulas $\varepsilon_i(x, y)$, $i \in \{0, 1, 2\}$, are defined in* (6.15.1)–(6.15.3).

6.16. Mechanism for Computing Formulas

The initial messages are transported for further processing in the mechanism for computing formulas. This mechanism is located over a line of type $D(3)$ in the second quarter. A general scheme of the location of informational lines is shown in Fig. 6.7.2.

We consider singular points on the K-line of the computing block $\mathcal{B}$. To these points (ordered from right to left) we ascribe the variables x_1,

x_2, ... as in Fig. 6.7.2. Let L be a Δ-line in the block $\mathcal{B}$ and let a be a 2-point of type $B\Delta$ such that it is the initial point of the line L. We ascribe the pair of variables (x_k, x_m), $k < m$, to L. We will obtain one of these variables if we drop down from the point a to the K-line along the D-line and obtain another variable if we drop down along the $V-B$-line (cf. Fig. 6.7.2).

The messages on the Δ-lines of the standard block are characterized by the following assertion.

Lemma 6.16.1. *If we ascribe the pair of variables (x_k, x_m), $k < m$, to a Δ-line L of the computing block $\mathcal{B}$, then for any 3-point $a \in L$ defining the sequence $\varkappa(a) = \langle u_1, \ldots, u_n \rangle$ we have $k < m \leqslant n$ and $\Delta_i(a) \leftrightarrow \varepsilon_i(u_k, u_m)$ for every $i \in \{0, 1, 2\}$, where ε_i are defined in (6.15.1)–(6.15.3).*

PROOF. If the point a is at the beginning of the line L, then the required assertions follow from Lemma 6.15.1. We now move the point a along the 3-layer inside the line L from its initial point. Then, within the framework of one region, the value of $\Delta_s(a)$ is preserved and the corresponding sequence $\varkappa(a)$ remains unchanged. After passage of the point a through the $V-B$-line, the value $\Delta_s(a)$ is expanded to all branchings and the sequence $\varkappa(a)$ is extended by adding a new element at the end while all other components remain unchanged. In any case, the truth of the relation under consideration remains constant under the above displacement of the 3-point a along the line L. □

We now characterize other informational lines of the mechanism for computing formulas. We ascribe a formula of the signature $\sigma_0 = \{\Gamma^2\}$ to each segment of the Σ-line, Ω-line, and to each Ω-point of the display of the block $\mathcal{B}$.

Further, we ascribe formulas in n free variables x_1, ..., x_n to lines of the nth region W_n. Since Ω-points of the display Z are related to the zero region W_0, we ascribe closed formulas of the signature σ_0 to these points.

The ascription operation for formulas is defined by induction as follows:

1. To the segment of the Σ-line starting at a point of type BD in the nth region (from the chain θ_{66} for $n = 1$), we ascribe the identically true formula True$(x_1, \ldots, x_n)$.

2. Let lines of types Δ and Σ intersect at a point of type $\Sigma\Delta$ so that the pair of variables (x_k, x_m) is ascribed to the Δ-line and the formula $\varphi(x_1, \ldots, x_n)$ is ascribed to the Σ-line. To three generated upward Σ-lines Σ', Σ'', and Σ''' (from left to right) we ascribe the following formulas:

to the line Σ' we ascribe the formula $\varphi(x_1, \ldots, x_n)$ & $\varepsilon_0(x_k, x_m)$

to the line Σ'' we ascribe the formula $\varphi(x_1, \ldots, x_n)$ & $\varepsilon_1(x_k, x_m)$

to the line Σ''' we ascribe the formula $\varphi(x_1, \ldots, x_n)$ & $\varepsilon_2(x_k, x_m)$

3. To the segment of the Ω-line starting at a point of type $B\Delta(3^*)$ in the nth region (from the chain θ_{66} for $n = 1$), we ascribe the identically false formula $\mathrm{False}(x_1, \ldots, x_n)$.

4. We consider a point of type $\Sigma\Omega$. To the Σ-line incoming from the bottom, we ascribe the formula $\varphi(x_1, \ldots, x_n)$, and to the Ω-line we ascribe the formula $\psi(x_1, \ldots, x_n)$. In the case of an extension of the Σ-line, to this extension we ascribe the same formula $\varphi(x_1, \ldots, x_n)$ and to the new Ω-lines generated upward (denoted by Ω', Ω'' from left to right) we ascribe formulas as follows:

to the line Ω' we ascribe the formula $\psi(x_1, \ldots, x_n)$

to the line Ω'' we ascribe the formula $\psi(x_1, \ldots, x_n) \vee \varphi(x_1, \ldots, x_n)$

5. We consider a point of type $Z\Omega$. To the Ω-line incoming from the bottom, we ascribe the formula $\psi(x_1, \ldots, x_n)$. Then to the generated upward Ω-lines Ω' and Ω'' (from left to right) we assign formulas as follows:

to the line Ω' we ascribe the formula $(\forall x_n)\, \psi(x_1, \ldots, x_n)$

to the line Ω'' we ascribe the formula $(\exists x_n)\, \psi(x_1, \ldots, x_n)$

6. We consider a point of type $Z\Omega$. To the Ω-line incoming from the bottom, we ascribe the formula $\psi(x_1)$. Then to the Ω-points of the display (denoted by Ω', Ω'' downward) we ascribe formulas as follows:

to the point Ω' we ascribe the formula $(\forall x_1)\, \psi(x_1)$

to the point Ω'' we ascribe the formula $(\exists x_1)\, \psi(x_1)$

The definition of the ascription operation is complete. The meaning of this procedure is explained in the following lemma.

Lemma 6.16.2. *Let Π be any of two symbols Σ and Ω. Let the formula $\psi(x_1, \ldots, x_n)$ be ascribed to the segment L of a line of type Π in the computing block B. Let a be a 3-point of L (or a 2-point of L on the display) and let $\varkappa(a) = \langle u_1, \ldots, u_n \rangle$ be the corresponding sequence of elements of the model kernel $\mathfrak{M}$. Then $\Pi(a) \leftrightarrow \mathfrak{M} \vDash \psi(u_1, \ldots, u_n)$.*

PROOF. We apply induction on the order of ascribing formulas to segments of lines and use Lemma 6.16.1 and the axioms in (CLC.10) and (CLC.14), which determine the initial instruction, transportation, and processing of messages on lines of types Σ and Ω.   □

We comment on the proof of Lemma 6.16.2. We discuss the change of messages along the Σ-line. The latter begins at points of type $BD(3)$ or from the chain θ_{66}, where Axiom 75° or Axiom 76° gives its initial value Σ_1 that presents the identically true formula. In the case considered, the Σ-line has the meaning of the empty conjunction of atomic formulas of the form (6.15.1)–(6.15.3). As is indicated in Fig. 6.7.2, exactly one Σ-line starts in every region, beginning from the first one. Intersecting Δ-lines, every Σ-line is divided and generates three new Σ-lines.

The change of messages on Σ-lines after their contact with Δ-lines is defined, by Axiom 79°, in Catal 6.2(20). We consider a contact point of the Σ-line and Δ-line which is associated with the pair of variables (x_k, x_m). At this point, three new lines Σ', Σ'', and Σ''' (from left to right) are generated. Analyzing Catal 6.2(20), we see that a message on the line Σ' is obtained by adding a new conjunctive term $\varepsilon_0(x_k, x_m)$ to the matter of the line Σ. In the same way, a message on the lines Σ'' and Σ''' is obtained by adding the conjunctive terms $\varepsilon_1(x_k, x_m)$ and $\varepsilon_2(x_k, x_m)$ to Σ.

Thus, starting with the identically true Σ-line and passing all the Δ-lines (the number of such lines is $s = n(n-1)/2$ in the case of the nth region), we obtain 3^s lines of type Σ, which present all possible formulas of the form

$$C_i = \varepsilon_{k_1}(x_1, x_2) \,\&\, \varepsilon_{k_2}(x_1, x_3) \,\&\, \dots \,\&\, \varepsilon_{k_s}(x_{n-1}, x_n) \qquad (6.16.1)$$

$$k_1, k_2, \dots, k_s \in \{0, 1, 2\}$$

$$i = k_1 + 3k_2 + 3^2 k_3 + \dots + 3^{s-1} k_s, \quad 0 \leqslant i \leqslant 3^s - 1$$

The obtained Σ-lines (located from left to right) are associated with the formulas C_i ordered by increasing i from 0 to $3^s - 1$.

We consider the change of messages along the Ω-line. The latter starts from a point of type $B\Delta(3^*)$ or from the chain θ_{66} so that Axiom 76° or Axiom 78° prescribes the identically false value presented by Ω_0. At this point, the Ω-line has the meaning of the empty disjunction of formulas of the form (6.16.1). Such a point of the initial start-up of the Ω-line is in each region, beginning with the first one.

Further change of messages on the Ω-line is defined by two different mechanisms: the mechanism for forming the matrix and the mechanism for realizing quantifiers. The first mechanism works in the region (for definiteness, in the nth region) in which the Ω-line has been started. A quantifier-free matrix is obtained as a result of division of the Ω-lines meeting the Σ-lines at points of the type $\Sigma\Omega$. By the scheme presented in Fig. 6.7.2, at this moment, the Σ-line is formed by elementary conjunctions of the form (6.16.1).

At a contact point of the Ω-line and the Σ-line, two lines Ω' and Ω'' are generated. It follows from Axiom 80° and Catal 6.2(21) that Ω', disregarding a message on the Σ-line, inherits a message that was on the arriving Ω-line, whereas the lower line Ω'' is defined by the disjunction rule for messages on the Σ-line and the Ω-line. Thus, after passing the pencil of Σ-lines of the nth region (the number of such lines is $t = 3^s$, $s = n(n-1)/2$), 2^t lines of type Ω appear which present all possible quantifier-free formulas of the form

$$D_j = C_0^{l_0} \vee C_1^{l_1} \vee \ldots \vee C_{t-1}^{l_{t-1}} \tag{6.16.2}$$
$$l_0, l_1, \ldots, l_{t-1} \in \{0, 1\}, \quad C^0 = \text{False}, \quad C^1 = C$$
$$j = l_0 + 2l_1 + 2^2 l_2 + \ldots + 2^{t-1} l_{t-1}, \quad 0 \leqslant j \leqslant 2^t - 1$$

where C_i are defined in (6.16.1). With the Ω-lines obtained (ordered from bottom to top in the given region) we associate the formulas D_j ordered by decreasing index j from $2^t - 1$ to 0.

The mechanism for realizing quantifiers begins to act after the line of quantifier-free matrices in n variables that are formed in the nth region after passage of the pencil of Σ-lines begins to cross the boundaries of the regions, approaching the display. After every passage of the Ω-line through the branching line at a point of type $B\Omega$, two lines Ω' and Ω'' appear. In accordance with Axiom 81°, the message is obtained from the message on the Ω-line by the rule of realizing the universal quantifier on the upper line Ω', and by the rule of realizing the existential quantifier on the lower line Ω''. Passing through all regions, the Ω-lines reach the V–Z-line, on which, by Axiom 82°, the last quantifier is realized and the truth values of the corresponding sentences of the signature σ_0 appear on the display.

Thus, in the nth region, the initial start-up of lines of types Σ and Ω lead to the appearance on the display Z of the computing block $\mathcal{B}$ of the truth values of the following sentences:

$$Z_r^{(n)} = (Q^{p_1} x_1)(Q^{p_2} x_2) \ldots (Q^{p_n} x_n) D_j(x_1, x_2, \ldots, x_n) \tag{6.16.3}$$
$$p_1, p_2, \ldots, p_n \in \{0, 1\}, \quad 0 \leqslant j \leqslant 2^t - 1, \quad Q^0 = \forall, \quad Q^1 = \exists$$
$$r = p_1 + 2p_2 + 2^2 p_3 + \ldots + 2^{n-1} p_n + 2^n j, \quad 0 \leqslant r \leqslant 2^{n+t} - 1$$

where D_j are defined in (6.16.2). To the sequence of the obtained Ω-points on the display (from bottom to top) we assign the sequence of formulas $Z_r^{(n)}$ in order of decreasing r from the maximal value to 0.

Lemma 6.16.3. *Let $\varphi(x_1, \ldots, x_k)$ be a formula of the signature $\sigma_0 = \{\Gamma^2\}$. Then it is possible to construct two formulas $\Psi_0, \Psi_1 \in FL_n(\sigma_1)$ such that Ψ_0 is an $\exists$-formula and Ψ_1 is a $\forall$-formula; moreover,* *for any model $\mathfrak{N}$ of the theory $F_0(\mathcal{M}, m)$ and sequence $x_1, \ldots, x_n$ of elements of $U(\mathfrak{N})$, the following relation holds:*

$$\mathbb{K}(\mathfrak{N}) \vDash \varphi(x_1, \ldots, x_k) \leftrightarrow \Psi_i(x_1, \ldots, x_k), \quad i = 0, 1$$

PROOF. It suffices to construct a formula Ψ_0. Using the axioms in (FRM.1), a given formula φ can be reduced to the form θ' that is similar to (6.16.3), but does not contain quantifiers in the variables $x_1, \ldots, x_k$. In this case, in the computing block, it is possible to indicate a 2-point a on the Ω-line to which the formula θ' is ascribed. For integers p and q, the point a can be characterized by a formula of the form $O_{p,q}(x)$ that asserts that x is the pth R_2-predecessor of the qth S_2-successor of the central point of the block. Hence the following formula is the required one:

$$\Psi_0(x_1, \ldots, x_k) = (\exists z)\big[O_{p,q}(z) \;\&\; \Omega_1(z) \;\&\; \varkappa(z) = \langle x_1, \ldots, x_k\rangle\big]$$

The latter is easily formalized in the form of $\forall$-formulas as well as in the form of $\exists$-formulas. $\square$

Lemma 6.16.3 immediately implies the following lemma.

Lemma 6.16.4. *For any models $\mathfrak{N}$ and $\mathfrak{N}'$ of the theory $F_0(\mathcal{M}, m)$ from $\mathfrak{N} \subseteq \mathfrak{N}'$, it follows that $\mathbb{K}(\mathfrak{N}) \preceq \mathbb{K}(\mathfrak{N}')$.*

6.17. Computing Controller and Programming

Our purpose is to define the role of the computing controller of the theory $F_0(\mathcal{M}, m)$ and make the choice of a Turing machine $\mathcal{M}$ and an input parameter m. First, we choose the value of the integer-valued parameter m depending on the initial axiomatizable theory T. Second, we construct a Turing machine. We use the notation $T = [S]_\sigma$ if the collection of sentences $S \subseteq SL(\sigma)$ is a system of axioms of the theory T of the signature σ.

We now choose the value of the parameter m. In accordance with Sec. 0.1, there is a fixed Gödel numbering

$$\Phi_i, \quad i \in \mathbb{N} \tag{6.17.1}$$

of all sentences of the signature $\sigma_0 = \{\Gamma^2\}$; moreover, the initial axiomatizable theory T extending the theory GRE is defined by the recursively enumerable index m_0 with respect to this numbering, i.e.,

$$T = \big[\{GRE^*\} \cup \{\Phi_i \mid i \in W_{m_0}\}\big]_{\sigma_0} \tag{6.17.2}$$

where GRE^* denotes the conjunction of axioms of the theory GRE.
Of particular interest are formulas of the signature σ_0 that take the
form (6.16.3) and appear in the computing mechanism. We call them *basic
sentences*. The set of basic sentences is recursive in the Gödel numbering
(6.17.1). Every closed formula of the signature σ_0 is equivalent to some
basic sentence in the theory GRE. There is a natural Gödel numbering of
the family of all basic sentences

$$\Psi_i, \quad i \in \mathbb{N} \qquad (6.17.3)$$

defined by the ordering of the corresponding Ω-points on the display of a
standard block.

We pass to a new axiomatics of the theory T in terms of basic propositions. Since the theory T is axiomatizable, the set S of all basic sentences
which is provable in T is a recursively enumerable set, hence it can be represented in the form $S = \{\Psi_j \mid j \in W_m\}$ for some m. We replace the
sentences Φ_i, $i \in W_{m_0}$, in (6.17.2) by basic sentences from S. As a result,
we obtain a new axiomatics of the theory T in the form

$$T = \left[\{GRE^*\} \cup \{\Psi_j \mid j \in W_m\}\right]_{\sigma_0} \qquad (6.17.4)$$

The number m constructed from the initial axiomatizable theory T plays
the role of a parameter in Axiom 29° and is called the *canonical recursively enumerable index* of the theory T. We emphasize the following
important fact: the passage to canonical indices is effective, i.e., the following lemma holds. This lemma can be proved with the help of the
s-m-n-theorem [56].

Lemma 6.17.1. *There exists a general recursive function $f(x)$
such that it takes only positive values and $m = f(m_0)$ is the canonical
recursively enumerable index of the theory T, where T is an axiomatizable theory of the signature σ_0 and m_0 is the recursively enumerable
index of T.*

We construct the computing controller. To provide the required conditions, we include the description of the Turing machine in Axiom 83°.
The latter controls the truth of all axioms of the initial axiomatizable theory T in the model kernel. We base this on the axiomatics of the theory
T in the form (6.17.4), where the canonical recursively enumerable index
m is used. Since the axioms of GRE^* in (6.17.4) are already taken into
account by the axioms in (FRM.1), it suffices that $\mathcal{M}$ controls the truth
of the sequence of basic sentences in the model kernel

$$\Psi_i, \quad i \in W_m \qquad (6.17.5)$$

For the initial data we take the input parameter m and information on the display.

The external alphabet of the Turing machine $\mathcal{M}$ contains five symbols a_0, a_1, a_2, a_3, and a_4, which are identified with the symbols

$$0, \ 1, \ B, \ 0', \ 1' \tag{6.17.6}$$

and are coded by the signature predicates A_i, $i < 5$. The situation on the tape, the location of the head, the internal state of the Turing machine $\mathcal{M}$, which is determined by Axioms 29° and 84°–87°, and the properties of Z generating tape cells are shown in Fig. 6.17.1(a). The messages ω_i, $i < \omega$, on the left-hand side of the tape codes the truth of the basic sentences (6.17.3) in the model kernel $\mathfrak{M}$ as follows:

$$\omega_i = 1 \text{ means } \mathfrak{M} \vDash \Psi_i \tag{6.17.7}$$

$$\omega_i = 0 \text{ means } \mathfrak{M} \vDash {}^\neg \Psi_i \tag{6.17.8}$$

In fact, at the start-up of the Turing machine $\mathcal{M}$, there exists only the right-hand side of the tape, including the field INDEX and the adjacent cell with a blank. The rest of the tape, i.e., the field CODES with messages ω_i, $i < \omega$, does not yet exist, as is seen in Fig. 6.17.1. But this part of the tape is generated by Ω-points of the display, and it is realized faster than the head can move. Therefore, every A-line with messages (6.17.7)–(6.17.8) will be obtained when the machine turns to this cell.

Let $\mathcal{M}$ be a Turing machine with symbols (6.17.6) and let $\mathfrak{M}$ be a model of the theory GRE. By the $(\mathfrak{M}, m)$-*computing*, we mean the work of the Turing machine $\mathcal{M}$ that starts from the initial state presented in Fig. 6.17.1 (a), where information of the field CODES is defined via the model $\mathfrak{M}$ by the relations (6.17.7)–(6.17.8).

Lemma 6.17.2. *Let* $\mathfrak{M} \in \mathrm{Mod}(GRE)$, $m < \omega$, *and let* $\mathcal{M}$ *be a Turing machine with the external alphabet* (6.17.6). *The following assertions are equivalent*:

(a) *the* $(\mathfrak{M}, m)$-*computing by the Turing machine* $\mathcal{M}$ *does not halt*,

(b) *there exists a model* $\mathfrak{N}$ *of the theory* $F_0(\mathcal{M}, m)$ *with the model kernel* $\mathbb{K}(\mathfrak{N}) = \mathfrak{M}$.

PROOF. We assume that (b) holds. We consider the computing block $\mathcal{B}$ of the model $\mathfrak{N}$. In accordance with axioms of the theory $F_0(\mathcal{M}, m)$, the first quarter of the block $\mathcal{B}$ is $(\mathfrak{M}, m)$-computing by the Turing machine $\mathcal{M}$ which cannot halt in view of Axiom 83°.

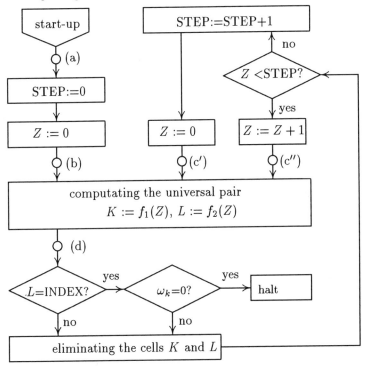

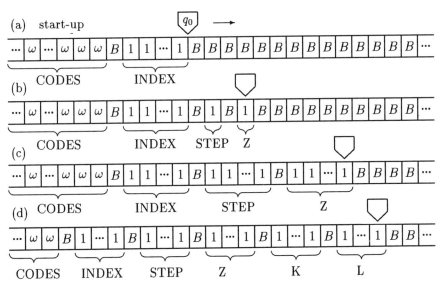

Fig. 6.17.1. Flow diagram of the Turing machine $\mathcal{M}$ and its intermediate states.

Let (a) hold. We construct a model $\mathfrak{N}$ of the theory $F_0(\mathcal{M}, m)$ with the model kernel $\mathbb{K}(\mathfrak{N}) = \mathfrak{M}$, performing sequentially the following stages.

STAGE 1. We construct a skeleton of the computing block $O(m, *)$, i.e., the structure of the computing block B_{25} at the abstraction level X, the first quarter remaining unfinished.

STAGE 2. In the second quarter of the block $O(m, *)$, we construct the structure of the tree of sequences, connecting it with the model kernel $\mathfrak{M}$ by the numbering predicate N on a subdivision line.

STAGE 3. We construct a commutative stratification of the tree of sequences in the domain Y by the fourth dimension. In accordance with the axioms in (CLC.12), we define the subdivision equivalence H. We take the number of H-classes equal to the cardinality $\|\mathfrak{M}\| + \omega$.

STAGE 4. We realize the subdivision of the model kernel and fill all the informational lines of the mechanism for processing the initial messages.

STAGE 5. Moving in the direction of the S_1-succession inductively, we fill all the informational lines of the mechanism for computing formulas. As a result, at the Ω-points of the display, we obtain the truth-value codes of all basic sentences in the model kernel $\mathfrak{M}$.

STAGE 6. On the line I, we introduce the initial messages, i.e., a blank, a series of $m + 1$ units, and an infinite series of blanks in the rest of the right-hand side of the tape. At the beginning of the Q-line directed to R, we set the start-up state Q_0 with the direction of motion of R.

STAGE 7. Moving through E_1-classes in the direction of the S_1-succession, we realize, in the first quarter of the block $O(m, *)$, the process of working the Turing machine $\mathcal{M}$, which is the $(\mathfrak{M}, m)$-computing and, in view of (a), does not halt. As a result, we construct the skeleton of the structure of the computing block.

STAGE 8. Take a countable number of copies of the computing block constructed at Stage 7. Putting their K-lines with shifts in the direction of the S_1-succession, we construct a skeleton of a zone.

STAGE 9. We thread the skeleton obtained of the zone by models of the successor theory through places indicated by the frame axioms and obtain the desired model $\mathfrak{N}$ of the theory $F_0(\mathcal{M}, m)$ with the model kernel $\mathbb{K}(\mathfrak{N}) = \mathfrak{M}$. $\square$

A Turing machine $\mathcal{M}$ with the external alphabet (6.17.6) is called *admissible* if for any model $\mathfrak{M}$ of the theory GRE and $m < \omega$ the following two conditions are equivalent:

(a) the $(\mathfrak{M}, m)$-computing by the Turing machine $\mathcal{M}$ does not halt,

(b) $\mathfrak{M} \vDash \Psi_i$ for all $i \in W_m$.

By Lemma 6.17.2, an admissible Turing machine can be taken as a computing controller and, thereby, provides the requirements of Catal 6.2(25). However, provision must be made for the work of the Turing machine to exert no negative effect on model-theoretic properties of the theory obtained. We consider an admissible Turing machine $\mathcal{M}$, a model $\mathfrak{M} \in \text{Mod}(GRE)$, and $m < \omega$ such that the $(\mathfrak{M}, m)$-computing by Turing machine $\mathcal{M}$ does not halt. By Lemma 6.17.2, there exists a model $\mathfrak{N}$ of the theory $F_0(\mathcal{M}, m)$ with the model kernel $\mathfrak{M}$. We consider the computing block $\mathcal{B}$ of the model $\mathfrak{N}$ and introduce the following quantities with values in $\mathbb{N} \cup \{\omega\}$:

$\|A(\nu, i)\|_{\mathfrak{M}, m}^{\mathcal{M}}$ is the number of 2-points $a \in \mathcal{B}$ of type $A(\nu, i)$

$\|Q(\nu, j)\|_{\mathfrak{M}, m}^{\mathcal{M}}$ is the number of 2-points $a \in \mathcal{B}$ of type $Q(\nu, j)$

$\|AQ(\nu, \nu', i, j)\|_{\mathfrak{M}, m}^{\mathcal{M}}$ is the number of 2-points $a \in \mathcal{B}$ of type $AQ(\nu, \nu', i, j)$

We consider an arbitrary $\nu, \nu' \in \{L, R\}$, $i < d$, $j < e$, $m < \omega$, $\mathfrak{M} \in \text{Mod}(GRE)$. A Turing machine $\mathcal{M}$ is called *normal* if it is admissible and, under the condition that the $(\mathfrak{M}, m)$-computing by Turing machine $\mathcal{M}$ does not halt, the values of the numbers $\|A(\nu, i)\|_{\mathfrak{M}, m}^{\mathcal{M}}$, $\|Q(\nu, j)\|_{\mathfrak{M}, m}^{\mathcal{M}}$, $\|AQ(\nu, \nu', i, j)\|_{\mathfrak{M}, m}^{\mathcal{M}}$ are independent of $\mathfrak{M}$ and m. For a normal Turing machine $\mathcal{M}$, we write $\|A(\nu, i)\|$, $\|Q(\nu, j)\|$, and $\|AQ(\nu, \nu', i, j)\|$ for brevity.

We give an exact formulation of the computing controller problem in the form of the problem on constructing an admissible normal Turing machine.

Lemma 6.17.3. *There exists an admissible normal Turing machine $\mathcal{M}$ with five external symbols* (6.17.6).

PROOF. The flow diagram is presented in Fig 6.17.1. In particular, intermediate states of the machine are shown in Fig. 6.17.1(a)–(d), as indicated by points. The names of the main domains and tape cells participating in the computation are also given here. As already mentioned, the domain CODES is virtual. The fields INDEX, STEP, Z, K, and L are numerical cells. The cell INDEX constantly keeps the canonical recursively enumerable index m of the theory T. As usual, the value of k in numerical cells is represented as a series of $k + 1$ units bounded from both sides by blanks.

Functions f_1 and f_2 appearing in the flow diagram are general recursive functions, and they must be taken so as to satisfy the relation

$$\{\langle k, l \rangle \mid k \in W_l\} = \{\langle f_1(t), f_2(t) \rangle \mid t \in \mathbb{N}\} \qquad (6.17.9)$$

i.e., the functions f_1 and f_2 present a set that is universal for the class of all recursively enumerable sets.

According to the flow diagram, the work of the Turing machine $\mathcal{M}$ is organized as two cycles embedded into each other. The external cycle is controlled by the integer-valued parameter STEP. At the next turn of the external cycle, the cell STEP takes the value t and the internal cycle starts. Note that the internal cycle is connected with the cell Z taking the values $0, 1, \ldots, t$. After that the cell STEP takes the value $t + 1$ and the internal cycle again starts for the values of Z from 0 to $t + 1$ and so on.

We consider the step of the external cycle STEP $= t$. By the flow diagram, at the following turn of the internal cycle $Z = s$, $0 \leqslant s \leqslant t$, the values $k = f_1(s)$ and $l = f_2(s)$ are computed and are written in two special cells K and L. The pair $\langle k, l \rangle$ obtained presents a single element of the universal set (6.17.9). But our purpose is to control the truth values of all the formulas (6.17.5); moreover, the truth values of basic sentences lie in the field CODES, and the value of the parameter m is contained in the cell INDEX. Therefore, among the pairs generated in such a way, only those pairs will be used for which $l = m$, i.e., the pairs relating to the set W_m. Thus, calculating by the flow diagram, we control the truth values of all the formulas in (6.17.5) in the model kernel, since the falsity of at least one of these formulas leads to a halt of the Turing machine $\mathcal{M}$.

It remains to show that, if we act in accordance with the flow diagram presented in Fig. 6.17.1, then we construct a Turing machine $\mathcal{M}$ with the desired properties.

Admissibility condition. The flow diagram itself stipulates the verification of the truth values of the formulas from (6.17.5) in the model kernel. It is necessary only to provide that the content of the fields CODES and INDEX remains unchanged because we must turn to them many times.

Normality condition. We consider a subprogram calculating elements of the universal set (6.17.9). By the flow diagram, every natural number s appears as the input data for the subprogram infinitely many times. Therefore, every instruction relating to the subprogram either is not executed or is executed infinitely many times. Regarding the remaining part of the block, it is not hard to organize elementary actions as the shuttle motion of the head, by using the special notation $0'$ and $1'$ for markers. Moreover, it is not hard to achieve that every used instruction is fulfilled infinitely many times. The sole exception is those instructions that, at the start-up of the Turing machine, form the initial values STEP $= 0$ and $Z = 0$. Unlike others, each such instruction holds once.

We note that if, checking the condition $INDEX = L$, we obtain a negative answer, then two different cases $INDEX < L$ and $INDEX > L$ appear. By Lemma 6.17.1, the value of m is nonzero, which provides an infinite number of repetitions for each of these two cases.

We have analyzed only the number of points of execution of commands. However, it is not hard to see that the values of $\|A(\nu, i)\|$ and $\|Q(\nu, j)\|$ are also independent of $\mathfrak{M}$ and m. Thus, acting in accordance with the flow diagram, we can construct an admissible normal Turing machine. □

From now on, we fix the admissible normal Turing machine $\mathcal{M}^*$ whose program is described in Axiom 83°. Our further purpose is to study the theory $F_0(\mathcal{M}^*, m)$, $m = f(m_0)$, where m_0 is the recursively enumerable index of the theory T.

6.18. Nonstandard Blocks and Normalization Axioms

To clarify the role of the normalization axioms, we consider an example of the proposition of Axiom 89°, which is denoted by Φ. In general, the formula Φ is not a consequence of axioms of the theory $F_0(\mathcal{M}^*, m)$. It is not hard to construct a nonstandard block of type D and, consequently, a model satisfying Axioms 1°–87° on which this formula is false. On the other hand, the assertion of the formula Φ is true with respect to all D-lines of the computing block, which follows from Axioms 4° and 5°. The normalization axioms listed below are assertions of the formula Φ which does not follow from axioms of the theory $F_0(\mathcal{M}^*, m)$, but expresses some special properties of informational lines and singular points of the computing block. Adding such assertions to axioms, we do not affect the work of the computing block, but all singularities of the latter will be transferred to nonstandard blocks, and we see that nonstandard blocks are embedded into standard blocks. Together with the normalization axioms, we give several axioms necessary for the model completeness.

(NOR.16) Normalization axioms for the computing mechanism

88°. In every E_3-class of dimension 4 located on the V-line, there is an element t satisfying $V_i(t)$ for any $i \in \{0, 1\}$.

89°. In every E_3-class located on the D-line, there are D_i-elements for any $i \in \{0, 1, 2\}$.

90°. In every E_2-class located on the Δ-line, there are Δ_i-elements for any $i \in \{0, 1, 2\}$.

91°. $\Sigma 0(x) \leftrightarrow \Sigma(x) \,\&\, (\forall y)\big[E_2(x, y) \,\&\, X_3(y) \rightarrow \Sigma_0(y)\big]$.

92°. $\Sigma 1(x) \leftrightarrow \Sigma(x) \,\&\, (\forall y)\big[E_2(x, y) \,\&\, X_3(y) \rightarrow \Sigma_1(y)\big]$.

93°. In every E_2-class on the Ω-line marked by the marker Ω^*, there is no Ω_1-element.

94°. $\Omega 0(x) \leftrightarrow \Omega(x) \,\&\, (\forall y)\big[E_2(x, y) \,\&\, X_3(y) \rightarrow \Omega_0(y)\big]$.

95°. $\Omega 1(x) \leftrightarrow \Omega(x) \,\&\, (\forall y)\big[E_2(x, y) \,\&\, X_3(y) \rightarrow \Omega_1(y)\big]$.

96°. In every E_1-class, there is a unique 2-point satisfying $\Sigma_1(x) \,\&\, \Omega^*(x)$.

97°. In every E_1-class, there is a unique 2-point satisfying $\Sigma_1(x) \,\&\, \Delta^*(x)$.

(NOR.17) Normalization axioms for the computing controller

98°. Let $\|A(\nu, i)\| = n$, $n < \omega$. In every E_1-class, there are exactly n copies of 2-points of type $A(\nu, i)$.

99°. Let $\|Q(\nu, j)\| = n$, $n < \omega$. In every E_1-class, there are exactly n copies of 2-points of type $Q(\nu, j)$.

100°. Let $\|AQ(\nu, \nu', i, j)\| = n$, $n < \omega$. In every E_1-class, there are exactly n copies of 2-points of type $AQ(\nu, \nu', i, j)$.

We have described the basic part of the axiomatics of the theory (6.1.2). It remains to describe the rigidity mechanism.

6.19. Axiomatics. Rigidity Mechanism

In view of the internal automorphisms, any model $\mathfrak{N}$ satisfying Axioms $1°$–$100°$ has nontrivial automorphisms even if its kernel is a rigid model. By indices of branching in H-classes, it is possible to code arbitrary binary predicates on the model kernel, which shows the maximality of the spectrum function for any completion of the theory under consideration. To avoid these undesirable properties, we introduce and describe below the rigidity mechanism. It consists of two parts. The first part operates as a rigidity function over the kernel suppressing internal automorphisms of the superstructure. The second part operates as a coordination function for different fragments of the superstructure in a similar way as Γ-connections for the construction IG (cf. Chapter 5).

We pass to axioms.

(RIG.18) The rigidity of the skeleton mechanism

101°. Every quasisuccession passing through X_1-elements along the S_1-succession is a model of the theory QSR in which the predicate Λ distinguishes a unique $\sim$-class.

102°. Every quasisuccession passing through X_2-elements in E_1-classes is a model of the theory QSR in which the predicate Λ distinguishes $\sim$-classes satisfying the formula $G(x) \vee Z(x)$.

103°. Every quasisuccession passing through X_3-elements in E_2-classes satisfying $KVD(x)$ is a model of the theory QSR in which the predicate Λ distinguishes one $\sim$-class in each EN-class inside the given E_2-class.

104°. Let a and t be elements such that $R_2(a,t)$, $OZ(t)$, $X_3(a)$, $\Lambda(a)$. Then the quasisuccession passing through X_4-elements in the E_3-class $[a]_{E_3}$ is a model of the theory QSR in which the predicate Λ distinguishes one $\sim$-class in each H-class inside a given E_1-class.

105°. In the remaining cases, except the cases indicated in Axioms 101°–104°, the crossing of the superstructure is made by the quasisuccession QS' without the rigidity mechanism; moreover, the predicates Λ, SP, and SN are assumed to be identically false.

REMARK 6.19.1. The scheme of axioms BR.41° of the theory QSR is not used in Axioms 101°–104°. Its requirements are provided by neighbors of the quasisuccession in the superstructure.

(RIG.19) Local isomorphism of transferring rigidity

106°. The predicates LR_3, LS_3, and LG are defined on the set X and are false outside this set.

107°. $LS_3(x, y) \rightarrow S_3(x, y) \& V(x) \& V(y)$.

108°. The predicate LS_3 defines an isomorphism (in the signature of the theory QS') between and two quasisuccessions passing through X_3-elements in S_2-neighboring E_2-classes on the V-line.

$109°$. $LR_3(x, y) \rightarrow R_3(x, y) \& \neg V(x)$.

$110°$. The predicate LR_3 defines an isomorphism between two quasisuccessions passing through X_3-elements in R_2-adjacent E_2-classes that are not separated by branchings.

$111°$. $LG(x, y) \rightarrow G(x) \& (\exists u)(\exists v)\big[S_2(x, u) \& R_2(u, v) \& R_2(v, y)\big]$.

$112°$. The predicate LG defines an isomorphism between two quasisuccessions passing through X_3-elements in neighboring E_2-classes located on a common G-line, including the point KVD placed on the extension of this line.

$113°$. The predicate $L\Pi_4$ is defined on the set X_4 and is false outside this set for every $\Pi \in \{R, S, P\}$.

$114°$. $L\Pi_4(x, y) \rightarrow \Pi_4(x, y)$, for every $\Pi \in \{R, S, P\}$.

$115°$. If $\Pi_3(x, y)$, $D4(x)$, and $D4(y)$ are satisfied, then $L\Pi_4$ defines an isomorphism between any quasisuccessions passing through X_4-elements in E_3-classes $[x]_{E_3}$ and $[y]_{E_3}$ for $\Pi \in \{R, S, P\}$.

A quantifier-free formula φ in n free variables $x_1, \ldots, x_n$ is called an $LRSP_4$-*formula* if φ has the form $\varphi_1 \& \ldots \& \varphi_{n-1}$, where every conjunctive term φ_i is an atomic formula of the form $L\Pi_4(x_i, x_{i+1})$ or $L\Pi_4(x_{i+1}, x_i)$ for some $\Pi \in \{R, S, P\}$.

$116°$. For any $3 \leqslant n \leqslant 8$ and $LRSP_4$-formulas $\varphi(x_1, \ldots, x_n)$ in n variables, the relation $\varphi(x_1, \ldots, x_n) \& E_3(x_1, x_n) \rightarrow x_1 = x_n$ holds.

The following axioms coordinate different fragments of the superstructure.

(RIG.20) Coordination of indices of numbering branchings

$117°$. The predicate CN is defined on the set $X_3 \cup X_4$ and is well defined on E_3-classes.

$118°$. $CN(x, y) \rightarrow KVD(x) \& KVD(y)$.

$119°$. The predicate CN is symmetric and transitive.

$120°$. The predicate CN defines a P_3-isomorphism between 3-points of any two EN-classes.

121°. $CN(x,y)$ & $X_3(x)$ & $X_3(y)$ $\rightarrow$ $\big[\Lambda(x) \leftrightarrow \Lambda(y)\big]$.

(RIG.21) Coordination of the branching indices

122°. The predicate CV is defined on the set $X_3 \cup X_4$ and is well defined on E_3-classes.

123°. $CV(x,y)$ $\rightarrow$ $E_2(x,y)$ & $V(x)$ & $V(y)$.

124°. The predicate CV is symmetric and transitive.

125°. The predicate CV defines a P_3-isomorphism between points of any two EV-classes in a common E_2-class on the line V.

126°. $CV(x,y)$ & $CV(u,v)$ $\rightarrow$ $\big[S_3(x,u) \leftrightarrow S_3(y,v)\big]$.

127°. $N(u,x)$ & $N(u,y)$ & $CN(x,y)$ $\rightarrow$ $CV(x,y)$.

(RIG.22) Unification of subdivision lines

128°. The predicate CK is defined on the set $X_2 \cup X_3 \cup X_4$ and is well defined on E_3-classes.

129°. $CK(x,y)$ $\rightarrow$ $K(x)$ & $K(y)$.

130°. The predicate CK is symmetric and transitive.

131°. The predicate CK establishes an R_2-isomorphism between the K-lines of any two E_1-classes.

132°. $CK(x,y)$ $\rightarrow$ $\big[B(x) \leftrightarrow B(y)\big]$.

(RIG.23) Coordination of classes of subdivision equivalences

133°. The predicate CH is defined on the set X_4 and is well defined on E_4-classes.

134°. $CH(x,y)$ $\rightarrow$ $CK(x,y)$ & $K(x)$ & $K(y)$.

135°. The predicate CH is symmetric and transitive.

136°. The predicate CH establishes a T_4-isomorphism between any two equivalence classes $\eta(x,y) = E_3(x,y)$ & $H(x,y)$.

137°. $K(x)$ & $R_4(x,y)$ $\rightarrow$ $CH(x,y)$.

138°. $K(x)$ & $P_4(x,y)$ $\rightarrow$ $CH(x,y)$.

(RIG.24) Coordination of the length of the fourth dimension

$139°$. The predicate CD is defined on the set X_4 and is well defined on E_4-classes.

$140°$. $CD(x, y) \to E_2(x, y)$ & $\dim(x) = \dim(y) = 4$.

$141°$. The predicate CD is symmetric and transitive.

$142°$. The predicate CD establishes a T_4-isomorphism between quasisuccessions passing through X_4-elements in any two E_3-classes placed within an E_2-class.

$143°$. $R_4(x, y)$ & $R_4(u, v) \to \left[CD(x, u) \leftrightarrow CD(y, v) \right]$.

$144°$. $S_4(x, y)$ & $S_4(u, v) \to \left[CD(x, u) \leftrightarrow CD(y, v) \right]$.

$145°$. $P_4(x, y) \to CD(x, y)$.

$146°$. $CH(x, y)$ & $E_2(x, y) \to CD(x, y)$.

We have described the axioms of the finitely axiomatizable theory (6.1.2) and the rigidity mechanism.

6.20. Description of Models and Isomorphisms

In this section, we characterize the structure of models of the theory $F = F(\mathcal{M}_0, f(m))$ and isomorphic embeddings between these models.

In fact, a general description of models of the theory F was given in Sec. 6.8 and remains in force. The existence of a model of the theory F is proved in Lemma 6.17.2. To obtain an arbitrary model of the theory F, we must, in addition to the construction carried out in the lemma, take an arbitrary number of zones and include in the zones, together with standard blocks, any number of those nonstandard blocks that are admissible from the standpoint of the normalization axioms. It is possible to vary blocks with respect to their third and fourth dimensions. Acting in such a way, we can construct an arbitrary model of the theory F. In view of the rigidity mechanism, the arbitrariness of the choice of parts of the superstructure is limited because we need to provide the validity of the requirements of the axioms in Sec. 6.19.

We now describe an isomorphic embedding $f\colon \mathfrak{N}_0 \to \mathfrak{N}_1$ of models $\mathfrak{N}_0$ and $\mathfrak{N}_1$ of the theory F. We sketch a construction of such an embedding. However, the choice of an intermediate correspondence between fragments of the models $\mathfrak{N}_0$ and $\mathfrak{N}_1$ is arbitrary.

We pass to constructing the isomorphic embedding $f\colon \mathfrak{N}_0 \to \mathfrak{N}_1$, where $\mathfrak{N}_0$ and $\mathfrak{N}_1$ are models of the theory F. For the initial data we take any elementary embedding of the kernels $\mu\colon \mathbb{K}(\mathfrak{N}_0) \to \mathbb{K}(\mathfrak{N}_1)$. The construction of the required embedding f is divided into several steps.

STEP 1. We map the zones of the model $\mathfrak{N}_0$ into the zones of the model $\mathfrak{N}_1$.

STEP 2. We spread the above mapping of zones to a mapping of E_1-classes preserving the S_1-succession.

STEP 3. We transform blocks of the model $\mathfrak{N}_0$ into blocks of the model $\mathfrak{N}_1$ in the corresponding zone. The blocks have the same description at the abstraction level E and their singular points, if they exist, must be located in the corresponding E_1-classes in accordance with item 2.

STEP 4. We spread the above mapping to an isomorphism between the geometric R_2–S_2-structures of blocks.

STEP 5. We extend the above mapping to an isomorphism between the geometric R_3–S_3–P_3-structure of blocks with preservation of values of messages on informational lines. In the case of standard blocks, to choose such a restriction we must take into account coordination between the numbering predicates in the models $\mathfrak{N}_0$ and $\mathfrak{N}_1$.

STEP 6. We establish a correspondence between H-classes of K-lines of the models $\mathfrak{N}_0$ and $\mathfrak{N}_1$, taking into account the coordination between H-classes of the models $\mathfrak{N}_0$ and $\mathfrak{N}_1$ with the model kernel.

STEP 7. We continue the above mapping to an isomorphic embedding of the geometric R_4–S_4–P_4–T_4-structure preserving messages on the V-lines and D-lines.

STEP 8. We establish a $\{\lhd, \sim\}$-isomorphism between the crossings by quasisuccessions at the corresponding places of the superstructures of the models $\mathfrak{N}_0$ and $\mathfrak{N}_1$. In view of the properties of the basic theories mentioned in Lemma 4.7.1, we restrict the external isomorphic embedding obtained to a complete isomorphic embedding of the corresponding models of the quasisuccessor theory.

At steps 5–8, we specify relations taking into account the coordination predicates and local isomorphisms acting in the rigidity mechanism. As a result, we obtain the desired isomorphic embedding $f\colon \mathfrak{N}_0 \to \mathfrak{N}_1$, which is kernel-elementary by definition. The description of the elementary transformation GF is complete.

6.21. Model Completeness and Quasiexactness

We show that the theory F is model-complete and the interpretation I of the theory T in the theory F is quasiexact. We first study the structure of models without the rigidity mechanism. To obtain such a simplified variant, we eliminate Axioms $101°$–$146°$ from the list of axioms and eliminate the corresponding rigidity predicates from the signature. For the simplified variant an assertion similar to Theorem 0.6.1 is valid but with the list of properties less than in the full variant (cf. Exercise 6). We describe a primitive enveloping for the simplified variant.

Let $\mathfrak{N}$ be a model of the theory F with the kernel $\mathfrak{M}$ and let $Y \subseteq |\mathfrak{N}|$. We can assume that $Y \subseteq |\mathfrak{N}| \setminus |\mathfrak{M}|$ since elements of $|\mathfrak{M}|$ do not affect the process of constructing envelopings. We describe how to construct a primitive enveloping over Y. Each element $a \in Y$ is located in some model $\mathfrak{A}$ of the theory QS' passing through one of the sets X_n, $n \in \{1, 2, 3, 4\}$, corresponding to the nth dimension of the superstructure. An element a is characterized by three coordinates in $\mathfrak{A}$. Therefore, a generates three n-points a_1, a_2, and a_3 in the superstructure. Moreover, a is located in a_1.

Thus, the set Y defines the family W of n-points for different $n \in \{1, 2, 3, 4\}$ in the superstructure. The construction of a primitive enveloping over Y consists in the construction of the closure of W by adding points of different depth to W.

STEP 1. If W is empty, then we add one 1-point to W.

STEP 2. We add extensions of points of W as follows: if a is an m-point, b is an n-point, $a \subset b$, $m > n$, and a belongs to W, then we add b to W.

STEP 3. If a 1-point a belongs to W, then we add all S_1-neighbors of a to W.

STEP 4. If a 1-point a belongs to W, then we add the 2-point 0 contained in a to W.

STEP 5. We obtain the closure of a 2-point to a block as follows: if a 2-point a belongs to W, then to W we add all R_2-neighbors of the point a, all S_2-neighbors of the added points, all R_2-neighbors of the added points, and so on. This action is called the *addition of $R_2 S_2$-neighbors*.

STEP 6. If a 3-point a belongs to W, then we add all $P_3S_3R_3$-neighbors of a passing no branching line (within a region).

STEP 7. If a 3-point a of W satisfies $V(a)$, then we add the R_3-successor of the point a to W.

STEP 8. We repeat Steps 6 and 7 up to stabilization (ω steps can be required) and pass to Step 9.

STEP 9. If a 3-point a of W satisfies $B(a)$ and none of its R_3-successors (through branching) belongs to W, then we add a 3-point b satisfying $R_3(b, a)$ to W.

STEP 10. We repeat Steps 6–9 up to stabilization and pass to Step 11.

STEP 11. If a 3-point a of W satisfies $K(a)\&B(a)$ and, in some numerating branching of a there are no 3-points of W, then we choose one such point and add it to W.

STEP 12. We repeat Steps 6–11 up to stabilization and pass to Step 13.

STEP 13. If a 4-point a belongs to W, then to W we add all 4-points that are $T_4P_4R_4S_4$-neighbors of a and are located inside of 3-points of W.

STEP 14. If there are two 3-points a and b in W such that a and b are located in the same E_1-class and satisfy $K(a)$ and $K(b)$ and, moreover, some H-class A contains a 4-point of W in a and b does not contain such points, then to W we add a 4-point a of A that is located in b.

STEP 15. We repeat Steps 13 and 14 up to stabilization and pass to Step 16.

STEP 16. If a 2-point a, $\dim(a) > 2$, of W contains no 3-points of W, then we add a 3-point b located inside a to W.

STEP 17. We repeat Steps 6–16 up to stabilization and pass to Step 18.

STEP 18. If a 3-point a, $\dim(a) = 4$, of W contains no 4-points of W, then we add a 4-point b located in a to W.

STEP 19. We repeat Steps 6–18 up to stabilization.

We have obtained the closure of the set W. The process is complete.

Some (determinate) steps are uniquely defined, whereas at other (nondeterminate) steps, we must make a certain choice. The construction is subject to the determination principle. This means that after a nondeterminate choice of an n-point the previous steps are repeated, which leads to a unique choice of added points as long as possible. Then a nondeterminate choice of a point must be made and determinate steps are again repeated and so on.

The procedure described is not an operator because of the nondeterminate steps of construction. However, analyzing the construction and using Axioms $1°-100°$, we can prove the following lemma.

Lemma 6.21.1. *If sets W_1 and W_2 have been constructed by the above procedure from the same set Y in a model $\mathfrak{N}$, then the following conditions hold:*

(a) *the superstructure of $\mathfrak{N}$ bounded on W_1 is isomorphic to the superstructure of $\mathfrak{N}$ bounded on W_2 (at the abstraction level $F4$),*

(b) $W_1 \subseteq W_2 \to W_1 = W_2$.

Thus, the procedure $Y \mapsto [Y]$ represents the closure quasioperator, where Y is a set of elements of the model $\mathfrak{N}$ and $[Y]$ is the obtained set of points of different depth in its superstructure. We also can define the quasioperator $W \mapsto [W]$, where W is the initial set of points of different depth and $[W]$ is the obtained set of points.

By construction, $[[W]] = [W]$ and $[W'] \subseteq [W]$ for $W' \subseteq [W]$ for any sets W and W'. A set W of points of the model $\mathfrak{N}$ is called *closed* if $W = [W]$. In view of Step 1, the closed set of points is nonempty.

Lemma 6.21.2. *Let $\mathfrak{N}$ be a model with the kernel $\mathfrak{M}$, let $\mathfrak{N}$ satisfy Axioms $1°-100°$, and let W be a set of points of the model $\mathfrak{N}$. On W, there exists a submodel $\mathfrak{N}' \subseteq \mathfrak{N}$ with the same kernel $\mathfrak{M}$ if and only if the set W is closed.*

PROOF. If a model $\mathfrak{N}'$ exists, then the corresponding set of points of $\mathfrak{N}$ is closed. Conversely, let W be a closed set of points of the model $\mathfrak{N}$. Then, at each depth, the set W is closed under the successor and predecessor relations. In $\mathfrak{N}$, we consider a set on which the axioms define a model $\mathfrak{A}$ of the theory QS'. We consider the set $\mathfrak{A}'$ formed by those elements of $\mathfrak{A}$ for which the first three coordinates are in W. If the model $\mathfrak{N}'$ exists, then $\mathfrak{A}'$ is nonempty; then $\mathfrak{A}'$ is a model of QS' and $\mathfrak{A}' \subseteq \mathfrak{A}$. Using this fact, it is not hard to construct the required model $\mathfrak{N}'$.   □

We consider the theory T defined by the complete system of Axioms $1°-146°$. The construction is the same, but some additional aspects appear owing to the rigidity predicates. Sometimes (at Steps 1, 11, and 14) the nondeterminate choice becomes determinate because of the marks of the rigidity of quasisuccession. However, the rigidity mechanism cannot transform Steps 16 and 18 to determinate ones. The difference is also in the fact that the rigidity predicates establish a correspondence between numbering branchings in the same way as Γ-connections act in IG. This remark is relative to the predicates CN, CV, CK, CH, and CD. Therefore, the addition of a point to some block can bring about the appearance of new points in other blocks and even in other zones.

For a set $Y \subseteq |\mathfrak{N}|$, we denote by $R[Y]$ the result of the procedure with the rigidity predicates. For a set W of points of the model $\mathfrak{N}$, we denote by $R[W]$ the obtained result of the above procedure. We have $R[R[W]] = R[W]$ for any set of points W. A set W of points of the model $\mathfrak{N}$ is called R-closed if $W = R[W]$. For a given $R[W]$ it is possible to construct $[W]$ so that $[W] \subseteq R[W]$. Qwing to the action of the rigidity predicates, nonstandard fragments in the complete theory F have the form of connected complexes of nonstandard fragments of the simpler case of Axioms $1°-100°$ (as IG).

We note that, in general, the procedure $W \mapsto R[W]$ is not an operator because of the nondeterminate steps. But the procedure is a quasioperator. For the quasioperator $R[W]$, assertions similar to Lemmas 6.21.1 and 6.21.2 hold.

Lemma 6.21.3. *If sets W_1 and W_2 are constructed from the same set Y in the model $\mathfrak{N}$ of the theory F in accordance with the above procedure (with the rigidity mechanism), then the following assertions hold:*

(a) *the superstructure of $\mathfrak{N}$ bounded on W_1 is isomorphic to the superstructure of $\mathfrak{N}$ bounded on W_2 (at the abstraction level $F4$),*

(b) *$W_1 \subseteq W_2 \to W_1 = W_2$.*

Moreover, if W is a set of points of the model $\mathfrak{N}$, then, on W, there exists a submodel $\mathfrak{N}' \subseteq \mathfrak{N}$ with the same model kernel if and only if the set W is R-closed.

We characterize standard and nonstandard elements of models.

Lemma 6.21.4. *Let $\mathfrak{N}$ be a model of the theory F with the model kernel $\mathfrak{M}$. Then the following assertions hold:*

(a) *all the steps of the procedure of the construction of the set $R[\emptyset]$ in $\mathfrak{N}$ are determinate,*

(b) *for any point a of $\mathfrak{N}$ such that $a \notin R[\emptyset]$ there exists an automorphism μ of the model $\mathfrak{N}$ which is identical on $\mathfrak{M}$ and $\mu(a) \neq a$,*

(c) *if $\mathfrak{N}'$ is a submodel of $\mathfrak{N}$ defined by the set $R[\emptyset]$, then $\mathrm{Stan}\,(\mathfrak{N}) = |\mathfrak{N}'|$ and $\mathrm{Nost}\,(\mathfrak{N}) = |\mathfrak{N}| \setminus |\mathfrak{N}'|$.*

We characterize primitive and perfect envelopings.

Lemma 6.21.5. *Let $\mathfrak{N}$ be a model of the theory F. A submodel $\mathfrak{N}' \preccurlyeq^\circ \mathfrak{N}$ is primitive over the set $X \subseteq \mathfrak{N}'$ if and only if the closure $R[X]$ in $\mathfrak{N}'$ contains all points of $\mathfrak{N}'$.*

PROOF. Let $\mathfrak{N}''$ be a model of the theory F and let $f\colon U(\mathfrak{N}') \cup X \to \mathfrak{N}''$ be a mapping subject to the conditions in Sec. 5.5. Since f is elementary on the kernel $\mathbb{K}(\mathfrak{N}')$, the computing mechanism identically works on the corresponding sequences of the model kernel, and the first quarters of the computing blocks of the models $\mathfrak{N}'$ and $\mathfrak{N}''$ are isomorphic. The position of the set X in the superstructure of the model $\mathfrak{N}'$ and the connections between X and the kernel are described by $\exists$-formulas that are preserved by the mapping f. Following the scheme of the construction of isomorphisms in Sec. 6.20, we can extend f to a kernel-elementary embedding of $\mathfrak{N}'$ into $\mathfrak{N}''$, which means that $\mathfrak{N}'$ is primitive over X.   □

Lemma 6.21.6. *A model $\mathfrak{N}$ of the theory F of cardinality α is perfect over a set $X \subseteq |\mathfrak{N}|$ if and only if the following conditions hold:*

(a) *the model $\mathfrak{N}$ contains α different zones that do not intersect the set X,*

(b) *in every zone, the model $\mathfrak{N}$ contains α different nonstandard blocks of every type at the abstraction level $F4$ that do not intersect the set X,*

(c) *in every E_2-class of dimension $\dim \geqslant 3$, there are α 3-layers that do not intersect X,*

(d) *in every branching of blocks with the $V\!-\!B$-line, there are α 3-layers which, together with their extensions to the left and to the right through branching, do not intersect X,*

(e) *in every numbering branching in blocks with the KVB-point, there are α 3-layers which, together with their extensions to the left and to the right through branching, do not intersect X,*

(f) *in every E_3-class of dimension 4, there are α 4-layers that do not intersect X,*

(g) *in every E_3-class on K and in every H-class, there are α 4-layers that do not intersect X.*

PROOF. The necessity of conditions (a)–(g) is obvious. The sufficiency is proved by constructing kernel-elementary embeddings in accordance with the scheme in Sec. 6.20. □

Lemma 6.21.7. *The interpretation I of the theory T in the constructed theory F is prequasiexact.*

PROOF. The validity of condition (a) from the definition of a prequasi-exact interpretation (cf. Sec. 5.5) is obvious. Conditions (b) and (c) are verified by constructing kernel-elementary embeddings in accordance with the scheme in Sec. 6.20. Condition (d) is proved by constructing a countable chain of elementary extensions such that, at each step, the diagram realizes new nonstandard configurations of different forms. The local consistency of such diagram follows from the description of the theory and the normalization axiom. Condition (e) is verified by a standard construction of a chain of models. □

Lemmas 6.16.4 and 6.21.7 imply the following lemma.

Lemma 6.21.8. *The theory F is model-complete.*

Lemma 6.21.9. *The interpretation I is quasiexact.*

PROOF. We verify the conditions from the definition of a quasiexact interpretation. We give the proof of the most complex conditions.

(Q.1) The assertion is proved in Lemma 6.21.7.

(Q.2)–(Q.4) The assertions are guaranteed by the properties of the closure quasioperator considered above.

(Q.5), (Q.6) It is easy to see that the type $p(x)$ of the theory F is a type of a nonstandard element provided that it describes neighborhoods of the element x as some nonstandard configuration. On the other hand, the standard type $p(x)$ is a type of either an element of the kernel or an element located on some standard configuration of one of the types. This allows us to establish the required properties of standard and nonstandard types.

(Q.7) The assertion follows from the fact that in an α^+-homogeneous model of the theory F of uncountable cardinality α, there are only a finite number of nonstandard configurations for any type.

(Q.8) The assertion is proved by constructing models and numberings with the help of the description of primitive envelopings.

(Q.9) The assertion is established by constructing models and numberings.

(Q.10) To obtain $\mathfrak{N}'$ it suffices to eliminate one of the nonstandard configurations from $\mathfrak{N}$.

(Q.11) All types of nonstandard configurations possess nontrivial automorphisms (cf. Lemma 6.21.4). Therefore, it is easy to obtain the required automorphism of a nonprimitive model by an isomorphic permutation inside a nonstandard configuration of this model.

(Q.12) The assertion follows from Lemma 6.21.4(a).

(Q.13) The required relation between isomorphism types of models of the theory F and their kernels is described in Lemma 6.21.10 below.

All the conditions from the definition of a quasiexact interpretation are verified. Thus, the interpretation I constructed is quasiexact. $\square$

The interpretation I is $\exists$-representable (cf. Table 5.8.1) in view of the description of LG. Indeed, the domain of the interpretation is distinguished by the quantifier-free formula $U(x)$. With an atomic formula $P_k(x_1, x_2, \ldots, x_{m_k})$ of the theory T, we associate the $\exists$-formula of the theory F indicated in Lemma 6.13.6. Thereby, the elementary transformation GF is constructed and Theorem 5.8.1 is proved. Thus, the proof of the main theorem (cf. Theorem 0.6.1) is complete.

Lemma 6.21.10. *Let the theory F and the interpretation I be obtained by GF from the axiomatizable theory T. Let $\mathfrak{M}$ be a model of the theory T_0 of cardinality $\|\mathfrak{M}\| \leqslant \omega_\alpha$. Then the number of isomorphism types of the model $\mathfrak{N}$ of the theory T_1 of cardinality $\|\mathfrak{N}\| = \omega_\alpha$ with the kernel $\mathbb{K}(\mathfrak{N}) \cong \mathfrak{M}$ is equal to $\varepsilon = \min(2^{\omega_\alpha}, J_2(|\omega + \alpha|))$. In other words, the spectral index of the interpretation I is equal to 2.*

PROOF. We compute the number of blocks that are intersected by a subdivision line and the number of the remaining blocks.

Let $\mathfrak{N}$ be a model satisfying Axioms $1°$–$146°$ of cardinality ω_α with the kernel $\mathbb{K}(\mathfrak{N}) \cong \mathfrak{M}$. In $\mathfrak{N}$, we consider the collection of blocks of types K, KM, and O that are intersected by the K-line of an E_1-class. In view of the axioms in (RIG.21)–(RIG.24), the collection of blocks is uniquely characterized up to an isomorphism between their geometric structure by the collection of ordinals

$$\delta_0,\ \delta_1,\ \delta_2[\beta', \beta''],\ \delta_3[\beta', \beta''],\ 0 \leqslant \beta', \beta'' \leqslant \omega + \alpha$$

$$0 \leqslant \delta_i \leqslant \omega + \alpha,\ i = 0, 1,\ 0 \leqslant \delta_j[\beta', \beta''] \leqslant \omega + \alpha,\ j = 2, 3$$

where

δ_0 characterizes the index of numbering branchings,

δ_1 characterizes the index of H-classes in a block of type O,

$\delta_2[\beta', \beta'']$ characterizes the number of blocks of type R with the index of the third dimension and the index of H-classes defined by the parameters β' and β'' respectively.

$\delta_3[\beta', \beta'']$ characterizes the number of blocks of type KM with the index of the third dimension on the right-hand side and the index of H-classes defined by the parameters β' and β'' respectively.

In view of the axioms in (RIG.24), the same values of the parameters characterize subdivision lines in all other E_1-classes of the model $\mathfrak{N}$. We introduce the addition parameter δ_4, which characterizes the number of S_1-chains of the model $\mathfrak{N}$. Computing the number of variants, we obtain exactly $J_1(|\omega + \alpha|)$ types of isomorphisms of the models $\mathfrak{N}$ of the theory F, which is required by the lemma, in which only blocks of types O, K, and KM are presented.

We consider blocks of the remaining types. In view of the axioms in (RIG.21)–(RIG.23), the isomorphism type of the geometric structure of such a block is characterized by at most three parameters. Therefore, there are at most $|\omega + \alpha|$ types of isomorphisms for the indicated blocks. Distributing these blocks in the superstructure in various ways, we obtain $J_1(|\omega+\alpha|)$ different types of isomorphisms for zones. As a result, we obtain the same number of types of isomorphisms of the model $\mathfrak{N}$ of cardinality ω_α with the kernel $\mathbb{K}(\mathfrak{N}) \cong \mathfrak{M}$ as in the formulation of the lemma. $\square$

An exact relation between spectrum functions of the theories indicated in the main theorem is of interest.

Lemma 6.21.11. *Let T be an axiomatizable theory without finite models and let F be a finitely axiomatizable theory obtained from T in accordance with Theorem 0.6.1. Then the spectrum functions of the corresponding completions T^* and F^* of the theories T and F are connected by the relation*

$$IM(\omega_\alpha, F^*) = J_2(|\omega + \alpha|) \circ \sum_{\beta \leqslant \gamma \leqslant \alpha} |\omega + \gamma| \circ IM(\omega_\beta, T^*)$$

PROOF. Considering the elementary transformation in Table 5.8.1, we see that the transformations AI, FG, and GL do not change the spectrum function, whereas the elementary transformations IG and GF transform the spectrum function in accordance with the relation (5.7.4) and Lemma 6.21.10, which leads to the required formula. $\square$

REMARK 6.21.1. Using the Shelah result that spectrum functions are monotone, it is possible to simplify the formula in Lemma 6.21.2 as follows: $IM(\omega_\alpha, F^*) = IM(\omega_\alpha, T^*) + J_2(|\omega + \alpha|)$. The same formula can be obtained under the assumption that the continuum-hypothesis holds.

Exercises

1. Write a complete program of a Turing machine under the condition that there is a subprogram of the universal block with some known form of access.

2. Write the following basic propositions: $\Psi_0, \Psi_1, \Psi_2, \Psi_3, \Psi_{11}, \Psi_{40}, \Psi_{100}, \Psi_{909}$.

3. Give lower and upper estimates for the distance from a point on the display Z corresponding to the sentence Ψ_{11} along the S-succession to the origin O. The same question for the proposition Ψ_n.

4. Explain the reasons for the fact that, in the universal construction, the Q-line cannot be as simple (without nonhomogeneity) as in the intermediate construction.

   HINT: in the universal construction there is no axiom concerning the uniqueness of a Q-point in each P-class.

5. What are the consequences of the elimination of the predicates L and R?

6. Show that, in the universal construction, we can do away with the predicates L and R, but the realized functions must be realized by a special programming of the Turing machine (double number of symbols of states of the Turing machine).

7. Let F^* be a subtheory of F defined by Axioms $1°-134°$ and omitting the predicates CD, CH, and CK from the signature. Let the assumptions of Lemma 6.21.7 hold. Show that for F^* a similar assertion holds with the estimate $\varepsilon = \min\left(2^{\omega\alpha}, J_3(|\omega + \alpha|)\right)$.

Catalogue Catal 6.1

Elementary Chains

$\theta_n,\ 1 \leqslant n \leqslant 72$

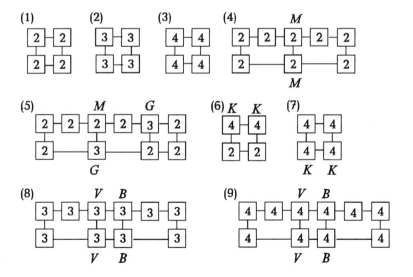

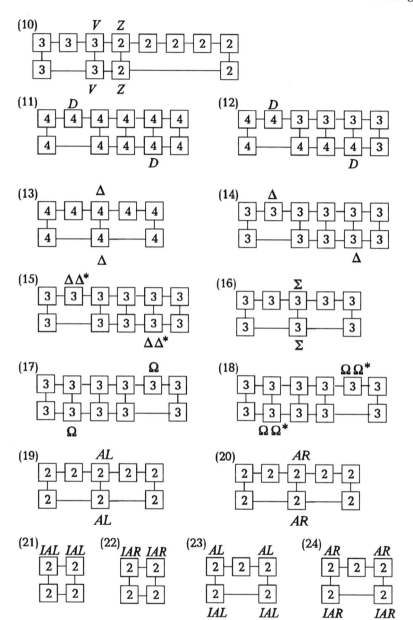

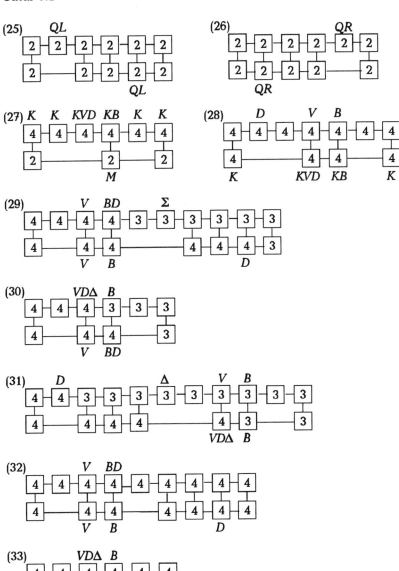

(34)
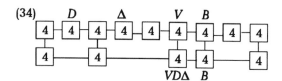

(35)
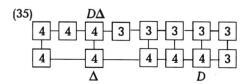

(36)
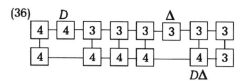

(37)
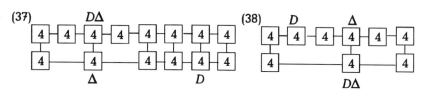

(38)

(39)
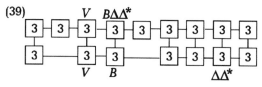

(40)
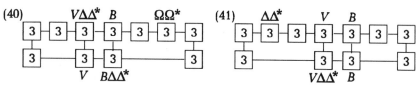

(41)

(42)
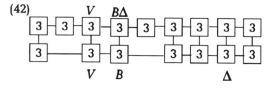

(43)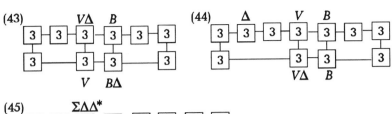

(44)

(45)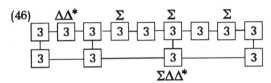

(46)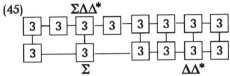

(47)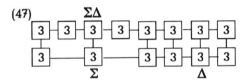

(48)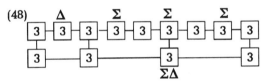

(49)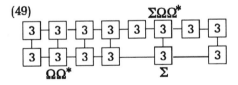

(50)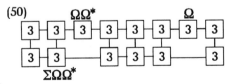

(51)

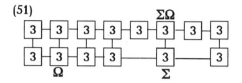

(52)

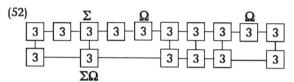

(53)

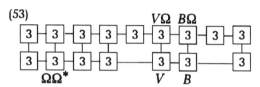

(54)

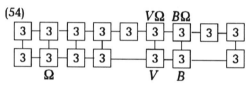

(55)   (56)

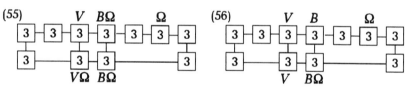

(57)

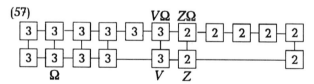

(58)

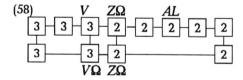

(59)
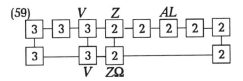

(60)
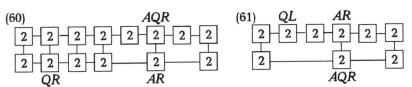

(61)

(62)
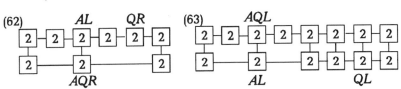

(63)

(64)
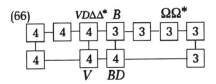

(65)

(66)
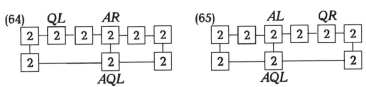

(67)
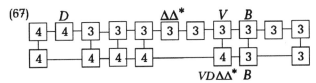

(68)
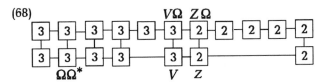

(69)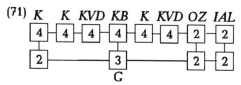

(70)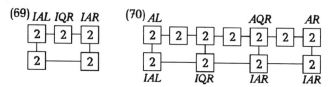

(71)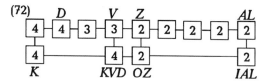

(72)

Catalogue Catal 6.2

Blocks
of the Basic Construction

Catal 6.2(1)–Catal 6.2(3) are pure blocks.
Catal 6.2(4)–Catal 6.2(15) are blocks with singular lines.
Catal 6.2(16)–Catal 6.2(24) are blocks with singular points.
Catal 6.2(25) is a functional block.

Catal 6.2(1)

$D2$ is a pure 2-dimensional net
dim = 2

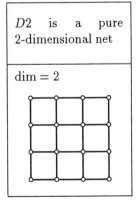

Catal 6.2(2)

$D3(\tau)$ is a pure 3-dimensional net
dim = 3

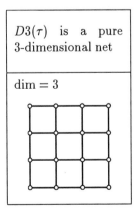

Catal 6.2(3)

$D4(\tau,\pi)$ is a pure 4-dimensional net
dim = 4   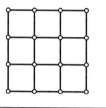

Catal 6.2(4)

M is the marking-out line
M is the (1,1)-line
dim = 2   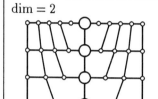

Catal 6.2(5)

G is the line generating the marking-out	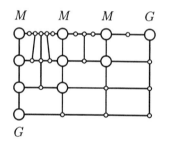
G is the ((1,1),1)-line, dim = 3 at the points of the G-line and dim = 2 at the rest of the 2-points; every G-point generates upward a line of type M	

Catal 6.2(6)

$K(\tau,\pi)$ is the subdivision line of the model kernel	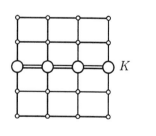
K is the horizontal line, dim = 2 below K and dim = 4 on K and above K	

Catal 6.2(7)

$B(\alpha, \tau, \pi(\text{if } \alpha = 4)), \alpha \in (3, 4)$ is the branching line	
B, V is the double line, $\dim = \alpha$; the branching of R_3–S_3-planes passing from B to V; for $\alpha = 4$, $V = V_0 \sqcup V_1$ holds and in every E_3-class in V there are V_0 and V_1	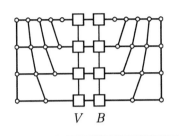

Catal 6.2(8)

$Z(\tau)$ is the screen of the values of formulas	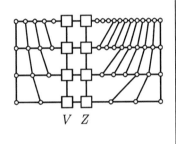
V, Z is the double (1,3)-line, $\dim = 3$ on V and to the left of V; $\dim = 2$ on Z and to the right of Z	

Catal 6.2(9)

$D(\alpha, \tau, \pi)$, $\alpha \in \{3, 4\}$ the line of transmission of messages	
D is the $(-2, 3)$-line, $\dim = 4$ on D and to the left of D; $\dim = \alpha$ to the right of D; $D = D_0 \sqcup D_1 \sqcup D_2$; there are D_0, D_1, D_2	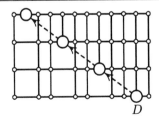

Catal 6.2(10)

$\Delta(\alpha, \tau, \pi)$, $\alpha \in \{3^*, 3, 4\}$, is the line of the transmission and lift of elementary messages	(a) the transmission line $\Delta(\alpha)$, $\alpha \neq 4$
Δ is the $(1, 1)$-line for $\alpha = 4$; Δ is the $(-2, 3)$-line for $\alpha \neq 4$; dim $= 4$ for $\alpha = 4$; dim $= 3$ for $\alpha \neq 4$; $\Delta = \Delta_0 \sqcup \Delta_1 \sqcup \Delta_2$; there are Δ_0, Δ_1, Δ_2; for $\alpha = 3^*$, the marker-predicate Δ^* is true along the Δ-line	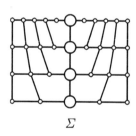 (b) the lift line $\Delta(4)$

Catal 6.2(11)

$\Sigma(\varepsilon, \tau)$, $\varepsilon \in \{0, 10, 1\}$ is the elementary conjunction line	
Σ is the $(1, 1)$-line; dim $= 3$; $\Sigma = \Sigma_0 \sqcup \Sigma_1$; for $\varepsilon = 0$, there is Σ_0 and there is no Σ_1; for $\varepsilon = 10$, there are Σ_0 and Σ_1; for $\varepsilon = 1$, there is no Σ_0 and there is Σ_1	Σ

Catal 6.2(12)

$\Omega(\alpha, \varepsilon, \tau)$, $\alpha \in \{3^*, 3\}$, $\varepsilon \in \{0, 10, 1\}$, $\alpha = 3^* \Rightarrow \varepsilon = 0$ the matrix line	
Ω is the $(3, -2)$-line; $\dim = 3$; $\Omega = \Omega_0 \sqcup \Omega_1$; for $\varepsilon = 0$, there is no Ω_0 and there is Ω_1; for $\varepsilon = 10$, there are Ω_0 and Ω_1; for $\varepsilon = 1$, there is Ω_0 and there is no Ω_1; for $\alpha = 3^*$, the marker-predicate Ω^* is true along the Ω-line	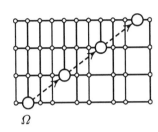

Catal 6.2(13)

$A(\nu, i)$, $\nu \in \{L, R\}$, $i \in \{0, 1, \ldots, d-1\}$ the line of the tape cell	
A is the $(1, 1)$-line; $\dim = 2$; $A = A_0 \sqcup A_1 \sqcup \ldots \sqcup A_{d-1}$; the line satisfies A_i; the polarizator-predicate ν is true along the A-line	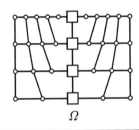

Catal 6.2(14)

I the line of generation of tape cells	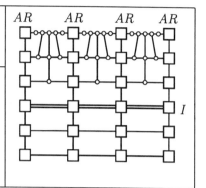
I the horizontal line; dim $= 2$; every I-point generates upward a line of type $A(R, 2)$	

Catal 6.2(15)

$Q(\nu, j)$, $\nu \in \{L, R\}$, $j \in \{0, 1, \dots, e-1\}$, the line of motion of the head	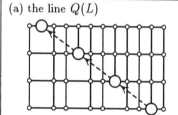
Q is the $(-2, 3)$-line for $\nu = L$; Q is the $(3, -2)$-line for $\nu = R$; dim $= 2$; $Q = Q_0 \sqcup Q_1 \sqcup \dots \sqcup Q_{e-1}$; the line satisfies Q_j; the polarizator-predicate ν is true along the Q-line	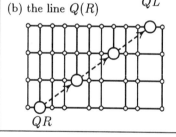

Catal 6.2(16)

$KM(\tau, \pi)$ is the subdivision pole of the model kernel

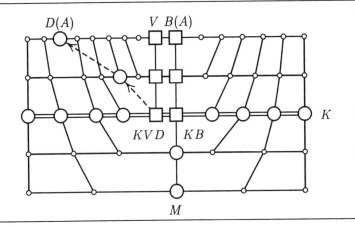

at the point KVD some initial message about the model kernel is introduced and transferred along the D-line and V-line

Catal 6.2(17)

$BD(\alpha, \tau, \pi)$, $\alpha \in \{3, 4\}$, an initial point of elementary messages
(a) — $BD(3)$, (b) — $BD(4)$.

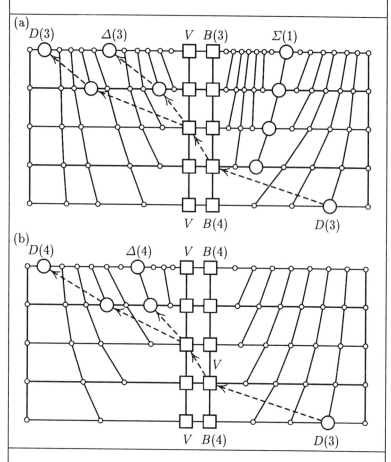

1. The meeting of the V-line and the D-line gets elementary messages which is transferred along the Δ-line.
2. The message on the D-line is reproduced after passing through branching.
3. In the case $BD(3)$, the elementary conjunction line Σ appears with the identically true message.

Catal 6.2(18)

$D\Delta(\alpha, \tau, \pi)$, $\alpha \in \{3, 4\}$, is a passive intersection
(a) — $D\Delta(3)$, (b) — $D\Delta(4)$.

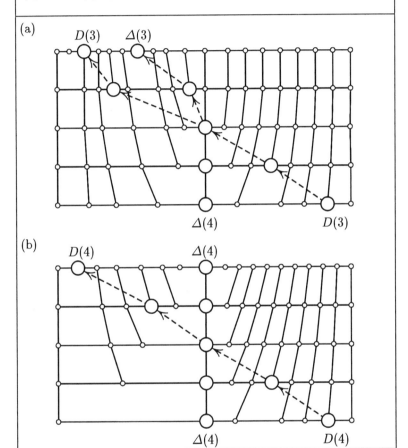

1. Messages pass through an intersection point along lines without interaction.
2. In the case $D\Delta(3)$, the lift is complete and the transmission of elementary messages starts (for use at a point of type $\Sigma\Delta$).

Catal 6.2(19)

$B\Delta(\alpha, \tau)$, $\alpha \in \{3^*, 3\}$, the point of branching of elementary messages: (a) — $B\Delta(3^*)$, (b) — $B\Delta(3)$.

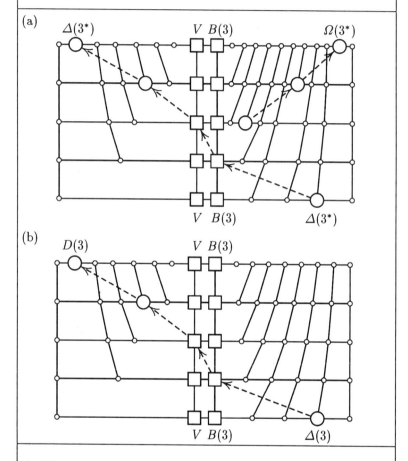

1. The message on the Δ-line is produced passing through branching.
2. In the case $B\Delta(3^*)$, the matrix line appears with the identically false message; it is marked by the marker-predicate Ω^*.

Catal 6.2(20)

$\Sigma\Delta(\alpha,\varepsilon,\tau)$, $\alpha \in \{3^*,3\}$, $\varepsilon \in \{10,1\}$, $\alpha = 3^* \Rightarrow \varepsilon = 10$, the point of formation of the elementary conjunction

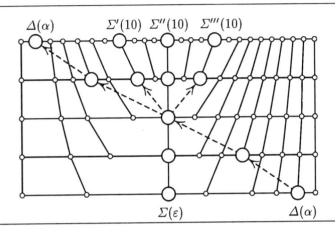

$\Delta(\alpha)$ $\Sigma'(10)$ $\Sigma''(10)$ $\Sigma'''(10)$

$\Sigma(\varepsilon)$ $\Delta(\alpha)$

1. The meeting of the Σ-line and Λ-line gets three new Σ-lines.

2. The message on the Λ-line passes through the meeting point without changes.

3. The interaction of messages is defined by the table to the right.

Σ	Λ	Σ'	Σ''	Σ'''
Σ_0	Λ_0	Σ_0	Σ_0	Σ_0
Σ_0	Λ_1	Σ_0	Σ_0	Σ_0
Σ_0	Λ_2	Σ_0	Σ_0	Σ_0
Σ_1	Λ_0	Σ_1	Σ_0	Σ_0
Σ_1	Λ_1	Σ_0	Σ_1	Σ_0
Σ_1	Λ_2	Σ_0	Σ_0	Σ_1

Catal 6.2(21)

$\Sigma\Omega(\alpha, \varepsilon, \delta, \tau)$, $\alpha \in \{3^*, 3\}$, $\varepsilon, \delta \in \{0, 10, 1\}$, $\varepsilon \leqslant \delta$ $(0 < 10 < 1)$, $\alpha = 3^* \Rightarrow \varepsilon = 0$ & $\delta = 10$, the point of formation of the matrix (a) — $\Sigma\Omega(3^*)$, (b) — $\Sigma\Omega(3)$.

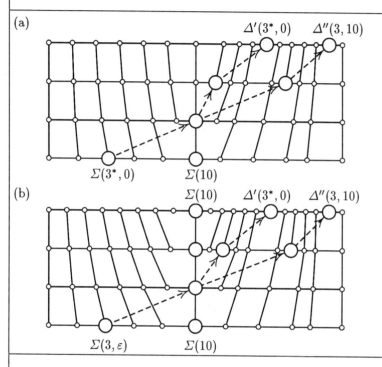

(a)

$\Delta'(3^*, 0)$ $\Delta''(3, 10)$

$\Sigma(3^*, 0)$ $\Sigma(10)$

(b)

$\Sigma(10)$ $\Delta'(3^*, 0)$ $\Delta''(3, 10)$

$\Sigma(3, \varepsilon)$ $\Sigma(10)$

1. $\Omega(\alpha, \varepsilon)$ meets $\Sigma(10)$.
2. $\Omega'(\alpha, \varepsilon)$ and $\Omega''(3, \delta)$ appear.
3. The Σ-line breaks if it meets the marker-line $\Omega(3^*)$.
4. The message on the Σ-line (if it is extended) passes through the meeting point without changes.
5. The interaction of messages is defined by the table to the right.

Ω	Σ	Ω'	Ω''
Ω_0	Σ_0	Ω_0	Ω_0
Ω_0	Σ_1	Ω_0	Ω_1
Ω_1	Σ_0	Ω_1	Ω_1
Ω_1	Σ_1	Ω_1	Ω_1

Catal 6.2(22)

$B\Omega(\alpha, \varepsilon, \delta', \delta'', \tau)$, $\alpha \in \{3^*, 3\}$, $\varepsilon, \delta', \delta'' \in \{0, 10, 1\}$, $\delta' \leqslant \varepsilon \leqslant \delta''$ $(0 < 10 < 1)$, $\alpha = 3^* \Rightarrow \varepsilon = 0$ & δ'', $\varepsilon = 1 \Rightarrow \delta' = 1$, is the point of the realization of quantifiers

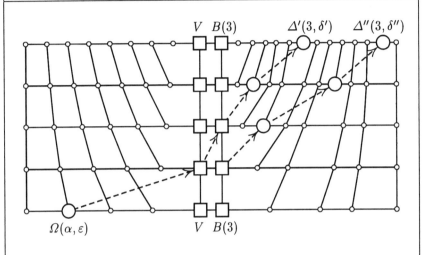

1. In the case $\alpha = 3^*$, the marker-predicate Ω^* of the line Ω breaks approaching the V-line.
2. At the point a, the message is computed by the rule of realization of the quantifier $\forall$ and is transmitted along the line Ω'.
3. At the point e, the message is computed by the rule of realization of the quantifier $\exists$ and is transmitted along the line Ω''.

Catal 6.2(23)

$Z\Omega(\varepsilon, \delta', \delta'', \tau)$, $\varepsilon \in \{0, 10, 1\}$, $\delta', \delta'' \in \{0, 1\}$, $\delta' \leqslant \varepsilon \leqslant \delta''$ $(0 < 10 < 1)$, $\varepsilon = 0 \Rightarrow \delta'' = 0$, $\varepsilon = 1 \Rightarrow \delta' = 1$, is the point of values of formulas appearing in the screen

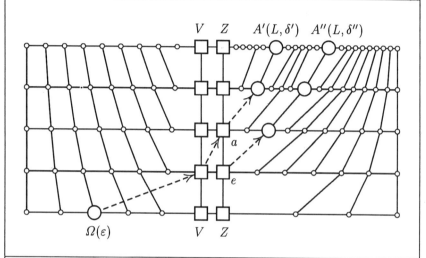

1. At the point a, the message is computed by the rule of realization of the quantifier $\forall$ and is transferred along the line A'.
2. At the point e, the message is computed by the rule of realization of the quantifier $\exists$ and is transferred along the line A''.

Catal 6.2(24)

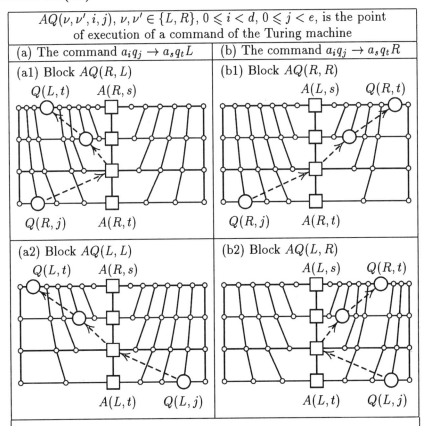

$AQ(\nu, \nu', i, j)$, $\nu, \nu' \in \{L, R\}$, $0 \leqslant i < d$, $0 \leqslant j < e$, is the point of execution of a command of the Turing machine

(a) The command $a_i q_j \to a_s q_t L$	(b) The command $a_i q_j \to a_s q_t R$
(a1) Block $AQ(R, L)$	(b1) Block $AQ(R, R)$
$Q(L, t)$ $A(R, s)$ $Q(R, j)$ $A(R, t)$	$A(L, s)$ $Q(R, t)$ $Q(R, j)$ $A(R, t)$
(a2) Block $AQ(L, L)$	(b2) Block $AQ(L, R)$
$Q(L, t)$ $A(R, s)$ $A(L, t)$ $Q(L, j)$	$A(L, s)$ $Q(R, t)$ $A(L, t)$ $Q(L, j)$

1. A block of type $A(\nu, \nu', i, j)$ is defined only if the Turing machine $\mathcal{M}$ has the command $a_i q_j \to a_s q_t \nu'$. Two blocks that differ in their lower parts correspond to one command.

2. The line $A(\nu, i)$ meets the line $Q(\nu, j)$ and two new lines $A(\nu'', s)$ and $Q(\nu', i)$, $\nu'' \in \{L, R\} \setminus \{\nu'\}$, appear; they are defined by the corresponding command of the Turing machine $\mathcal{M}$ of the form $a_i q_j \to a_s q_t \nu'$.

Catal 6.2(25)

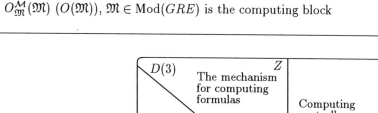

$O_{\mathfrak{M}}^{\mathcal{M}}(\mathfrak{M})$ $(O(\mathfrak{M}))$, $\mathfrak{M} \in \mathrm{Mod}(GRE)$ is the computing block

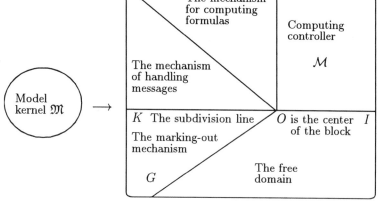

1. Lines of types G, K, $D(3)$, $\Delta(3^*)$, Z, I, and Q start in neighborhoods of the central point O. They generate the system of informational-marking-out lines (cf. Figs. 6.7.1 and 6.7.2) that is the basis of the computing mechanism controlling the elementary theory of the model kernel of $\mathfrak{M}$.

2. The Turing machine $\mathcal{M}$ and the input parameter m included in the axiomatics of the theory F are taken so that after the work of the computing mechanism, the following assertion holds:

the block $O(\mathfrak{M})$ is defined $\Leftrightarrow \mathfrak{M} \in \mathrm{Mod}(T)$.

Moreover, the Turing machine $\mathcal{M}$ is independent of T and the parameter m is effectively defined by the recursively enumerable index of the theory T.

3. At the abstraction level X, the block can be represented in the form $O(m, \lambda)$, where m is the parameter in Axiom 29°, $\lambda = \langle \lambda_i ; i < \omega \rangle$, $\lambda_i \in \{L, R\}$, where the components λ_i of the sequence λ define the choice of direction of motion of the head after passage through points of execution of command in the domain of the computer-controller.

Chapter 7

Existence Theorems

Most of the results of this chapter follow from the intermediate construction. The method of reducing signatures presented allows us to pass to any finite rich signature. We emphasize that one of the obvious advantages of the construction 𝔽𝕄 is the possibility of constructing a finitely axiomatizable theory by constructing a recursively enumerable tree and omitting the intermediate construction of a recursively axiomatizable theory.

7.1. Lindenbaum Algebras and the Hanf Problem

In the sequel, we assume that every Lindenbaum algebra is enumerated and is equipped with the natural (Gödel) numbering. From this standpoint, the Lindenbaum algebra of any recursively axiomatizable theory is positively enumerated. On the other hand, for any positively enumerated Boolean algebra $(\mathcal{B}, \nu)$ it is possible to construct a recursively axiomatizable theory the Lindenbaum algebra of which is equivalent to $(\mathcal{B}, \nu)$. Using the intermediate construction, we can characterize the Lindenbaum algebras of finitely axiomatizable theories in a similar way.

Theorem 7.1.1. *Let σ be a finite rich signature. An enumerated Boolean algebra $(\mathcal{B}, \nu)$ is equivalent to the Lindenbaum algebra of some finitely axiomatizable theory F of a signature σ if and only if $(\mathcal{B}, \nu)$ is a positively enumerated Boolean algebra.*

PROOF. If $(\mathcal{B}, \nu)$ is equivalent to the Lindenbaum algebra of some finitely axiomatizable theory, then $(\mathcal{B}, \nu)$ is a positively enumerated algebra. Therefore, it suffices to develop a method of constructing F from a given positively enumerated Boolean algebra $(\mathcal{B}, \nu)$.

Let $\mathcal{B}$ denote the Boolean algebra whose elements are classes of equivalent truth-table conditions and whose operations are induced by the corresponding propositional connectives. It is obvious that $\mathcal{B}$ is a countable atomless Boolean algebra generated freely by the elementary truth-table conditions ε_k, $k < \omega$. The algebra $\mathcal{B}$ has a natural constructivization μ defined by the Gödel numbering of the truth-table conditions. By assertion (a) of Theorem 3.1.1, the Lindenbaum algebra of the theory $\mathbb{F}(m, s)$ is recursively isomorphic to an enumerated quotient algebra

$$\left(\mathcal{B}/\mathcal{F}_m, \mu^*\right) \tag{7.1.1}$$

where $\mathcal{F}_m$ denotes the filter generated by the set $\{\tau_k \mid k \in W_m\}$ and μ^* is the numbering of the quotient algebra induced by the numbering μ. It is easy to see that every recursively enumerable filter of the algebra $\mathcal{B}$ coincides with $\mathcal{F}_m$ for some $m < \omega$. Since any positively enumerated Boolean algebra can be represented up to an isomorphism in the form (7.1.1), we arrive at the required assertion. □

Theorem 7.1.2. *The following assertions hold:*

(a) *if the theory $PC(\sigma)$ of a finite rich signature σ is defined by an identically true formula, then the Lindenbaum algebra is not constructivizable,*

(b) *there exists a decidable finitely axiomatizable theory whose Lindenbaum algebra is constructivizable but not strongly constructivizable.*

PROOF. (a). Feiner [14] constructed a positively enumerated Boolean algebra that is nonconstructivizable. By Theorem 7.1.1, there exists a finitely axiomatizable theory F of a finite rich signature σ such that the Lindenbaum algebra is nonconstructivizable. Since F is finitely axiomatizable, the algebra $\mathcal{B}(F)$ is isomorphic to the quotient algebra $\mathcal{B}(PC(\sigma))/\mathcal{F}$ by some principal filter $\mathcal{F}$. Consequently, the algebra $\mathcal{B}(PC(\sigma))$ is not constructivizable.

(b). Goncharov [20] has constructed constructivizable but not strongly constructivizable Boolean algebras of each elementary type except the categorical type in the cardinality $\alpha \leqslant \omega$. With the help of this result, we can obtain the required theory in accordance with Theorem 7.1.1.   □

The following theorem solves the well-known Hanf problem [24] (see, for example, the Friedman list [15]). It is equivalent to Theorem 7.1.1.

Theorem 7.1.3. *For any recursively axiomatizable theory T there exists a finitely axiomatizable theory F of a finite rich signature σ such that the Lindenbaum algebras of the theories T and F are recursively isomorphic. The system of axioms for the theory F as well as a recursive function presenting an isomorphism between the Lindenbaum algebras can be found effectively in the signature σ and the recursively enumerable index of the system of axioms of the theory T.*

PROOF. Let Φ_k, $k < \omega$, be the Gödel numbering of sentences of the signature σ' of the theory T. We define f acting from the set of truth-table conditions into the set of sentences of the signature σ' by induction:

$$f(\varepsilon_i) = \Phi_i, \quad f(\tau'|\tau'') = f(\tau')|f(\tau'') = \neg[f(\tau') \,\&\, f(\tau'')]$$

Let m be a recursively enumerable index of the set of numbers of the family of truth-table conditions $M = \{\tau : \vdash_T f(\tau)\}$. The number m is found effectively in the recursively enumerable index of the system of axioms of the theory T. It is easy to show that the theory $F = F_\sigma(m, s)$ is the required finitely axiomatizable theory for an arbitrary s. Moreover, the recursive isomorphism μ from $\mathcal{L}(T)$ into $\mathcal{L}(F)$ is defined by the formula $\mu(\Phi_i) = \Psi_i$, $i \in \mathbb{N}$. Let us show that F and μ satisfy the imposed conditions.

If $\vdash_T (\Phi_i \leftrightarrow \Phi_j)$, then the truth-table condition $\varepsilon_i \leftrightarrow \varepsilon_j$ occurs in M. By the definition of $\mathcal{R}_m$, for any $A \in \mathcal{R}_m$ we have $i \in A \leftrightarrow j \in A$. Consequently, $\mathbb{F}_\sigma(m, s)[A] \vdash (\Psi_i \leftrightarrow \Psi_j)$ is true for any such A. Since the formula $\Psi_i \leftrightarrow \Psi_j$ is true in each of the completions of F, it is deducible in F. Thereby, the mapping μ is well defined.

If, in the theory T, the formulas $\Phi_i \leftrightarrow (\Phi_j \vee \Phi_k)$ and $\Phi_r \leftrightarrow (\neg\Phi_t)$ are provable, then the truth-table conditions $\varepsilon_i \leftrightarrow (\varepsilon_j \vee \varepsilon_k)$ and $\varepsilon_r \leftrightarrow (\neg\varepsilon_t)$ occur in M. The aforesaid means that, in the theory F, the formulas $\Psi_i \leftrightarrow (\Psi_j \vee \Psi_k)$ and $\Psi_r \leftrightarrow (\neg\Psi_t)$ are provable. Consequently, μ is a homomorphism from the Boolean algebra $\mathcal{L}(T)$ into the Boolean algebra $\mathcal{L}(F)$.

If $(T \vdash \Phi_i)$ is not true, then the truth-table condition ε_i does not occur in M. Based on the properties of the set M, it is possible to find a set $A \in \mathcal{R}_m$ such that $i \notin A$. Then $\mathbb{F}_\sigma(m, s)[A] \vdash \neg\Psi_i$. Consequently, Ψ_i

cannot be provable in the theory T. Thus, μ is injective. Theorem 3.1.1 implies that μ is surjective.   □

Using Theorem 7.1.3, it is possible to construct finitely axiomatizable theories with a rather complete structure of the Lindenbaum algebra by a uniform method. First, we construct a recursively axiomatizable theory with given properties of the Lindenbaum algebra. After that, with the help of Theorem 7.1.3, we pass to a finitely axiomatizable theory with the same Lindenbaum algebra. For example, it is possible to prove the existence of finitely axiomatizable theories possessing the property indicated in the following theorem.

Theorem 7.1.4. *There exists a finitely axiomatizable theory $F_{\alpha,n}$ such that it has any given recursively enumerable truth-table degree of undecidability of A and its Lindenbaum algebra is a superatomic Boolean algebra of a given ordinal type (α, n), $\omega \leqslant \alpha < \lambda$, $1 \leqslant n < \omega$.*

PROOF. Feferman (see, for example, [56]) has constructed a recursively axiomatizable theory that has an arbitrary recursively enumerable truth-table degree; moreover, the Lindenbaum algebra is superatomic and has the ordinal type $(\omega, 1)$. It is not hard to increase the ordinal type to the fixed value (α, n) so that the truth-table degree of undecidability of the theory is unchanged. It remains to apply Theorem 7.1.3.   □

7.2. Complete Theories

The following partial case of the main theorem is often used for the study of complete finitely axiomatizable theories.

Theorem 7.2.1. *Let σ be a finite rich signature and let $\mathcal{D}$ be a recursively enumerable tree. We can construct, effectively in the signature σ and the recursively enumerable index of the tree $\mathcal{D}$, a complete, model-complete, and finitely axiomatizable theory $\mathbb{F}_\sigma(\mathcal{D})$ of the signature σ such that*

(a) *the theory $\mathbb{F}_\sigma(\mathcal{D})$ has a prime model if and only if the tree $\mathcal{D}$ is atomic,*

(b) *a prime model (if it exists) of the theory $\mathbb{F}_\sigma(\mathcal{D})$ is strongly constructivizable if and only if the family of chains $\Pi^{\mathrm{fin}}(\mathcal{D})$ is computable,*

(c) *a prime strongly constructivizable model (if it exists) of the theory* $\mathbb{F}_\sigma(\mathcal{D})$ *is autostable under strong constructivizations if and only if the tree* $\mathcal{D}$ *is recursive,*

(d) *the theory* $\mathbb{F}_\sigma(\mathcal{D})$ *has a countable saturated model if and only if the tree* $\mathcal{D}$ *is superatomic,*

(e) *a countable saturated model of the theory* $\mathbb{F}_\sigma(\mathcal{D})$ *is strongly constructivizable if and only if the family of chains* $\Pi(\mathcal{D})$ *is computable,*

(f) *the theory* $\mathbb{F}_\sigma(\mathcal{D})$ *is totally transcendental if and only if the tree* $\mathcal{D}$ *is superatomic,*

(g) *the Morley rank of the theory* $\mathbb{F}_\sigma(\mathcal{D})$ *is equal to* $\max\{15,\ 1 + \mathrm{Rank}(\mathcal{D}) + \gamma\}$*, where* $\gamma = 2$ *if the tree* $\mathcal{D}$ *is superatomic and* $\gamma = 0$ *otherwise.*

PROOF. Let m be a recursively enumerable index of the set of numbers of tt-conditions of the form $\neg\varepsilon_k$, $k \in \mathbb{N}$. By definition, $\mathcal{R}_m = \{\varnothing\}$. By Theorem 3.1.1, the theory $F_\sigma(m, s)$ is complete for any s since $F_\sigma(m, s)[\varnothing]$ is a unique completion of $F_\sigma(m, s)$. We choose s so that the tree $\mathcal{D}_s^\varnothing$ coincides with a given tree $\mathcal{D}$. For $\mathbb{F}_\sigma(\mathcal{D})$, we take the theory $F_\sigma(m, s)$. Theorem 3.1.1 provides all the necessary properties of this theory. $\square$

Theorem 7.2.1 is a finitely axiomatizable version of the construction $\mathbb{T}(\mathcal{D})$ (cf. Sec. 2.4). We pass to applications of this theorem.

Theorem 7.2.2. *The following assertions hold:*

(a) *there exists a complete, finitely axiomatizable, totally transcendental theory whose prime model and countable saturated model are not constructivizable,*

(b) *there exists a complete, finitely axiomatizable, totally transcendental theory whose prime model is strongly constructivizable but is nonautostable under strong constructivizations.*

PROOF. (a). We consider the recursively enumerable superatomic tree $\mathcal{D}$ constructed in [17] such that the families of chains $\Pi(\mathcal{D})$ and $\Pi^{\mathrm{fin}}(\mathcal{D})$ are not computable. Let σ be a finite rich signature. Then the theory $\mathbb{F}_\sigma(\mathcal{D})$ possesses the required properties because, by the model completeness, any constructive model of this theory is a strongly constructive model.

(b). The theory $\mathbb{F}_\sigma(\mathcal{D})$ has the above-mentioned properties, where $\mathcal{D}$ is a recursively enumerable and not recursively superatomic tree such that $\Pi(\mathcal{D})$ is computable. The construction of such a tree is trivial. $\square$

Theorem 7.2.2 solves two of Harrington's problems stated in [26].

Theorem 7.2.3. *There exists a complete, finitely axiomatizable, ω-stable theory T that has no constructive homogeneous models.*

PROOF. An example of an axiomatizable theory with the indicated properties has been constructed in [18]. It remains to apply the universal construction (cf. Theorem 0.6.1). $\square$

Let us show that a prime model and a countable saturated model of a complete finitely axiomatizable theory can have rather complicated structures provided that they are nonconstructivizable. We denote by $FD(\mathfrak{M}, \nu)$ the complete first-order definable diagram of the enumerated model $(\mathfrak{M}, \nu)$.

Theorem 7.2.4. *The following assertions hold:*

(a) *there exists a numbering ν of a prime model of a complete decidable theory T such that $FD(\mathfrak{M}, \nu) \in \Delta_2^0$,*

(b) *for any $A \in \Delta_2^0$ there exists a complete, finitely axiomatizable, ω-stable theory F such that for any numbering ν of a prime model $\mathfrak{M}$ of the theory F the set $FD(\mathfrak{M}, \nu)$ is not m-reducible to A.*

PROOF. (a). A standard procedure of constructing a prime model of a complete decidable theory has the same complexity as that of a recursive set with respect to a creative set $K \subseteq \mathbb{N}$ since, in order to verify the atomicity of formulas, it suffices to turn to the oracle K. Consequently, this procedure leads to the model $(\mathfrak{M}, \nu)$ such that $FD(\mathfrak{M}, \nu)$ is recursive with respect to K, i.e., in Δ_2^0.

(b). We consider the superatomic recursively enumerable tree $\mathcal{D}$ constructed in [10] from a given $A \in \Delta_2^0$ such that the family of chains $\Pi^{\text{fin}}(\mathcal{D})$ is not recursively enumerable with respect to A. We show that the theory $\mathbb{F}(\mathcal{D})$ from Theorem 7.2.1 is the required one. If the complete diagram $FD(\mathfrak{M}, \nu)$ is m-reducible to the set A, then it is recursive with respect to A. In this case, we can enumerate components of the model $\mathfrak{M}$, starting from U, and pass to the corresponding finite chains of the tree $\mathcal{D}$; thereby, we enumerate all the family $\Pi^{\text{fin}}(\mathcal{D})$ by a recursive (with respect to A) function, which contradicts the properties of the tree $\mathcal{D}$. $\square$

Theorem 7.2.5. *The following assertions hold:*

(a) *there exists a numbering ν of a countable saturated model $\mathfrak{M}$ of a complete decidable theory T such that $FD(\mathfrak{M}, \nu) \in \Delta_1^1$,*

(b) *for any $A \in \Delta_1^1$ there exists a complete, finitely axiomatizable, ω-stable theory F such that for any numbering ν of a countable saturated model $\mathfrak{M}$ of the theory F the reducibility $F \leqslant_m FD(\mathfrak{M}, \nu)$ holds.*

PROOF. (a). A standard procedure of constructing a saturated model of a complete decidable theory has the complexity of a hyperarithmetic set, which leads to assertion (a).

(b). As in [16], given $A \in \Delta_1^1$, we construct a recursively enumerable superatomic tree $\mathcal{D}$ such that the reducibility $A \leqslant_m \pi^*$ holds for some chain π^* of this tree. Then the theory $\mathbb{F}(\mathcal{D})$ defined by Theorem 7.2.1 is the required one. Indeed, let $c \in U(\mathfrak{M})$ be an element of $\mathfrak{M}$ such that it generates the component corresponding to the chain π^*. Enumerating this component in the enumerated model $(\mathfrak{M}, \nu)$ and passing to a tree, it is possible to enumerate, with respect to $FD(\mathfrak{M}, \nu)$, the chain π^* as well as its complement. Then the chain is recursive with respect to $FD(\mathfrak{M}, \nu)$. Thus, $A \leqslant_m FD(\mathfrak{M}, \nu)$.   □

7.3. Value of the Morley Rank

The following two theorems characterize possible values of the Morley rank in the case of finitely axiomatizable theories.

Theorem 7.3.1. *For any constructive ordinal $\beta \geqslant 0$ there exists a complete, finitely axiomatizable, totally transcendental theory of Morley rank $\alpha_T = \beta + 3$.*

PROOF. In the case $\beta < \omega$, the theory T can be taken to be ω_1-categorical. It is not hard to construct it from the quasisuccessor theory QS of Morley rank 3 (cf. Chapter 1). In the case $\beta \geqslant \omega$, the theory $\mathbb{F}(\mathcal{D})$ is the required one, where $\mathcal{D}$ is a recursive superatomic tree of rank $\beta + 1$. Such a tree exists in view of the results of Sec. 2.2.   □

Let λ denote the first nonconstructive ordinal.

Theorem 7.3.2. *There exists a complete finitely axiomatizable theory F of Morley rank $\alpha_T = \lambda$.*

PROOF. We consider the direct sum of the following sequence of trees: $\mathcal{D} = \oplus \langle \mathcal{D}_i \mid i \in \mathbb{N} \rangle$, where the terms $\mathcal{D}_i = [W_i]_{\mathcal{D}}$ are defined in Sec. 2.3. It is easy to show that $\mathcal{D}$ is recursively enumerable. Since there exists a super-atomic recursive enumerable tree of constructive rank as large as desired, we have Rank$(\mathcal{D}) = \lambda$. Consequently, the theory $\mathbb{F}(\mathcal{D})$, which is defined in accordance with Theorem 7.2.1, possesses the required properties.   □

We present one more result concerning the value of rank.

Theorem 7.3.3. *There exists a complete finitely axiomatizable theory F of Morley rank $\alpha_T = \omega_1$.*

PROOF. Lachlan [32] has constructed a complete decidable theory T of Morley rank ω_1 (the decidability of this theory was not considered in [32]). Using the universal construction, we transform T into a finitely axiomatizable theory $F = \mathbb{F}(T)$. Then the rank of the theory F must be at least ω_1. By the well-known result of Lachlan, ω_1 is the maximal value of all possible ranks of the theory. Therefore, $\alpha_F = \omega_1$.   □

Theorem 7.3.3 strengthens the result of Lachlan [32], who has constructed a complete decidable theory of a finite signature such that $\alpha_T = \omega_1$. We note that the Morley rank of a totally transcendental theory is not a limit ordinal. Hence the theories mentioned in Theorems 7.3.2 and 7.3.3 cannot be totally transcendental. It was shown by Sacks [58] that the Morley rank of a totally transcendental decidable theory is a constructive ordinal. Thus, Theorems 7.3.1–7.3.3 describe complete finitely axiomatizable theories for the majority of values of the Morley rank.

Exercises

1. Prove that for a complete decidable theory T there exists a homogeneous model $\mathfrak{M}$ of the theory T and a numbering ν such that $FD(\mathfrak{M}, \nu) \in \Delta_2^0$.

2. Prove that for any $A \in \Delta_2^0$ there exists a complete finitely axiomatizable ω-stable theory F such that for any numbering ν of any homogeneous model $\mathfrak{M}$ of the theory F the diagram $FD(\mathfrak{M}, \nu)$ is not m-reducible to A.

Chapter 8

Complexity of Semantic
Classes of Sentences

In this chapter, we present the most interesting results obtained with the
help of the constructions described in the previous chapters. In particular,
we consider exact estimates for the algorithmic complexity of those semantic classes of sentences that are often used in model theory and logic. In
many cases in which the existence theorem for a finitely axiomatizable theory is obvious, the deduction of exact estimates for the complexity turns
out to be nontrivial. Most of the results presented in this chapter are
proved by means of the intermediate construction and the standard sets
constructed in Sec. 4.2. Furthermore, we use the fact that the construction
of finitely axiomatizable theories by the above-mentioned constructions is
effective. Throughout the chapter, we adopt the following notation: σ is
a finite rich signature, Φ_i, $i \in \mathbb{N}$, is the Gödel numbering of all sentences
of the signature σ, SC is the class of strongly constructivizable models, D
is the class of models with a decidable elementary theory, F is the class of
models with a finitely axiomatizable theory, P is the class of prime models,
S is the class of countable saturated models, T is the class of models of an

ω-stable theory. For a set S of sentences of the signature σ, we denote by Nom S the set of Gödel numbers of sentences from S.

8.1. General Questions

We begin by proving estimates for the class of complete sentences.

Theorem 8.1.1. $\{n \mid \Phi_n$ defines a complete theory$\} \approx \Pi_2^0$.

PROOF. The upper estimate $\forall \exists$ is obvious. To derive the lower estimate we take the standard Π_2^0-set $I = \{n \mid W_n$ is infinite$\}$ introduced in [56]. We consider the signature σ' consisting of only the equality predicate. We denote by Δ_k the sentence asserting that there are at least k different elements and consider the theory T_n of the signature σ' with axioms $\{\Delta_k \mid k \in W_n\}$. It is easy to see that the theory T_n is complete if and only if $n \in I$. Using Theorem 0.6.1, given n, we construct effectively a finitely axiomatizable theory F_n of the signature σ such that $\mathcal{L}(F_n) \cong \mathcal{L}(T_n)$. By construction,

$$n \in I \Leftrightarrow T_n \text{ is complete} \Leftrightarrow F_n \text{ is complete}$$

Given T_n, the theory F_n is effectively constructed. Hence there exists a general recursive function $f(n)$ such that the sentence $\Phi_{f(n)}$ is an axiom of this theory. Finally, we obtain

$$n \in I \Leftrightarrow \Phi_{f(n)} \text{ defines a complete theory}$$

which leads to the required lower estimate.   □

We note that the set in Theorem 8.1.1 is not recursively enumerable. This fact can be regarded as an answer to the question from [23].

Theorem 8.1.2. $\{n \mid \Phi_n$ defines a stable complete theory$\} \approx \Pi_2^0$.

PROOF. The number n belongs to the above-mentioned set if and only if Φ_n defines a complete theory and the theory $\{\Phi_n\}$ has no first-order definable order on an infinite set of sequences. Each of these conditions is described by a prefix of the form $\forall \exists$.

To obtain the lower estimate, we use the set I and the function $f(x)$ constructed in the proof of Theorem 8.1.1. By Lemma 3.13.5,

$$n \in I \Rightarrow \Phi_{f(n)} \text{ is superstable} \Rightarrow \Phi_{f(n)} \text{ is stable}$$
$$n \notin I \Rightarrow \Phi_{f(n)} \text{ is not a complete theory}$$

We obtain the lower estimate: $n \in I \Leftrightarrow \Phi_{f(n)}$ is complete and stable. $\square$

It is worth pointing out that the estimate for the complexity of the class of complete stable theories is based on the intermediate construction whose list of properties does not contain the stability property.

Theorem 8.1.3. $\{n \mid \Phi_n$ has the finite Lindenbaum algebra$\} \approx \Sigma_3^0$.

PROOF. The upper estimate is obvious. To obtain the lower estimate we use the standard Σ_3^0-set $J = \{n \mid$ the set $\mathbb{N} \setminus W_n$ is finite$\}$ introduced in [56]. The signature σ' and sentences Δ_n, $n \in \mathbb{N}$, are defined in the same way as in the proof of Theorem 8.1.1. We consider the theory T_n of the signature σ' given by the set of axioms $\{\Delta_k \to \Delta_{k+1} \mid k \in W_n\}$. We transform the theory T_n into a finitely axiomatizable theory F_n such that $\mathcal{L}(F_n) \cong \mathcal{L}(T_n)$. Let $f(n)$ be a general recursive function such that $\Phi_{f(n)}$ is an axiom of the theory F_n. By construction,

$$n \in J \Leftrightarrow \mathcal{L}(T_n) \text{ is finite } \Leftrightarrow \mathcal{L}(\Phi_{f(n)}) \text{ is finite}$$

which provides the required lower estimate. $\square$

We proceed to estimates for the algorithmic complexity of concrete semantic classes of sentences. Note that these classes are important in logic.

Theorem 8.1.4 [6]. $\{n \mid \Phi_n$ is tautology$\} \approx \Sigma_1^0$.

PROOF. The upper estimate follows from the existence of the axiomatization for the classical predicate logic. The lower estimate can be obtained with the help of the Church construction. $\square$

Theorem 8.1.5 [62]. $\{n \mid \Phi_n$ is true in finite models$\} \approx \Pi_1^0$.

PROOF. The upper estimate is obtained immediately. To deduce the lower estimate we use the Trakhtenbrot construction. $\square$

In the following theorem, we denote by $M_{\text{fin}}(\sigma)$ the class of all finite models and by $M_{r.e}(\sigma)$ the class of all recursively enumerable models of the signature σ. This theorem follows from the Vaught construction.

Theorem 8.1.6 [63]. Let K be the class of models of a finite rich signature σ such that $M_{\text{fin}}(\sigma) \subseteq K \subseteq M_{r.e}(\sigma)$. Then $\Pi_1^0 \leqslant_m \text{Th}(K)$.

Theorem 8.1.6 has a fundamental meaning because it shows the impossibility of logic that is based on traditional principles of axiomatization, but its semantics is based on some version of the notion of a constructive model.

In the most important cases of constructivizability and strong constructivizability, there are exact estimates for complexity, which will be discussed below.

Theorem 8.1.7 [38]. *Let $M_c(\sigma)$ be the class of all constructiviz-able models of a finite rich signature σ. Then $\mathrm{Th}\,(M_c(\sigma)) \approx \varnothing^{(\omega)}$, i.e., the theory of the above-mentioned class is recursively isomorphic to the elementary theory of a standard model of arithmetics.*

PROOF. The upper estimate can be obtained by standard methods from the arithmetic representation of constructive models. To deduce the lower estimate it suffices to consider the signature $\sigma = \{+, -, 0, 1\}$ because the remaining cases can be obtained by the reduction of signatures. In [38], the formula Ψ of the signature σ is constructed such that it has a unique constructive model $\mathfrak{N}$ that is isomorphic to the standard model of arithmetics. Therefore, $\mathfrak{N} \vDash \Phi_i \Leftrightarrow M_c(\sigma) \vDash \Psi \to \Phi_i$, which provides the lower estimate for $\mathrm{Th}\,(M_c(\sigma))$. □

Let $M_{\mathrm{dec}}(\sigma)$ be the class of models with a decidable elementary theory and let $M_{s.c}(\sigma)$ denote the class of all strongly constructivizable models of the signature σ.

Theorem 8.1.8. *Let σ be a finite rich signature. Then*

$$\mathrm{Th}\,(M_{\mathrm{dec}}(\sigma)) = \mathrm{Th}\,(M_{s.c}(\sigma)), \quad \mathrm{Th}\,(M_{\mathrm{dec}}(\sigma)) \approx \Pi_3^0, \quad \mathrm{Th}\,(M_{s.c}(\sigma)) \approx \Pi_3^0$$

PROOF. The estimates coincide because each decidable theory has a strongly constructive model. Hence it suffices to consider the estimate for the complexity of $\mathrm{Th}\,(M_{\mathrm{dec}}(\sigma))$. It suffices to deduce an auxiliary estimate

$$\{n \mid \Phi_n \text{ has a decidable extension}\} \approx \Sigma_3^0$$

It is not hard to obtain the upper estimate. To deduce the lower estimate we take the standard Σ_3^0-set (cf. [56])

$$A = \{n \mid W_{l(n)} \text{ and } W_{r(n)} \text{ are recursively inseparable}\}$$

Given n, we can effectively construct an axiomatizable theory T_n such that

$$n \in A \Rightarrow T_n \text{ has a decidable extension}$$
$$n \notin A \Rightarrow T_n \text{ has no decidable extensions}$$

As a result, we obtain a general recursive function $f(n)$ such that

$$n \in A \Rightarrow \Phi_{f(n)} \text{ has a decidable extension}$$
$$n \notin A \Rightarrow \Phi_{f(n)} \text{ has no decidable extensions}$$

The function $f(n)$ realizes the reduction. $\square$

8.2. Prime Models

Theorem 8.2.1. *The following assertions hold:*

(a) $\{n \mid \Phi_n \text{ is complete and has a prime model}\} \approx \Pi_3^0$,

(b) $\{n \mid \Phi_n \text{ is complete and has no prime models}\} \approx \Sigma_3^0$,

PROOF. As is known, a complete theory T has a prime model if and only if every Lindenbaum algebra $\mathcal{L}_n(T)$, $n \in \mathbb{N}$, is atomic. Using this fact, it is easy to derive the upper estimate. To deduce the lower estimate, we take the standard Π_3^0-set A introduced in Lemma 2.3.1. By Theorem 7.2.1,

$$n \in A \Leftrightarrow \mathbb{F}(\mathcal{D}_n) \text{ is complete and has a prime model}$$
$$n \notin A \Leftrightarrow \mathbb{F}(\mathcal{D}_n) \text{ is complete and has no prime model}$$

Given n, the finitely axiomatizable theory $\mathbb{F}(\mathcal{D}_n)$ is effectively constructed. Let $f(n)$ be a general recursive function such that $\Phi_{f(n)}$ is an axiom of this theory. Then

$$n \in A \Leftrightarrow \Phi_{f(n)} \text{ is complete and has a prime model}$$
$$n \notin A \Leftrightarrow \Phi_{f(n)} \text{ is complete and has no prime models}$$

The last relations give the required lower estimates. $\square$

Theorem 8.2.2. *The following relations hold:*

(a) $\{n \mid \Phi_n \text{ is complete and has a strongly constructivizable prime model}\} \approx \Sigma_4^0$,

(b) $\{n \mid \Phi_n \text{ is complete and has no strongly constructivizable prime model}\} \approx \Pi_4^0$.

PROOF. The upper estimate can be obtained with the help of the Goncharov–Harrington theorem. To derive the lower estimate, we use the standard sets A' and A'' defined in Lemma 2.3.1. By Theorem 7.2.1,

$n \in A' \Leftrightarrow \mathbb{F}(\mathcal{D}_n)$ is complete and has a strongly constructivizable prime model

$n \notin A'' \Leftrightarrow \mathbb{F}(\mathcal{D}_n)$ is complete and has no strongly constructivizable prime model

which provides the required lower estimate.   □

Theorem 8.2.3. *The following relations hold:*

(a) $\{n \mid \Phi_n \ has \ a \ prime \ model\} \approx \Sigma_1^1$,

(b) $\mathrm{Nom}\,(\mathrm{Th}\,(\Pi)) \approx \Pi_1^1$.

PROOF. (a). The existence of a prime model for Φ_n is equivalent to
the existence of a complete theory T such that $\Phi_n \in T$ and every algebra
$\mathcal{L}_k(T)$, $k \in \mathbb{N}$, is atomic. Formalizing, we arrive at the upper estimate Σ_1^1.
To obtain the lower estimate, we use the standard Σ_1^1-set A^* introduced
in Lemma 2.3.1. We choose a number m so as to satisfy the equality
$\mathcal{R}_m = \mathcal{P}(\mathbb{N})$. Then

$$s \in A^* \Leftrightarrow (\exists A \subseteq \mathbb{N})[\mathcal{D}_s^A \text{ is atomic}]$$
$$\Leftrightarrow (\exists A \subseteq \mathbb{N})[\mathbb{F}(m,s)[A] \text{ has a prime model}]$$
$$\Leftrightarrow F(m,s) \text{ has a prime model}$$

(b). It is obvious that the set $\mathrm{Nom}\,\mathrm{Th}\,(\Pi)$ is recursively isomorphic
to the complement to the set from (a), which leads to (b).   □

Theorem 8.2.4. *The following assertions hold:*

(a) $\{n \mid \Phi_n \ has \ a \ decidable \ completion \ with \ a \ prime \ model\} \approx \Sigma_4^0$.

(b) $\{n \mid \Phi_n \ has \ a \ finitely \ axiomatizable \ completion \ with \ a \ prime$
$model\} \approx \Sigma_4^0$.

(c) $\mathrm{Nom}\,\mathrm{Th}\,(\Pi \cap P) \approx \Pi_4^0$,

(d) $\mathrm{Nom}\,\mathrm{Th}\,(\Pi \cap F) \approx \Pi_4^0$.

PROOF. It suffices to prove (a) and (b) since (c) and (d) are direct
consequences of them.

A number n belongs to the set indicated in (b) if and only if there
exists a sentence Ψ of the signature σ such that Φ_n & Ψ is complete and
has a prime model. This condition is described by a prefix of the form $\forall\exists\forall$.

Thus, we obtain the estimate Σ_4^0. The upper estimate can be obtained in a similar way, but instead of a single sentence Ψ one must consider a pair of indices, one of which is a recursively enumerable index of the theory and the other of which is a recursively enumerable index of its complement (and a decidable theory is represented in such a way). To derive the lower estimate, we use the standard Σ_4^0-sets B_0 and B_1 introduced in Lemma 2.3.1. Let m be a recursively enumerable index of the collection of numbers of tt-conditions of the form $i \in A \to j \notin A$, $i, j \in \mathbb{N}$, $i \neq j$. Then $\mathcal{R}_m$ contains $\varnothing$ and all one-element subsets of $\mathbb{N}$. It is easy to see that the theory $F(m, s)[A]$ is finitely axiomatizable if A consists of a single element and is decidable but not finitely axiomatizable if A is the empty set. By Theorem 3.1.1,

$$s \in B_0 \Leftrightarrow (\exists A \in \mathcal{R}_m) \, \mathcal{D}_s^A \text{ is atomic}$$
$$\Leftrightarrow (\exists A \in \mathcal{R}_m)[\mathbb{F}(m, s)[A] \text{ has a prime model}]$$
$$\Leftrightarrow \mathbb{F}(m, s) \text{ has a decidable completion with a prime model}$$

Moreover,

$$s \in B_1 \Leftrightarrow (\exists A \in \mathcal{R}_m \setminus \{\varnothing\}) \, \mathcal{D}_s^A \text{ is atomic}$$
$$\Leftrightarrow (\exists A \in \mathcal{R}_m \setminus \{\varnothing\})[\mathbb{F}(m, s)[A] \text{ has a prime model}]$$
$$\Leftrightarrow \mathbb{F}(m, s) \text{ has a finitely axiomatizable completion with a prime model}$$

which gives the lower estimates in cases (a) and (b). $\square$

Theorem 8.2.5. *The following assertions hold:*

(a) $\{n \mid \Phi_n \text{ has a strongly constructivizable prime model}\} \approx \Sigma_4^0$,

(b) $\{n \mid \Phi_n \text{ has a finitely axiomatizable extension with a strongly constructivizable prime model}\} \approx \Sigma_4^0$,

(c) $\operatorname{Nom} \operatorname{Th}(\Pi \cap SC) \approx \Pi_4^0$,

(d) $\operatorname{Nom} \operatorname{Th}(\Pi \cap SC \cap F) \approx \Pi_4^0$.

PROOF. It suffices to prove (a) and (b) since the sets indicated in (c) and (d) are their complements. By the Goncharov–Harrington theorem, Φ_n has a strongly constructivizable prime model if and only if there exists a complete decidable theory T and a computable family S of types of T such that, in $\Phi_n \in T$, every type of S is principal and any formula satisfiable in T belongs to some type of S. A detailed formalization of these conditions gives a prefix of the form Σ_4^0. The upper estimate in case (b) is obtained in a similar way. To obtain the lower estimate, we use the set A' defined in Lemma 2.3.1. Applying Theorem 7.2.1, we find

$n \in A' \Leftrightarrow \mathbb{F}(\mathcal{D}_n)$ has a strongly constructivizable prime model

$n \in A' \Leftrightarrow \mathbb{F}(\mathcal{D}_n)$ has a finitely axiomatizable completion with a strongly constructivizable prime model

$\square$

8.3. Countable Saturated Models

In this section, we study the algorithmic complexity of some classes of sentences under some conditions on countable saturated models.

Theorem 8.3.1. *The following assertions hold:*

(a) $\{n \mid \Phi_n$ *is complete and has a countable saturated model*$\} \approx \Pi_1^1$,

(b) $\{n \mid \Phi_n$ *is complete and has no countable saturated models*$\} \approx \Pi_1^1$.

PROOF. A complete formula Φ_n has a countable saturated model if and only if every algebra $\mathcal{L}_k(\Phi_n)$ is superatomic. The upper estimates can be obtained if we take into account that the superatomicity of a Boolean algebra is described by a prefix of the form Π_1^1. To obtain the lower estimate, we use the set H introduced in Lemma 2.3.1. By Theorem 7.2.1,

$n \in H \Leftrightarrow \mathbb{F}(\mathcal{D}_n)$ is complete and has a countable saturated model

$n \notin H \Leftrightarrow \mathbb{F}(\mathcal{D}_n)$ is complete and has no countable saturated models

$\square$

Theorem 8.3.2. $\{n \mid \Phi_n$ *is complete and has a strongly construc-tivizable countable saturated model*$\} \approx \Pi_1^1$.

PROOF. By the Morley theorem [34], the number n belongs to the set indicated in the theorem if and only if Φ_n is complete, every algebra $\mathcal{L}_k(\Phi_n)$, $k \in \mathbb{N}$, is superatomic, and there exists a computable family of types S of the theory $[\Phi_n]$ such that every type of this theory is contained in S. A detailed formalization of these conditions leads to a prefix of the form Π_1^1. To obtain the lower estimate, we use the set H' defined in

Lemma 2.3.1. We have

$n \in H' \Rightarrow \mathbb{F}(\mathcal{D}_n)$ is complete and has a strongly constructivizable countable saturated model

□

Theorem 8.3.3. *The following assertions hold:*

(a) $\{n \mid \Phi_n \text{ has a countable saturated model}\} \approx \Sigma_2^1$,

(b) $\operatorname{Nom} \operatorname{Th}(H) \approx \Pi_2^1$.

PROOF. It suffices to prove (a). The upper estimate is immediate. To
obtain the lower estimate, we use the set H^* introduced in Lemma 2.3.1.
Let m be a number such that $\mathcal{R}_m = \mathcal{P}(\mathbb{N})$. Then

$$s \in H^* \Leftrightarrow (\exists A \subseteq \mathbb{N})[\mathcal{D}_s^A \text{ is superatomic}]$$
$$\Leftrightarrow (\exists A \subseteq \mathbb{N})[\mathbb{F}(m, s)[A] \text{ has a countable saturated model}]$$
$$\Leftrightarrow \mathbb{F}(m, s) \text{ has a countable saturated model}$$

Thus, we have obtained the lower estimate in the case (a). □

Theorem 8.3.4. *The following assertions hold:*

(a) $\{n \mid \Phi_n \text{ has a decidable completion with a countable saturated}$
$model\} \approx \Pi_1^1$,

(b) $\{n \mid \Phi_n \text{ has a finitely axiomatizable completion with a countable}$
$saturated model\} \approx \Pi_1^1$,

(c) $\operatorname{Nom} \operatorname{Th}(\Pi \cap P) \approx \Sigma_1^1$,

(d) $\operatorname{Nom} \operatorname{Th}(\Pi \cap F) \approx \Sigma_1^1$.

PROOF. It suffices to prove (a) and (b) since (c) and (d) are direct
consequences of them. It is not hard to deduce the upper estimates. To
obtain the lower estimates, we use the standard set H introduced in Lemma
2.3.1. In view of the completeness of the theory $\mathbb{F}(\mathcal{D})$, we find

$n \in H \Leftrightarrow \mathbb{F}(\mathcal{D}_n)$ has a decidable completion with a countable
saturated model

$n \in H \Leftrightarrow \mathbb{F}(\mathcal{D}_n)$ has a finitely axiomatizable completion with a
countable saturated model

□

Theorem 8.3.5. *The following assertions hold:*

(a) $\{n \mid \Phi_n$ *has a strongly constructivizable countable saturated model*$\}$ $\approx \Pi_1^1$,

(b) $\{n \mid \Phi_n$ *has a finitely axiomatizable extension with a strongly constructivizable and countable saturated model*$\} \approx \Pi_1^1$,

(c) $\mathrm{Nom}\,\mathrm{Th}\,(H \cap SC) \approx \Sigma_1^1$,

(d) $\mathrm{Nom}\,\mathrm{Th}\,(H \cap SC \cap F) \approx \Sigma_1^1$.

PROOF. It suffices to prove (a) and (b). By the Morley theorem [34], the sentence Φ_n has a strongly constructivizable countable saturated model if and only if there exists a complete decidable theory T and a computable family S of types of T such that $\Phi_n \in T$ and every type of T belongs to S. A detailed formalization of these conditions leads to a prefix with a single functional quantifier "every type." The upper estimate in case (b) can be derived in the same way. To obtain the lower estimate, we use the set H' introduced in Lemma 2.3.1. Applying Theorem 7.2.1, we obtain

$n \in H' \Leftrightarrow \mathbb{F}(\mathcal{D}_n)$ has a strongly constructivizable countable saturated model

$n \in H' \Leftrightarrow \mathbb{F}(\mathcal{D}_n)$ has a finitely axiomatizable completion with a countable saturated and strongly constructivizable model

$\square$

8.4. Totally Transcendental Theories

Theorem 8.4.1. *The following assertions hold:*

(a) $\{n \mid \Phi_n$ *is complete and is totally transcendental*$\} \approx \Pi_1^1$,

(b) $\{n \mid \Phi_n$ *is complete and is not totally transcendental*$\} \approx \Sigma_1^1$.

PROOF. As is known, the total transcendence, i.e., the function of the Morley rank defined everywhere, is equivalent to the stability of the theory in a countable cardinality. Therefore, the number n belongs to the set indicated in (a) if and only if the theory of Φ_n is complete and for every countable model $\mathfrak{M}$ of this theory the Lindenbaum algebra $\mathcal{L}_1\big(\mathrm{Th}\,(\mathfrak{M}, |\mathfrak{M}|)\big)$ is superatomic. This condition is described by a prefix of the form Π_1^1. A similar estimate can be obtained in case (b).

To obtain the lower estimate, we use the set H from Lemma 2.3.1. By Theorem 7.2.1, we have

$n \in H \Leftrightarrow \mathbb{F}(\mathcal{D}_n)$ is complete and is totally transcendental

$n \notin H \Leftrightarrow \mathbb{F}(\mathcal{D}_n)$ is complete and is not totally transcendental

$\square$

Theorem 8.4.2. *The following assertions hold:*

(a) $\{n \mid \Phi_n$ *has a totally transcendental completion*$\} \approx \Sigma_2^1$,

(b) $\operatorname{Nom Th}(T) \approx \Pi_2^1$.

PROOF. It suffices to prove (a). The deduction of the upper estimate is obvious. To obtain the lower estimate, we use the set H^* introduced in Lemma 2.3.1. Let m be such that $\mathcal{R}_m = \mathcal{P}(\mathbb{N})$. Then

$$s \in H^* \Leftrightarrow (\exists A \subseteq \mathbb{N})[\mathcal{D}_s^A \text{ is superatomic}]$$
$$\Leftrightarrow (\exists A \subseteq \mathbb{N})[\mathbb{F}(m,s)[A] \text{ is totally transcendental}]$$
$$\Leftrightarrow \mathbb{F}(m,s) \text{ has a totally transcendental completion.}$$

Thus, we have obtained the lower estimate in case (a). $\square$

Theorem 8.4.3. *The following assertions hold:*

(a) $\{n \mid \Phi_n$ *has a totally transcendental decidable completion*$\} \approx \Pi_1^1$,

(b) $\{n \mid \Phi_n$ *has a finitely axiomatizable totally transcendental completion*$\} \approx \Pi_1^1$,

(c) $\operatorname{Nom Th}(T \cap P) \approx \Sigma_1^1$,

(d) $\operatorname{Nom Th}(T \cap F) \approx \Sigma_1^1$.

PROOF. It suffices to prove (a) and (b) since (c) and (d) are direct consequences of them. The deduction of the upper estimates is obvious. To obtain the lower estimates, we use the standard set H introduced in Lemma 2.3.1. Since the theory $\mathbb{F}(\mathcal{D})$ is complete, we have

$n \in H \Leftrightarrow \mathbb{F}(\mathcal{D}_n)$ has a totally transcendental decidable completion

$n \in H \Leftrightarrow \mathbb{F}(\mathcal{D}_n)$ has a finitely axiomatizable totally transcendental completion $\square$

Theorem 8.4.4. *The following assertions hold:*

(a) $\{n \mid \Phi_n$ *has a strongly constructivizable model with a totally transcendental theory*$\} \approx \Pi_1^1$,

(b) $\{n \mid \Phi_n$ *has a finitely axiomatizable and totally transcendental extension with a strongly constructivizable model*$\} \approx \Pi_1^1$,

(c) $\mathrm{Nom\,Th}\,(T \cap SC) \approx \Sigma_1^1$,

(d) $\mathrm{Nom\,Th}\,(T \cap SC \cap F) \approx \Sigma_1^1$.

PROOF. It suffices to prove (a) and (b). As above, the formalization of conditions gives a prefix of the form Π_1^1. To establish the lower estimates, we use the set H' defined in Lemma 2.3.1. Applying Theorem 7.2.1, we find

$n \in H' \Leftrightarrow \mathbb{F}(\mathcal{D}_n)$ has a totally transcendental completion with
    a strongly constructivizable model

$n \in H' \Leftrightarrow \mathbb{F}(\mathcal{D}_n)$ has a finitely axiomatizable completion with a
    strongly constructivizable model

□

Appendix

Open Questions

QUESTION 1 [known]. *Does a countably categorical, complete, and finitely axiomatizable theory T for which the Lindenbaum function l is not recursive exist? ($l(n)$ is equal to the number of atoms of the Lindenbaum algebra $\mathcal{L}_n(T)$).*

There exists a similar example of an axiomatizable theory.

QUESTION 2. *Does a complete, finitely axiomatizable, and ω-stable theory T of Morley rank $\alpha_T = \omega + 1$ exist?*

QUESTION 3. *Does a complete finitely axiomatizable theory T such that the Morley rank α_T satisfies $\lambda < \alpha_T < \omega_1$, exist? ($\lambda$ denotes the first nonconstructive ordinal.)*

It suffices to study this question for axiomatizable theories and, if the answer is positive, apply the universal construction.

QUESTION 4 [known]. *Does a complete, finitely axiomatizable, and uncountably categorical theory exist such that it is strongly minimal?*

The following question (a weak version of Question 4) is also open for theories of rank 1, i.e., if models of the theory consist of a finite number of strongly minimal sets.

QUESTION 5 [known]. *Does a stable, complete, and finitely axiomatizable theory exist?*

For superstable theories, a similar question has been negatively solved (Zil'ber, Cherlin, Lachlan, and Harrington).

QUESTION 6. *Does a complete finitely axiomatizable theory possessing exactly n, $1 < n < \omega$, countable models exist?*

QUESTION 7. *Does an infinite group exist the elementary theory of which is finitely axiomatizable and ω-stable?*

QUESTION 8. *Let T_1 and T_2 denote group theory and graph theory respectively. Is it true that $T_1 \equiv_q T_2$?*

QUESTION 9. *Describe complete, countably categorical, and finitely axiomatizable theories in terms of the structure of Lindenbaum algebras.*

This description must be such that, based on it, one can easily solve other questions concerning this class of theories.

QUESTION 10. *Let Φ_n, $n \in \mathbb{N}$, be the Gödel numbering of all sentences of some finite rich signature σ. Find exact estimates of the algorithmic complexity for the following classes of sentences:*

(a) $\{n \mid \Phi_n$ *is countably categorical*$\}$,

(b) $\{n \mid \Phi_n$ *is uncountably categorical*$\}$,

(c) $\{n \mid \Phi_n$ *is totally categorical*$\}$,

(d) $\{n \mid \Phi_n$ *is complete and countably categorical*$\}$,

(e) $\{n \mid \Phi_n$ *is complete and uncountably categorical*$\}$,

(f) $\{n \mid \Phi_n$ *is model-complete*$\}$,

(g) $\{n \mid \Phi_n$ *is complete and model-complete*$\}$.

QUESTION 11. *Compute exact estimates of the algorithmic complexity (by Secs. 8.2–8.4) for the following model-theoretic properties:*

(a) *stability,*

(b) *superstability,*

(c) *atomic models,*

(d) *minimal models,*

(e) *rigid models,*

(f) *models with first-order definable and almost first-order definable elements,*

(g) *nonmaximality of the spectrum function.*

QUESTION 12. *Find exact estimates of the algorithmic complexity for the following classes of models of a fixed finite rich signature* ($R(\mathfrak{M})$ *denotes the Morley rank of* Th $(\mathfrak{M})$):

(a) $\{\mathfrak{M} \mid R(\mathfrak{M}) < \omega\}$,

(b) $\{\mathfrak{M} \mid R(\mathfrak{M}) < \omega \& \text{Th}(\mathfrak{M}) \text{ is } \omega\text{-stable}\}$,

(c) $\{\mathfrak{M} \mid R(\mathfrak{M}) < \lambda\}$,

(d) $\{\mathfrak{M} \mid R(\mathfrak{M}) < \lambda \& \text{Th}(\mathfrak{M}) \text{ is } \omega\text{-stable}\}$,

(e) $\{\mathfrak{M} \mid R(\mathfrak{M}) \leqslant \lambda\}$,

(f) $\{\mathfrak{M} \mid R(\mathfrak{M}) < \omega_1\}$,

(g) $\{\mathfrak{M} \mid R(\mathfrak{M}) = \omega_1\}$.

QUESTION 13. (a) *Find a maximal list* L_1 *for which an analog of the first-level Rice theorem for sentences is valid.*
(b) *Find a maximal list* L_2 *for which an analog of the second-level Rice theorem for sentences is valid.*

Tentative values of L_1 and L_2 are discussed in Sec. 11.4.
The following three questions differ from the previous ones. In particular, they are not too specific, but direct the way toward interesting new results.

QUESTION 14 [problem of a basic theory]. *Does a universal construction exist in which the rigidity mechanism is based on the quasisuccessor theory of rank 2 (QS from Chapter 1 or some other similar theory)?*

In this direction, some simplification of the universal construction is possible.

QUESTION 15 [problem of an algebraic version]. *Does a construction exist such that it is similar to the universal construction, but its domain of action is represented by the list AQL that is an algebraic extension of the model list MQL?*

The zero version of this question is to construct a construction with the list L^a which contains at least one algebraic (but not model) property, for example, the model completeness property L^a (AQL can consist only of this property). The complete list AQL can be defined as a list of those properties that are preserved by IG.

The following question is devoted to the noncompleteness of the general scheme of construction depicted in Fig. 0.7.1. It is shown that the branch of constructions for which an axiomatizable theory is taken as a parameter has developed and reached its peak in the form of the construction $\mathbb{FU}$. Simultaneously, another branch of constructions with a natural parameter has remained stationary after the Vaught construction in 1961. The above-mentioned constructions could be called *polar*. Their distinguishing feature is that the theory $F_n = \mathbb{VT}(n)$ is effectively constructed by n, but its properties are determined by the membership of n in the creative set K. In addition, the properties of the theory F_n are polar-opposite in two cases, $n \in K$ and $n \notin K$ (e.g., see Theorem 0.7.3).

The aforesaid leads to the following natural question.

QUESTION 16 [problem of a polar construction]. *Construct a polar construction such that it is as strong as possible and it strengthens the Vaught construction.*

As one of the possible applications of such a construction, we mention a solution of Question 13(a). In this case, the volume of the list L_1 plays the role of the measure of the strength of the construction.

References

1. BÜCHI, J. R., Turing machines and the Entscheidungsproblem, *Math. Ann.*, **148**, 201–213, 1962.
2. BOONE, W. W. and ROGERS, H. JR., On a problem of J. H. C. Whitehead and a problem of Alonzo Church, *Math. Scand.*, **19**,185–192, 1966.
3. CHANG, C. C. and KEISLER, H. J., *Model Theory*, Elsevier, New York, 1992.
4. CHERLIN, G., HARRINGTON, L. A., and LACHLAN, A. H., $\aleph_0$-categorical $\aleph_0$-stable structures, *Ann. Pure Appl. Logic*, **28**, No. 2, 103–135, 1985.
5. CHURCH, A., A note on the Entscheidungsproblem, *J. Symbolic Logic*, **1**, No. 1, 40–41, 1937. Correction: 101–102.
6. CHURCH, A., *Introduction to Mathematical Logic*, Vol. 1, Princeton Univ. Press, Princeton, 1956.
7. COBHAM, A., Some remarks concerning theories with recursively enumerable complement, *J. Symbolic Logic*, **28**, No. 1, 72–74, 1963.
8. CRAIG, W. and VAUGHT, R. L., Finite axiomatizability using additional predicates, *J. Symbolic Logic*, **23**, No. 3, 289–308, 1958.
9. DENISOV, S. D., Models of noncontradictory formulas and the Ershov hierarchy, *Algebra Logic*, **11**, No. 6, 359–362, 1972.

10. DROBOTUN, B. N., Enumerations of prime models, *Sib. Math. J.*, **18**, No. 5, 707–715, 1977.

11. ERSHOV, YU. L., Constructive models, *Selected Questions of Algebra and Logic*, Novosibirsk, 1973, 111–130.

12. ERSHOV, YU. L., On a hierarchy of sets. I, *Algebra Logika*, **7**, No. 1, 47–74, 1968.

13. ERSHOV, YU. L., *Decidability Problems and Constructible Models* [in Russian], Nauka, Moscow, 1980.

14. FEINER, L., Hierarchies of Boolean algebras, *J. Symbolic Logic*, **35**, No. 3, 305–373, 1970.

15. FRIEDMAN, H., One hundred and two problems in mathematical logic, *J. Symbolic Logic*, **40**, No. 2, 113–129, 1975.

16. GONCHAROV, S. S. and DROBOTUN B. N., Numeration of saturated and homogeneous models, *Sib. Math. J.*, **2**, No. 21, 164–175, 1980.

17. GONCHAROV, S. S. and NURTAZIN, A. T., Constructive models of complete solvable theories, *Algebra Logic*, **12**, No. 2, 67–77, 1973.

18. GONCHAROV, S. S., A totally transcendental decidable theory without constructivizable homogeneous models, *Algebra Logic*, **19**, No. 2, 85–93, 1980.

19. GONCHAROV, S. S., Constructivizability of superatomic Boolean algebras, *Algebra Logic*, **12**, No. 1, 17–22, 1973.

20. GONCHAROV, S. S., Some properties of constructivizations of Boolean algebras, *Sib. Math. J.*, **16**, No. 2, 264–278, 1975.

21. GONCHAROV, S. S., Strong constructivizability of homogeneous models, *Algebra Logic*, **17**, No. 4, 247–262, 1978.

22. GONCHAROV, S. S., *Countable Boolean Algebras and Decidability*, Plenum, New York, 1997.

23. GUREVICH, YU. SH., Problem on decidability for the logic of predicates and operations, *Algebra Logika*, **8**, No. 3, 284–308, 1969.

24. HANF, W., Model-theoretic methods in the study of elementary logic, *Symposium on the Theory of Models*, North-Holland, Amsterdam, 1965, pp. 33–46.

25. HANF, W., The Boolean algebra of logic, *Bull. Am. Math. Soc.*, **31**, 587–589, 1975.

26. HARRINGTON, L., Recursively presented prime models, *J. Symbolic Logic*, **39**, No. 2, 305–309, 1974.

27. HENSEL, G. and PUTNAM, H., Normal models and the field Σ_1^*, *Fundam. Math.*, **64**, No. 2, 231–248, 1969.

28. HRUSHOWSKI, E., Finitely axiomatizable $\aleph_1$-categorical theories, *J. Symbolic Logic*, **8**, No. 3, 838–844, 1994.

29. KARATABANOVA, S. ZH., A finitely realizable list of properties, *Algebra Logic*, **31**, No. 4, 245–247, 1992.

30. KLEENE, S. C., Finite axiomatizability of theories in the predicate logic using additional predicate symbols, *Mem. Am. Math. Soc.*, **10**, 27–68,

1952.

31. KREISEL, G., Note on arithmetic models for consistent formulae of the predicate calculus, *Fundam. Math.*, **37**, 265–285, 1950.

32. LACHLAN, A. H., The transcendental rank of a theory, *Pac. J. Math.*, **27**, No. 27, 119–122, 1971.

33. MAKOWSKY, J. A., On some conjectures connected with complete sentences, *Fundam. Math.*, **81**, No. 3, 193–202, 1974.

34. MORLEY, M., Decidable models, *Israel J. Math.*, **25**, No. 3–4, 233–240, 1976.

35. MOSTOWSKI, A., A formula with no recursively enumerable model, *Fundam. Math.*, **42**, 125–140, 1955.

36. MOSTOWSKI, A., Modern state of research on the foundations of mathematics, *Usp. Mat. Nauk*, **9**, No. 3, 3–38, 1954.

37. MOSTOWSKI, A., On a system of axioms which has no r.e. arithmetic model, *Fundam. Math.*, **40**, 56–61, 1953.

38. MOSTOWSKI, A., On recursive models of formalized arithmetic, *Bull. Acad. Pol. Sci.*, No. 7, 705–710, 1957.

39. NURTAZIN, A. T., Strong and weak constructivizations and computable families, *Algebra Logic*, **13**, No. 3, 177–184, 1974.

40. PERETYAT'KIN, M. G., Analogues of Rice's theorem for semantic classes of propositions, *Algebra Logic*, **30**, No. 5, 332–348, 1991.

41. PERETYAT'KIN, M. G., Finitely axiomatizable theories, *Proc. of the International Congress of Mathematicians*, Berkeley, California, 1986, **1**, 322–330; English translation: Amer. Math. Soc., Transl., (2), **147**, 1990, 11–19.

42. PERETYAT'KIN, M. G., Example of an ω_1-categorical complete finitely axiomatizable theory, *Algebra Logic*, **19**, No. 3, 202–229, 1980.

43. PERETYAT'KIN, M. G., Expressive power of finitely axiomatizable theories. I. Introduction, interpretations, reduction to graphs, *Sib. Adv. Math.*, **3**, No. 2, 153–197, 1993.

44. PERETYAT'KIN, M. G., Expressive power of finitely axiomatizable theories. II. Rigid quasisuccession, *Sib. Adv. Math.*, **3**, No. 3, 123–145, 1993.

45. PERETYAT'KIN, M. G., Expressive power of finitely axiomatizable theories. III. Main construction, *Sib. Adv. Math.*, **3**, No. 4, 131–201, 1993.

46. PERETYAT'KIN, M. G., Finitely axiomatizable totally transcendental theories, *Tr. Inst. Mat. SO Akad. Nauk SSSR*, 1982, **2**, pp. 88–135.

47. PERETYAT'KIN, M. G., Semantic universality of theories over a superlist, *Algebra Logic*, **31**, No. 1, 30–47, 1991.

48. PERETYAT'KIN, M. G., Semantical significances of formulas in the classical predicate logic, *Contemporary Math.*, 1992, Vol. 131, part 3, pp. 623–643.

49. PERETYAT'KIN, M. G., Semantically universal classes of models, *Algebra Logic*, **30**, No. 4, 271–281, 1991.

50. PERETYAT'KIN, M. G., Similarity of properties of recursively enumerable

and finitely axiomatizable theories, *Dokl. Akad. Nauk SSSR*, **308**, 788–791, 1989.

51. PERETYAT'KIN, M. G., Strongly constructive models and numerations of the Boolean algebra of recursive sets, *Algebra Logic*, **10**, No. 5, 332–345, 1971.

52. PERETYAT'KIN, M. G., Turing machine computations in finitely axiomatizable theories, *Algebra Logic*, **21**, No. 4, 272–295, 1982.

53. PERETYAT'KIN, M. G., Uncountably categorical quasisuccession of Morley rank 3, *Algebra Logic*, **30**, No. 1, 51–61, 1991.

54. PUTNAM, H., Trial and error predicates and the solution to a problem of Mostowski, *J. Symbolic Logic*, **30**, No. 1, 49–57, 1965.

55. RICE, H. G., Classes of recursively enumerable sets and their decision problems, *Trans. Am. Math. Soc.*, **74**, 358–366, 1953.

56. ROGERS, H. J., *Theory of Recursive Functions and Effective Computability*, McGraw-Hill, New York, 1967.

57. SACKS, G., *Saturated Model Theory*, Benjamin, Reading, Massachusetts, 1972.

58. SACKS, G., Effective bounds on Morley rank, *Fundam. Math.*, **103**, No. 2, 111–121, 1972.

59. SHELAH, S., *Classification Theory and the Number of Nonisomorphic Models*, North-Holland, Amsterdam, 1990.

60. SOARE, R. I., *Recursively Enumerable Sets and Degrees*, Springer-Verlag, Berlin–Heidelberg–New York, 1986.

61. SHOENFIELD, J. R., *Mathematical Logic*, Addison-Wesley, Reading, Massachusetts, 1967.

62. TRAKHTENBROT, B. A., Impossibility of algorithm for the decidability problem on finite classes, *Dokl. Akad. Nauk SSSR*, **70**, No. 4, 569–572, 1950.

63. VAUGHT, R. L., Sentences true in all constructive models, *J. Symbolic Logic*, **25**, No. 1, 39–58, 1961.

64. ZAURBEKOV, S. S., A restricted analog of the third level of Rice's theorem, *Algebra Logic*, **32**, No. 2, 71–74, 1993.

65. ZIL'BER, B. I., Totally categorical theories: structural properties and the nonfinite axiomatizability, in: *Model Theory of Algebra and Arithmetic. Proc. Conf. Karpacz*, 1979, Lecture Notes in Math., **834**, Springer Verlag, Berlin, 1980, pp. 381–410. Correction: Solution of the problem of finite axiomatizability for theories that are categorical in all infinite powers, in: *Investigations in Theoretical Programming*, Kaz. Univ., Alma Ata, 1981, pp. 69–74.

Subject Index

293